Non-logic Devices in Logic Processes

Yanjun Ma • Edwin Kan

Non-logic Devices in Logic Processes

Springer

Yanjun Ma
Invention Development Fund
Bellevue, WA, USA

Edwin Kan
School of Electrical and Computer Engineering
College of Engineering, Cornell University
Ithaca, NY, USA

ISBN 978-3-319-48337-5 ISBN 978-3-319-48339-9 (eBook)
DOI 10.1007/978-3-319-48339-9

Library of Congress Control Number: 2017934520

Printed on acid-free paper

This Springer imprint is published by Springer Nature
The registered company is Springer International Publishing AG
The registered company address is: Gewerbestrasse 11, 6330 Cham, Switzerland

Preface

For over 10 years, one of the authors (YM) has had the privilege to work in a small, growing semiconductor company that focused on producing integrated circuits (ICs) that consume microwatt power and cost less than a penny—chips for the passive radiofrequency identification (RFID) market. The goal of the RFID industry is to produce RFID tags cheap enough to replace the ubiquitous barcodes. As can be expected, since it costs almost nothing to print a barcode, the cost of the RFID chip is one of the most important factors for this lofty goal. Thus one of the driving forces is to squeeze every bit of cost, from the designing to manufacturing, of the RFID chips, which usually entails the IC design methods and manufacturing technologies of the lowest cost.

Integrated circuit technology is one of the most amazing engineering marvels in the history of mankind. It has brought forth the revolution in computing, control, and communication and realizes the platform of the information age. For over 50 years, this technology has taken an exponential growth path, following the famed Moore's law. While it has been projected to continue on this exponential path for at least another 10 years, it will slow down eventually (and may have already slowed down), as expected from laws of mathematics: that exponential growth cannot continue forever. Thus IC technology is destined to mature, not unlike the steam engine and automotive technologies which powered the previous industrial revolutions.

The slowing down and eventual stoppage of Moore's law lead to the question of why writing this book and who are the targeted readers. After all there are numerous CMOS device and technology books on the market already, including the encyclopedic *Physics of Semiconductor Devices* by Simon Sze. The corresponding series on CMOS processing by S. Wolf and coauthors are equally impressive and are textbook examples. Additionally, a number of semiconductor device books also have good process coverage.

The answer to the first question is in the title of the book, *Non-logic Devices in Logic Processes*. Our belief is that as IC technology scaling slows down, the users of the IC technology will be much more motivated to find ways to produce their non-logic chips using the most economical IC fabrication process, which usually means the logic CMOS process.

Therefore the main purpose of this book is to teach innovative use of the standard logic CMOS process to design non-logic devices that can reduce the cost of ICs. The contents of this book are based on practices that have been frequently used in the industry but, to the authors' knowledge, have not been systematically covered in a monogram. Indeed, one of the main sources for the materials in this book is from relevant patents. As such this book can serve as a complementary resource to standard textbooks on semiconductor devices, which in general tend to focus more on the theory and less on practical design of semiconductor devices, especially outside the standard logic and memory modules.

It is noteworthy that we have excluded the discussion of MEMS devices in this book. While many MEMS processes have been designed to be CMOS compatible, additional process steps such as release of the mechanical structures are necessary. Not to mention that there are many textbooks on MEMS already, such as by Kovacs and Senturia. We have mostly excluded the discussion of RF CMOS devices due to similar reasons.

Our targeted readers are semiconductor technology developers, device engineers, analog and memory designers, who are seeking unconventional uses of the standard CMOS devices. This book can also be used by graduate students in those areas.

Bellevue, WA Yanjun Ma
Ithaca, NY Edwin C. Kan

Organization of This Book

This book has three basic parts.

In Part I we review basic CMOS devices commonly used in logic CMOS circuits as well as the process for making these devices. Because of their simplicity and widespread use, logic CMOS devices have been well studied and their associated processes fully developed for economy of scale. After reviewing the history of semiconductor scaling, with emphasis on the Moore's law and its corollaries, we review the economics of semiconductor in Chap. 1. We then describe the basic device physics in Chap. 2 and briefly review a basic CMOS process flow in Chap. 3. These will form the foundation for designing non-logic CMOS devices to be described in Part II.

Part II (Chaps. 4–7) is the core of this book and describes some of the common practices for designing non-logic devices that can be manufactured in the most basic logic CMOS processes without any process modifications. In particular, Chap. 4 focuses on the art of designing non-logic MOSFETs, including those with nonstandard threshold voltage, and high-voltage MOSFETs in the logic CMOS process. In Chap. 5 we review several floating-gate devices. Floating-gate transistors have found widespread use in embedded nonvolatile memory and neuromorphic circuits applications. Chapters 6 and 7 discuss methods for obtaining diodes, varactors, and bipolar transistors in the CMOS process.

Part III (Chaps. 8–13) illustrates the use of the devices described in Part II in selected applications. Chapters 8–10 explore in depth how to design and construct several types of nonvolatile memories in the logic CMOS process. In Chap. 8 we review basic floating-gate memory cell designs and several methods for adding and removing static charge from the floating gate. We will then briefly discuss the basic circuit modules for embedded memory, including charge pumps and sense amplifiers. Chapter 9 focuses on one-time programmable (OTP) memories, including fuse, antifuse, and floating-gate OTP memories. Chapter 10 shows several multiple-time programmable (MTP) memories that have seen commercial deployment. In Chap. 11 we discuss some non-NVM applications of floating-gate transistors, including digital-to-analog converters, RFID, neuromorphic computing.

Chapter 12 will treat the CMOS biosensors in both Faradic (current sensing) and non-Faradic (field sensing) operations. Examples of isolation and compensation techniques to interface with the biological saline buffers will be presented. One of the main reasons of IC clean-room manufacturing is to eliminate the influence of alkali ions in the semiconductor wafers, but the CMOS biosensors have to be operated with close proximity to this electrolyte environment with high variability. In Chap. 13 we will describe the distributive waveform shaping circuits in high-frequency applications by employing embedded waveguides and varactors. Finally Chap. 14 presents a brief overview of recent development in CMOS processing, especially with regard to the emergence of FinFETs and universal substrates, as well as emerging nonvolatile memory technologies which may offer future opportunities.

Acknowledgements

YM would like to thank his former colleagues at Impinj—Chris Diorio, Todd Humes, Ron Paulsen, Bin Wang, Rui Deng, Eric Wong, Ron Koepp, John Quist, John Hyde, Alberto Pesavento, Wendy Malabuyo, Jay Kuhn, Charles Peach, David Pelke, and Jaideep Mavoori for sharing their knowledge in the areas of semiconductor physics and IC design. He especially thanks Eric Wong for reviewing many of the chapters in this book and for contributing materials for Sect. 11.2.

He also would like to thank Dr. Patrick Ennis for reading the manuscript and the encouragement to undertake this project.

YM is indebted to his former colleagues at Sharp Labs of America, especially Yoshi Ono, Dave Evans, and S.T. Hsu, for teaching him much of his knowledge on CMOS processing.

Finally he would like to thank his family: Michelle, Christine, and Kevin for the support during the long process of writing this book.

ECK would like to thank his Ph.D. students at Cornell over the last 19 years. His students have not only been his research contributors and collaborators, but have also become his lifelong friends. He feels honored and rewarding to have the opportunity to advise and work with these students during an important development period in their lives. He hopes to especially thank Yumin Shen (now at TSMC), Pingshan Wang (now at Clemson), Jinsook Kim (now at Spansion), Blake Jacquot (now at Google), Keith Lyon (now at KLA-Tencor), and Yingqiu Cao (still at Cornell), whose Ph.D. research works serve as the base for several chapters in this book.

Acknowledgements

[illegible]

Contents

Abbreviations

ADC	Analog to digital converter
AFM	Atomic force microscope
ALD	Atomic layer deposition
ANN	Artificial neutral network
BEOL	Backend of line
BG	Bandgap
BGR	Bandgap reference
BIT	Best industry practice
BJT	Bipolar junction transistor
BTBT	BAND to band tunneling
CMOS	Complementary metal-oxide-semiconductor
CMP	Chemical mechanical polish
CPLD	Complex programmable logic device
CPS	Coplanar stripline
CPW	Coplanar waveguide
CR	Coupling ratio
CVD	Chemical vapor deposition
CνMOS	Chemoreceptive neuron MOS
DAC	Digital-to-analog converter
DAHC	Drain avalanche hot carrier
DIBL	Drain-induced barrier lowering
DR	Design rules
DRAM	Dynamic random access memory
DRC	Design rule check
DRIR	Deep reactive ion etch
ECC	Error correction codes
EDL	Electrical double layer
EEPROM	Electrically EPROM
EKV	Enz-Krummenacher-Vittoz model

EOT	Equivalent oxide thickness
EPROM	Erasable PROM
ESD	Electrostatic discharge
FB	Flatband
FEOL	Front end of line
FET	Field effect transistor
FG	Floating gate
FN	Fowler-Nordheim
FPGA	Field programmable gate array
GAA	Gate all around
GIDL	Gate-induced drain leakage
I/O	Input/output
IC	Integrated circuit
ILD	Interlayer dielectric
IMD	Intermetal dielectric
IoT	Internet of things
IP	Intellectual property
ITRS	International technology roadmap for semiconductors
LDD	Light doped drain
LDMOS	Lateral diffused drain MOS
LNVM	Logic NVM
LoC	Lab-on-a-Chip
LOCOS	Local oxidation of silicon
MEA	Microelectrode array
MIM	Metal-insulator-metal
MLC	Multilevel cell
MMIC	Microwave monolithic integrated circuits
MO	Molecular orbital
MOM	Metal-oxide-metal
MOS	Metal oxide semiconductor
MOSCAP	MOS capacitor
MOSFET	Metal oxide semiconductor field effect transistor
MtM	More than Moore
MTP	Multiple times programmable
NFC	Near field communication
NLTL	Nonlinear transmission line
NRE	Nonrecurring engineering
NVM	Nonvolatile memory
OTP	One time programmable
PBS	Phosphate buffered saline
PCB	Printed circuit board
PCM	Phase change memory
PCM	Process control monitor

PLD	Programmable logic device
PMU	Power management unit
PROM	Programmable read only memory
Qbd	Charge to breakdown
RAM	Random access memory
RFID	Radiofrequency identification
RIE	Reactive ion etch
ROM	Read only memory
RRAM	Resistive RAM
SAM	Self-assembled monolayer
SCE	Short channel effect
SILC	Stress-induced leakage current
SIP	System in package
SOC	System on chip
SOI	Silicon on insulator
SRAM	Static RAM
SSD	Solid state drive
STI	Shallow trench isolation
STTRAM	Spin transfer torque RAM
TDDB	Time-dependent dielectric breakdown
TSV	Through silicon via
WAT	Wafer acceptance test

About the Authors

Yanjun Ma obtained his B.S. degree from the University of Science and Technology of China and Ph.D. degree in solid state physics from the University of Washington in Seattle and has since had wide research and development experiences in academic, industrial, and government research labs and in small and large commercial companies. He started his research career studying high temperature superconductors, quasicrystals, and X-ray physics which led to the discovery of a momentum conservation law in X-ray fluorescence spectroscopy. He later expanded to the area of semiconductor device physics, including as a leading researcher in high k gate dielectrics and nonvolatile memories.

Yanjun has over 37 issued US patents and a number of international patents and patent applications, with some very highly cited patents, including a few of the earliest patents on high k gate dielectrics. He also has over 80 publications.

Yanjun is currently a vice president of innovation and a Principal Hardware Architect at the Invention Development Fund and Xinova. He is also a founder and chief technical officer at MCLogic, an integrated circuit start-up company that is deploying some of the techniques detailed in this book. Prior to his current positions, Yanjun was a principal engineer and a director of Technology Development and Production Engineering at Impinj, Inc., an RFID company in Seattle, Washington. He also had held senior research and engineering positions at Lattice Semiconductors, Sharp Labs of America as well as research positions at Bell Labs, Brookhaven National Labs, and the University of Washington.

Edwin C. Kan received his B.S. degree from National Taiwan University in 1984 and his M.S. and Ph.D. degrees from the University of Illinois at Urbana-Champaign in 1988 and 1992, all in electrical engineering. In January 1992, he joined Dawn Technologies as a Principal CAD Engineer developing advanced electronic and optical device simulators and technology CAD framework. He was then with Stanford University as a Research Associate from 1994 to 1997. From

1997, he became an Assistant Professor in the School of Electrical and Computer Engineering, Cornell University, where he is now a Professor.

He has spent the summers of 2000 and 2001 at IBM Microelectronics, Yorktown Heights and Fishkill, NY, in the Faculty Partner Program. In 2004 and 2005, he has been a visiting researcher at Intel Research, Santa Clara, CA, and a visiting professor at Stanford University during his sabbatical leave. His main research areas include CMOS technologies, semiconductor device physics, flash memory, CMOS biosensors, RFID, RF indoor locating and tracking, and numerical methods for PDE and ODE.

Edwin has over 10 US patents, 80 journal publications, and 160 conference papers. He has graduated 25 Ph.D. students at Cornell and worked with many master's and undergraduate students. He received the Presidential Early Career Award for Scientists and Engineer (PECASE) in October 2000 from the White House. He also received several teaching awards from Cornell Engineering College for his CMOS and MEMS courses.

Part I
Basics

In the first part of this book, we review the basic CMOS devices commonly used in logic CMOS circuits, as well as the process for making these devices. Because of their simplicity and widespread use, logic CMOS devices have been well studied and their associated processes well developed to yield economy of scale. We briefly describe the basic device physics and a brief process flow. These will form the foundation for designing non-logic CMOS devices in Part II.

In Chap. 1 we provide a brief introduction and review the economics of the semiconductor industry, the Moore's law, and a number of its corollaries that may be used to derive the best practices in designing devices for use in integrated circuits.

Chapter 2 is an overview of the basic structure and physics of logic CMOS devices, including the basic metal-oxide-semiconductor (MOS) structures, MOSFETs, capacitors, and resistors.

In Chap. 3 we go over a brief process flow for the fabrication of the CMOS devices, and discuss some layout and manufacturing issues.

Chapter 1
Introduction

Abstract In this chapter, we provide a brief introduction and review the importance of semiconductor devices in supporting the whole electronics industry, the economics of semiconductor industry, the Moore's law, and a number of corollaries that may be used to derive the best practices in designing devices for use in integrated circuits.

Keywords Moore's law • Integrated circuit • CMOS • Scaling

1.1 The Semiconductor Industry

The global electronics industry, including cell phones, PCs, TVs, and automotive electronics, is a trillion-dollar market. Semiconductors are the brains inside virtually all modern electronics, serving as key enablers for the growth of the electronics market. The hierarchy of the electronics and semiconductor industry may be described by the inverted pyramid shown in Fig. 1.1.

At the bottom of the pyramid is the materials used in semiconductors. In particular, silicon-based integrated circuits (ICs), which power an overwhelming percentage of all electronics, are made with only a very limited selection of materials, mostly because of process compatibility and cost. For a long time, before around 2005, the only materials used in Si ICs were **silicon** as the substrate in Metal-oxide-semiconductor field effect transistors (MOSFETs), $\mathbf{SiO_2}$ for the gate and intermetal insulators, silicon nitrides ($\mathbf{Si_3N_4}$) for the process development and isolation, **polycrystalline silicon** (poly) for the local MOSFET gate, **aluminum** (doped with silicon or copper) for the interconnect between the devices, and **tungsten** and a handful of **silicides** ($TiSi_2$, $CoSi_2$, $NiSi_2$, and WSi) for contacts and gate extension. There were also a few other elements such as **boron** and **phosphorus** mainly used as the dopants to make the p–n junctions. More recently, **copper** has replaced aluminum as the interconnect, while high dielectric constant materials such as $\mathbf{HfO_2}$ are added to $\mathbf{SiO_2}$ as the gate dielectrics. Occasionally, more exotic materials are used in specialty ICs, e.g., $\mathbf{ZrO_2}$ and TiDyO have been used in the storage capacitors in dynamic random access memories (DRAMs) and $\mathbf{Al_2O_3}$ in the control dielectric of flash memories. We should mention that even more materials are used in the IC manufacturing process, including many types of

Y. Ma, E. Kan, *Non-logic Devices in Logic Processes*,
DOI 10.1007/978-3-319-48339-9_1

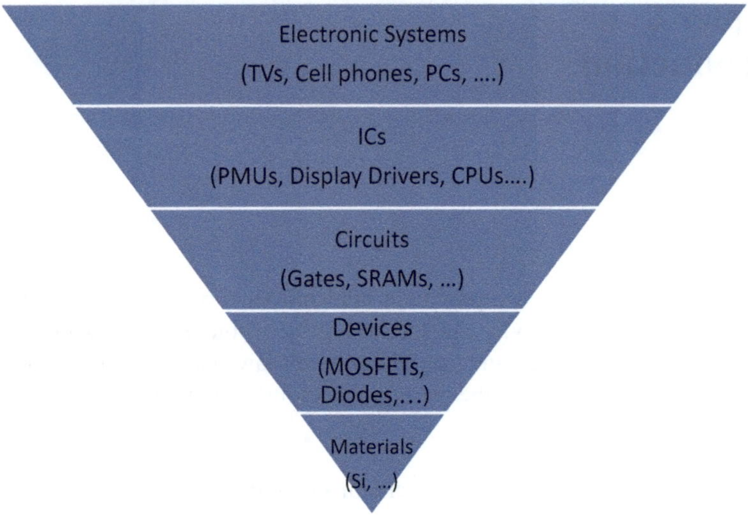

Fig. 1.1 The inverted pyramid where semiconductor materials and devices supporting the trillion-dollar electronics industry

organic photoresists for lithography as well as various gases and acids during the etch processes.

Using these materials, we can build a large number of semiconductor devices, which by category have historically included many types of **bipolar junction transistors** (**BJTs**), **field effect transistors** (**FETs**), **diodes**, **field emission devices**, and so on. Simon Sze's classic book on semiconductor devices described most of the semiconductor devices that had found commercial use (Sze, 1981). Even though the IC era started with BJTs, they are mostly replaced by MOSFETs and can only be found in niche ICs, such as the SiGe bipolar heterojunction transistors (HBT) for high-frequency circuits where linearity and dynamic range are important.

Because of their energy efficiency and simple isolation structures that are more amenable for feature size scaling, CMOS circuits dominate the present-day semiconductor market. For CMOS circuits, the number of devices commonly used in digital or mixed-signal ICs is even less. A few basic devices, diodes, capacitors, resistors, and MOSFETs, and a few BJTs here and there support >95 % of the $300 billion (in 2016) semiconductor IC market.

On the next level up, these devices are used to construct circuits that may number in the tens to thousands, including major types such as logic, volatile and nonvolatile memory, mixed signal, communication, timing, amplification, and power management circuits. Many of these circuits can be designed using only MOSFETs. For example, the digital circuits such as logic gates (**inverters**, **NAND**, **NOR**, and **XOR** gates), **flip-flops**, **adders** and **multipliers**, and **gate arrays** need only n- and p-type MOSFET. Other circuits, such as **oscillators**, may need a few capacitors and resistors. **Voltage reference circuits** and **electrostatic discharge**

(ESD) protection circuits typically make use of resistors, diodes, and BJTs. Radio-frequency circuits such as **power amplifiers** and **filters** employ on-chip inductors as part of the resonance circuits for frequency selection.

From these circuits, we can build ICs that perform the desired functions for end use, from cloud computing to consumer applications. Display driver ICs drive the visible pixels. **Microcontroller units** (**MCU**) control the fuel injection of an internal combustion engine. **Microprocessor units** (**MPUs**) perform the basic operations in a computer. **Analog-to-digital converter** (**ADC**) and **digital-to-analog converter** (**DAC**) interface our physical world with the digital world. Mixed-signal and RF circuits construct the **wired and wireless communication** functions. Eventually these ICs are put together in complete systems that we use to make phone calls, surf the internet, control the power generation plants, drive an autonomous car, and so on.

This book focuses on the second level in Fig. 1.1, the devices that enable the electronics industry. It is truly remarkable that only a handful devices form the bottom of the pyramid, supporting the trillion-dollar electronics industry. This is precisely the reason for the growth of the electronics industry—we only need to perfect the manufacturing of very few devices. The semiconductor industry, in particular, has been mastering their production. We can scale smaller and smaller and pack more and more of these devices into the same die area, year after year, all in working order—a process described by the Moore's Law.

Briefly stated, Moore's law states that the number of transistors in an integrated circuit doubles every 12 months (in early years of IC) to 24 months (more recently). The Moore's law has been shown to be valid for over 50 years now, during which the number of transistors on an IC has increased from a handful to over a billion.

1.2 Moore's Law and Its Corollaries

Moore's law, which has predicted the exponential growth of semiconductor technology development and economics since the 1960s, is slowing down (Simonite, 2016) and will undoubtedly stop in the future. The eventual ending of Moore's law is a certainty because physical limits will be reached due to the finite size of atoms, number of atoms to form electronic bands, and quantized nature of information process (see Chap. 14 for a more in-depth discussion). The minimal semiconductor device feature size is already around 10 nm in 2016, which makes "**More Moore**" (**MM**) very difficult due to both economical and technological reasons. We may ultimately be able to continue scaling the device size to 3–4 nm or about 10 atoms across, a goal that may be reached in 10 years. After that it is anticipated that new computing paradigms such as quantum computing will be needed to carry on the so-called **More than Moore** (**MtM**) scaling (see the international technology road map for semiconductors or ITRS, the latest of which can be found on their website at www.itrs.net). However, quantum computing by electronic, magnetic, or optical devices is still illusive at the present time and no experimental evidence has shown any competitive advantage in either density or power as compared to the current state-of-the-art CMOS technology.

In many practical aspects the scaling has already stopped. For example, for power consumption, power density cannot be increased much further than about 100 W/cm^2 due to the lack of practical cooling technology. The CPU clock has been stopped at around 4 GHz, and the motherboard clock has been stopped at around 1 GHz, due to switching power limits or long interconnect delays.

Moore's law is essentially an economics principle. By providing more function and higher performance at the same or lower cost, semiconductor industry rides on a virtuous cycle to continue expanding the end use and market, which in turn provides the income for more research and development that drive the next cycle. But leave no doubt, lower cost is the driving force of the IC industry. Improvement in cost per transistor is why the industry strives to adopt new process nodes at ever smaller device size.

From Moore's law, several economic corollaries are observed:

C. 1 ***"Transistors are cheap"*** [1]***—The incremental cost of manufacturing a transistor is essentially zero.***

The first commercial ICs in the early 1960s contained only a few transistors and sold for about $50, at an average cost of about $10 per transistor. We have had 50 years of scaling where the cost of a chip has been kept about the same, but the number of transistors on the chip doubled almost every year, following Moore's law. Nowadays, a chip with over a billion transistors can sell for less than a dollar, such as a gigabyte (GB) NAND flash chip with over 10^9 transistors. Thus, each transistor currently costs less than one billionth of a dollar. Adding hundreds or thousands of transistors to a chip usually does not add meaningful cost to the chip, except for the cost of designing the circuit using these transistors.

C.2 One-time vs. recurring cost trade-off is important.

While adding transistors costs essentially nothing, the cost of design and tooling for manufacturing (principally the cost of a mask set and associated testing setup cost) continues to grow quickly. This one-time, so-called **NRE** (**nonrecurring engineering**) cost is usually very high and is actually rapidly increasing for more advanced processes, reaching beyond millions of dollars. This cost needs to be amortized (averaged) over a large number of chips to make the final cost of each chip economical.

On the other hand, once the chip is designed and the manufacturing tooling is made, the recurring cost of fabricating wafers has been approximately constant, usually a few thousand dollars per wafer for the leading-edge processes. Since each wafer can contain thousands to tens of thousands of dies, the cost of fabricating each die is very inexpensive as long as the die is not excessively large.

[1] YM thank Chris Diorio for this quote and for impressing him this concept throughout his stay at Impinj.

The cost of each die is then,

$$\text{Die cost} = \text{wafer cost/number of sellable die per wafer}$$

The final cost of making each chip includes the cost of the die, cost of testing, packaging, and the amortized development cost. This can be calculated via:

$$\begin{aligned}\text{chip unit cost} = {} & \frac{\text{total NRE cost}}{\text{total number of chips to sell wafer fab cost}} \\ & + \frac{\text{wafer fab cost}}{\text{total yieldable die per wafer}} + \text{test cost per chip} \\ & + \text{packaging cost per chip}\end{aligned}$$

In the previous equation, the total NRE is measured in millions of dollars. Wafer fab cost per wafer is measured in thousands of dollars. Test and packaging cost is on the order of pennies to a few dollars per chip. Total yieldable die per wafer can range from hundreds for large ICs such as microprocessors or large FPGAs, to thousands for small dies such as RFID dies.

The total number of projected chips that can be sold is a strong function of the cost of the chip—more will be sold if the price is lower.

Therefore, the final cost of a chip depends on the trade-off between the nonrecurring and recurring costs in the design and manufacturing processes.

C.3 Process complexity adds cost

The dominant cost of wafer processing is the lithography steps (a recurring cost) and the cost of the mask set that is used to pattern the designs (an NRE item). As a result, process complexity and hence cost of fabricating a wafer is, to the first order, measured by the number of masks in the full process flow. More masks needed to fabricate the chip will translate into higher cost per wafer and higher unit cost for each chip.

The production yield, i.e., the percentage of defect-free chips made over total number of chips produced, also goes down with increasing number of lithography steps and with increased die size.

C.4 Economy of scale drives the use of commodity process technology

The cost of developing a production-grade semiconductor manufacturing process is high to the point where it is becoming nearly prohibitive for most semiconductor companies. For the most advanced processes, the development cost can be billions of dollars. Only a few large semiconductor companies such as Intel, Samsung, and Taiwan Semiconductor Manufacturing Company (TSMC) can afford to develop the 20 nm and beyond today. As a result, the process with the highest production volume will also be the cheapest, as the development cost can be amortized over higher wafer volume. In addition, the learning curve for these processes will go down faster as the production kinks can get ironed out faster with more production experience. The yield will normally be better than processes of equivalent complexity but at lower volume.

The previous corollaries translate into a **best industry practice** (**BIP**):

Focusing IC development effort on using the simplest, most basic, and highest volume process.
BIP.1

Unless the design of the chip requires other choices, the first choice of an IC designer will be to select the basic logic CMOS process as the baseline to start evaluation.

The basic CMOS process comes with its own limitations, however. For simplicity, most processes only offer a basic set of CMOS devices, including p and n MOSFETs. Often IC designs require additional devices. For example, even for a pure logic IC, the main core functional circuit cannot be designed with only the basic devices that are available from the logic process, especially considering interface with the outside world. Often another set of input/output devices will be needed for the I/O circuits. The bonding pads will also need **electrostatic discharge protection** (**ESD**), which often requires passive or bipolar devices.

A straightforward way to satisfy new device demands of the IC designers is to modify the process, a procedure that is often performed. However, process modifications will also come with a cost, often in the process development time, the need for extra masks, and in the cost of fabricating the ICs. As a result of these additions, the cost can often appear to be unacceptable, as the required process modification may be needed for a handful of devices among the millions of other devices. For example, in mixed-signal designs, high-value linear resistors, e.g., 300 kΩ, are often needed. Let's say the typical polysilicon resistor has sheet resistance of 300 Ω/square. Then each resistor will need 1000 squares, which can be a large area. On the other hand, one can modify the process by adding a high-value resistor mask that selectively reduces the doping in the polysilicon layer. This additional resistor will have incremental cost of the additional mask, which can run up to a few percent of the total cost of a wafer.

Thus, our basic Moore's law C.1 has to be modified to be,

C.1a. Incremental cost of manufacturing an additional device is negligible only if a process change is not needed. High cost penalty exists for using devices not available in the baseline process.
Corollary 1a

Thus, there is a strong motivation to use a given process without modification. How do we use the basic CMOS process to get these additional devices, without incurring these costs?

1.3 Aim and Scope of This Book

The aim of this book is to put the earlier economic principle to the practice of designing more functional, non-logic devices in logic CMOS processes.

Logic ICs, including MPUs and FPGAs, and memory ICs, including DRAM and flash memories, are the main drivers for the continued scaling of semiconductors.

Increasingly, more ICs are being fabricated by semiconductor foundries in the model of "horizontal integration" where multiple design companies have choices of multiple manufacturing sites to complete the entire production. This is in contrast with the "vertical integration" where one company starts from product design and ends all the way at in-house manufacturing. From the 1980s, foundries have garnered ever increasing share of the semiconductor fabrication market. Economies of scale dictate that the process offerings from foundries are for the most popular applications. It takes a lot of money, time, and development effort to customize a process. Such customization can only be justified for products that are guaranteed to have a very high volume. Some of these special processes include processes for making DRAM and flash memories. For the overwhelming number of small- to medium-sized fabless IC companies, as well as many university-led prototypes, it is most beneficial to stick to the use of logic CMOS processes for all their designs.

However, for the industry-standard logic CMOS process, the available types of devices are limited to the most common devices, including n- and p-channel MOSFETs. The purpose of this book is then to teach ways to design non-logic CMOS devices in logic CMOS processes and circuits, with the goal of producing specialized circuits in the low-cost CMOS processes.

This book is not a textbook about process technology, nor is it aimed as a comprehensive text on device physics. It touches on both subjects but at shallower levels. What it teaches is the use of standard, lowest cost logic process and innovative device design to obtain nonconventional devices and circuits. We will share innovative practices for using the basic, simplest processes to their maximum.

In the remainder of this chapter, we will survey the historical development of ICs and take a quick look at the state-of-the-art processes. We will then move on to a quick review of the basic CMOS device and the underlying physics in Chap. 2. In Chap. 3, we will review the process of making MOSFETs by going through basic logic CMOS processes, as well as briefly discussing the mask design and some economics of semiconductor manufacturing.

In Part II of this book (Chaps. 4–7), we present some methods for making new, non-logic devices using the basic logic CMOS process, without any process modifications. In particular, Chap. 4 focuses on the art of designing non-logic MOSFETs, including those with nonstandard threshold voltage and high-voltage MOSFETs in the logic CMOS process. In Chap. 5 we review several floating-gate devices. Floating-gate transistors have found widespread use in embedded nonvolatile memory, neuromorphic circuits, and biosensor applications. Chapters 6 and 7 discuss methods for obtaining diodes, varactors, and bipolar transistors in the CMOS process.

In Part III (Chaps. 8–13) we show some examples of applications using the devices showcased in Part II. In particular, in Chaps. 8–10 we will explore in depth how to design and construct several types of nonvolatile memories in the logic CMOS process. In Chap. 8 we will review the basic memory cell design and several methods for adding and removing charges from the floating gate. We then briefly discuss the basic circuit modules for embedded memory, including charge pumps and sense amplifiers. Chapter 9 focuses on one-time programmable (OTP) memories, including fuse, antifuse, and floating-gate OTP memories. Chapter 10 shows several multiple-time programmable (MTP) memories that have seen commercial deployment. In Chap. 11 we discuss some non-data-storage applications of floating-gate transistors, including digital-to-analog converters, RFID, neuromorphic computing, and field programmable gate arrays. Chapter 12 will treat the CMOS biosensors in both Faradic (current sensing) and non-Faradic (field sensing) operations. Examples of isolation and compensation techniques to interface with the biological saline buffers will be presented. One of the main reasons of IC clean-room manufacturing is to eliminate the influence of alkali ions in the semiconductor wafers, but the CMOS biosensors have to be operated with close proximity to this electrolyte environment with high variability. In Chap. 13 we will describe the distributive waveform shaping circuits in high-frequency applications by employing embedded waveguides and varactors. Finally Chap. 14 present a brief overview of recent development in CMOS processing, especially with regard to the emergence of FinFETs and universal substrates, as well as emerging nonvolatile memory technologies which may offer future opportunities.

In this book we only review the essential elements of the CMOS device physics and fabrication technology that are needed to understand the practices discussed in later chapters. The readers are assumed to know the basic CMOS device physics and some knowledge of semiconductor processing, similar to what is covered in an undergraduate level introductory CMOS course. Deeper knowledge of CMOS devices physics will be helpful but not necessary. There are many good CMOS device books that the readers can refer to learn or refresh the basic concepts.

In particular, the following books on semiconductor devices provided good introductory as well as comprehensive coverage:

"The Physics of Semiconductor Devices" by Simon Sze (Sze, 1981)—This is not a textbook but rather a comprehensive reference book. This is almost regarded as the "Bible" of virtually all semiconductor device physics, from p–n junctions, bipolar and FET transistors, to microwave devices, image sensors, and solar cells. The third edition, published in 2007, is completely updated with the addition of a second author, K.K. Ng.

"Fundamentals of Modern VLSI Devices" by Y. Taur and T. K. Ning (1998)—With in-depth coverage of MOSFET and Bipolar transistor devices, this is an upper undergraduate or a graduate level textbook on device physics. The second version has been published in 2013 with updates on recent device technology.

"Modern Semiconductor Devices for Integrated Circuits" by C. Hu (2009)—This is a textbook with a good collection of problems for an introductory device physics course.

"CMOS VLSI Design: A circuits and systems perspective" (Weste & Harris, 2011)—This book has broad coverage on both semiconductor devices, process, and circuits.

"Silicon VLSI Technology: Fundamentals, Practice and Modeling" by J. D. Plummer, M. Deal and P. D. Griffin (Plummer, 2001)—This is a standard textbook to teach the unit semiconductor processes and their integration. A revision is available in 2009.

"Fabrication Engineering at the Micro-and Nanoscale" by S. A. Campbell (2013)—the fourth Edition was published in 2013. This textbook covers the similar content with the Plummer's book but has more details on the equipment design and control parameters.

"The VLSI Handbook" (Chen, 2010)—This is a comprehensive handbook on all aspects of integrated circuits, including chapters on CMOS processing and flash memory.

On a historical note, the bipolar transistor was invented by Bardeen, Brattain and Shockley at Bell Labs in 1947. The initial invention was for a Ge transistor but it was soon realized that Si makes a better substrate in terms of process integration. Si has therefore taken over ever since the late 1950s. But what goes around often comes around; more recently, MOSFET with Ge channel on Si substrate has gained a lot of attention due to the superior hole mobility of Ge over Si.

In 1958, the integrated circuit was invented, principally by Kilby at Texas Instruments. See Sidebar on the history of IC patents and monetization of semiconductor patents. At nearly the same time, planar IC technology was invented by Noyce et al. at Fairchild Semiconductors. This invention arguably had more impact than the Kilby invention because the planar version of the IC was more adaptable to mass production.

For a long time, Si bipolar transistors dominated the IC industry, before FETs took over from 1980s. It is worth noting that the FET was actually conceived and patented much earlier than the bipolar transistors, all the way back in 1925 by physicist Lilienfeld. However, it was not until 1960 that the first MOSFET was actually made by Kahng at Bell Labs. Complementary MOSFET (CMOS) technology was invented by Wanlass and Sah at Fairchild in 1963. Due to the low power nature of the CMOS technology, CMOS circuits were initially used in low power applications such as digital watch. Only since the late 1980s has CMOS took over and become the dominant semiconductor technology, due to its advantage of superior power consumption and allowance of tight device pitch.

On a side note, the floating-gate memory was invented by Kahng and Sze at Bell labs in 1967. This invention became the foundation to the large class of solid-state memory that gives nonvolatile storage—in particular, flash memory and solid-state disk (SSD) that have found ubiquitous applications from consumer smartphones to hyper-scale data centers for cloud computing.

In 1965, Moore observed that the number of transistors in Fairchild's ICs doubled almost every year. In 1972 this so-called Moore's law was revised to the doubling of transistor counts every 2 years. Moore predicted that the trend would continue. Indeed, Moore's law has been proven true for 50 years and is still going as of this writing (2016)!

Once accepted by the semiconductor industry, Moore's law became a self-fulfilling prophecy and emerged as one of the driving principles of the semiconductor industry. Technology roadmaps were generated according to Moore's law and technology developers were challenged to deliver progress following the roadmap.

In 1972, Dennard at IBM proposed a set of rules for semiconductor device scaling which formally provided the physical foundation for the Moore's law—as device size scales down, the IC performance scales up while its cost and power consumption goes down, opening up more applications. As ICs found uses in more applications, semiconductor industry revenue grows and provides more funding to the next new innovations, thus ensuring the IC industry to remain in a virtuous cycle.

As discussed in the previous sections, after more than 50 years of Moore's law and scaling of device size to ever smaller dimensions, Moore's law has slowed down in recent years. The feature size of semiconductor devices is expected to hit the physical limit in less than 10 years. If the history is any guide, the focus of the semiconductor industry will soon be on more innovative designs and on ways to increasing the efficiency to reduce the cost of manufacturing, just like what has happened to a number of other industries, for example, the automotive industry, as the market saturates. Until another technology can be found, such as one of the numerous More than Moore (MtM) technologies that are currently being investigated, focus will be on squeezing any manufacturing efficiency possible out of IC design and manufacturing processes. The materials in the rest of this book are meant to teach some of the principles for achieving this increased efficiency.

Patent History and Patent Wars Involving the Invention of IC

The invention of integrated circuits resulted from the work of many pioneers of semiconductors. It is generally accepted that Jack Kilby of Texas Instrument was the first to propose integrating different circuit components in an integrated circuit. Indeed, Kilby was awarded the Nobel Prize for Physics in 2000 for this achievement.

Kilby obtained many patents related to the invention of the ICs, including his first U.S. Patent 3,138,743, titled "miniaturized electronic circuits," which was filed on Feb 6, 1959 and granted June 23, 1964. The patent states:

> "This invention relates to miniature electronic circuits, and more particularly to unique integrated electronic circuits fabricated from semiconductor material..." and further states,
>
> "The present invention has resulted from a new and totally different concept for miniaturization. Radically departing from the teachings of the art, it is proposed by the invention that miniaturization can best be obtained by use of as few materials and operations as possible. In accordance with the principles of the invention, the ultimate in circuit miniaturization is attained using only one material for all circuit elements"

The patent claims:

> 1. In an integrated circuit having a plurality of electrical circuit components in a wafer of single-crystal semiconductor material, a plurality of junction transistors defined in a wafer ....

However, Kilby's patents formed only part of the picture for making an IC. Kilby demonstrated that various circuit elements such as active components, resistors, and capacitors can be integrated on a single chip. But none of his patents dealt with isolation and interconnection—the components were separated by cutting grooves on the chip and connected by gold wires. Kilby mentioned the use of metallization layer. However, Kilby favored thick coating layers of different metals (aluminum, copper, or antimony-doped gold) and silicon monoxide instead of the dioxide. These ideas were not adopted in the commercial production of ICs.

This part of the work was done by Robert Noyce in 1959, then at Fairchild Semiconductors, solved another important problem, the problem of interconnections that hindered mass production of ICs. Noyce submitted his application on July 30, 1959, and on April 25, 1961 received U.S. Patent 2,981,877. According to the patent, the invention consisted of preserving the oxide layer, which separated the metallization layer from the chip (except for the contact window areas), and of depositing the metal layer so that it is firmly attached to the oxide. The deposition method proposed by Noyce included vacuum deposition of aluminum through a mask and deposition of a continuous layer, followed by photolithography and etching off the excess metal. Noyce has been credited as the coinventor of IC. Were not for the premature death of Noyce in 1990, he would likely have shared the Nobel prize with Kilby.

The claim of Noyce patent included:

> 1. A semiconductor device comprising a body of semiconductor having a surface, said body containing adjacent P-type and N-type regions with a junction there between extending to said surface, two closely spaced contacts adherent to said surface upon opposite sides of and adjacent to one portion of said junction, an insulating layer consisting essentially of oxide of said semiconductor on and adherent to said surface, said layer extending across a different portion of said junction, and an electrical connection to one of said contacts comprising a conductor adherent to said layer, said conductor extending from said one contact over said layer across said different portion of the junction, thereby providing electrical connections to both of the closely spaced contacts.

Note that this description is more nuanced and showed difference with the Kilby patent.

In addition to Kilby and Noyce, many others made contributions that were essential to the development of IC, including J.A. Hoerni who invented the planar manufacturing process, US Patent No. 3,025,589 "Method of Manufacturing Semiconductor Devices" filed May 1, 1959 and K. Lehovec who invented the concept of p–n junction isolation.

These ideas turned the relatively crude ideas of Kilby into reliable manufacturing processes what were used to produce trillions of ICs since.

The invention of the IC not only brought fame to Kilby but also brought substantial patent-related royalty income to his employer, Texas Instruments.

It is worth noting the invention of ICs and the patent disputes are made possible because Bell Labs decided to license the basic transistor invention and manufacturing technology to all. It is believed that Bell Labs did not anticipate the wide uses of transistors, and at that time was not interested in its consumer applications. As a result, anybody could license the transistor technology for a small sum of $25,000 and with no royalties.

The invention of p–n junction isolation is another case where substantial inventorship dispute occurred. Lehovec's 1962 patent was challenged by TI, which claimed that the p–n junction isolation was anticipated by one of Kilby's patents. However, Lehovec conclusively proved that Kilby did not mention isolation of components. His priority on the isolation patent was finally acknowledged in April 1966.

Because of growth of semiconductor, the monetization of the early IC patents has been notable. This in large part is due to the aggressive litigation strategy employed by TI. Since TI began aggressively asserting its patent rights in early 1980s, the semiconductor manufacturer has collected many billion dollars in royalty payments.

An example is Japan v. Texas Instruments. For almost 30 years, Japanese companies were producing ICs without paying royalties to Texas Instruments, and only in 1989 the Japanese court acknowledged the patent rights to the invention by Kilby. As a result, in the 1990s, all of Japanese IC manufacturers had to pay for the 30 years old patent or enter into cross-licensing agreements.

As an example, as reported in EE Times ("TI, Toshiba pact includes Kilby patents," Electronic Engineering Times, 17 December 1990), Toshiba and TI signed 10-year cross licensing agreement, plus Toshiba will pay royalty on a percentage of its worldwide sales, upward of $75 million a year for the next 10 years.

More history of the invention of IC can be found in, for example, "History of Semiconductor Engineering" (Lojek, 2007) and "Makers of the Microchip: A Documentary History of Fairchild Semiconductor" (Lecuyer & Brock, 2010).

References

Campbell, S. (2013). *Fabrication engineering at the micro-and nanoscale. Oxford: Oxford Press.*

Chen, W. (2010). *The VLSI handbook*. Boca Raton: CRC.

Hu, C. (2009). *Modern semiconductor devices for integrated circuits*. Englewood Cliffs: Prentice Hall.

Lecuyer, C., & Brock, D. (2010). *Makers of the microchip: A documentary history of fairchild semiconductor*. Cambridge: MIT Press.

Lojek, B. (2007). *History of semiconductor engineering*. New York: Springer.

Plummer, J.D., Deal, M., and Griffin, P.D. (2001). *Silicon VLSI technology: Fundamentals, practice and modeling. NJ: Prentice Hall.*

Simonite, T. (2016, March 24). *Intel puts the brakes on Moore's law*. Retrieved from https://www.technologyreview.com/s/601102/intel-puts-the-brakes-on-moores-law/#/set/id/601103/.

Sze, S. (1981). *Physics of semiconductor devices* (2nd ed.). Amsterdam: Wiley.

Taur, Y., & Ning, T. H. (1998). *Fundamentals of moden VLSI devices*. Cambridge: Cambridge University Press.

Weste, N., & Harris, D. (2011). *CMOS VLSI design: A circuits and systems perspective*. Boston: Addison-Wesley.

Chapter 2
Overview of Logic CMOS Devices

Abstract In this chapter, we provide a brief review of basic CMOS devices and device physics. We start by walking through the basic devices that are available in generic CMOS processes, covering rudimentary device physics that are of particular interests to the central theme of this book (including p–n junctions and gate oxide breakdown) and their effects on the design rules of non-logic active and passive devices. As power consumption has become the main barrier for IC design and applications, for example, in wearable ICs, it is useful to always keep in mind the fundamental factors affecting CMOS power consumption. We conclude with a brief analysis of power consumption of CMOS ICs.

Keywords CMOS • Scaling • Diodes • Field effect transistors • Bipolar junction transistors • MOSFET • MOSCAP • Hot carrier effect • Short channel effect • Threshold voltage • Inversion • Accumulation • Schottky diodes • Tunneling • Fowler–Nordheim tunneling • P–N junction • Junction breakdown • Zener Breakdown • Inverter • Leakage power • Dynamic power consumption

2.1 Basic CMOS Devices and Processes

ICs are fabricated using process technologies that have evolved over the last ~60 years. In Fig. 2.1 we show the relationship between the basic CMOS processes used for producing logic circuits and a number of derivative processes used to produce more specialized ICs.

We start with a generic logic CMOS process that makes basic n- and p-channel MOSFETs. This basic process is used to produce a large category of ICs—logic ICs including MPUs and FPGAs, as well as static random access memory (SRAM). This is the largest segment of the IC market, making up roughly half of the total semiconductors sold. It is also the most profitable market segment, thanks largely to the special functions that these ICs can perform. As a result, the logic CMOS process was usually the most advanced process to develop until around 2005. Since usually CMOS logic process is also the simplest, it is most often the lowest cost baseline process from which other more complex processes are built on for a given technology node. However, the design and verification process of such logic ICs can be long due to the functional complexity. Recently, flash memories have been at

Y. Ma, E. Kan, *Non-logic Devices in Logic Processes*,
DOI 10.1007/978-3-319-48339-9_2

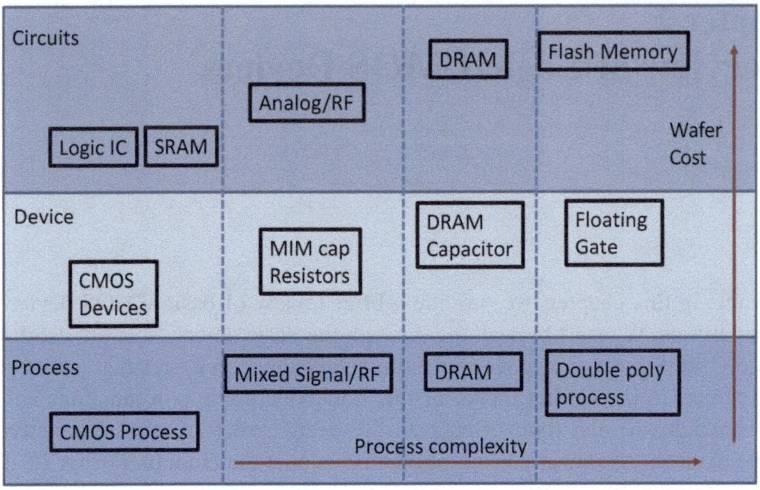

Fig. 2.1 Major silicon-based processes, devices, and IC categories. The cost indicated is the cost of processing one wafer

the technology development forefront, due to its regular array architecture that enables both short design/verification time and aggressive lithography such as double patterning.

By adding a few process steps to the baseline logic process for making—for example, metal–insulator–metal (MIM) capacitors, high-value resistors, and deep n-well for substrate isolation—analog, mixed signal, and radiofrequency ICs can be fabricated. Due to these extra processing steps, analog and mixed-signal processes are generally somewhat more expensive than the underlying logic CMOS process. Analog and mixed-signal ICs occupy a major market segment with about $60B annual sales in 2015.

By adding process steps for making trench or overlay capacitors, the baseline CMOS process can be modified to make dynamic random access memory (DRAM). DRAM market is also worth about $50B in annual sales in 2015. In reality, DRAM market is a commodity market with typically low profit margins. Extensive modifications of the process are done to obtain optimized DRAM cells to achieve higher packing density and lower cost. As a result, dedicated DRAM processes are fairly different from the baseline CMOS process.

Through process modifications, including adding a second polysilicon layer and adding process steps for making high-voltage transistors and floating-gate transistors, flash memory ICs can be made. The flash memory market is another ~$40B market in 2015.

Note that additional process steps add process complexity and cost to wafer processing, as indicated in Fig. 2.2. In addition, if the process involves making many types of devices it is often difficult to optimize each device that is important to the function of the IC. For example, in making high-voltage transistors for flash memory, we may need long annealing steps which can degrade the performance of

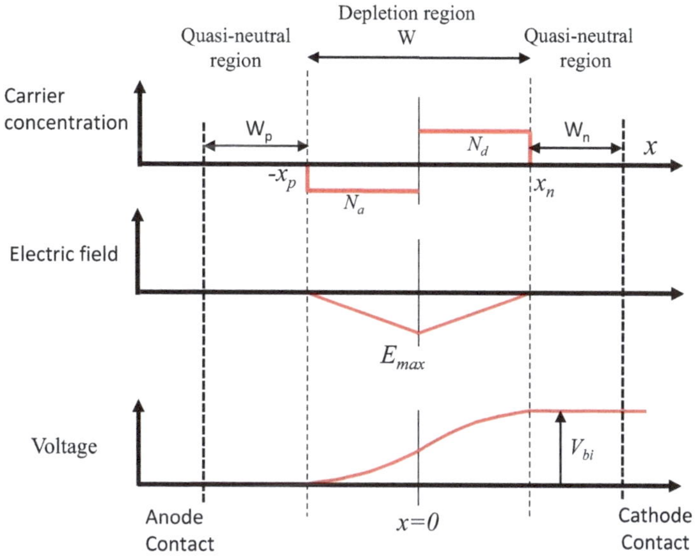

Fig. 2.2 Charge concentration (*top*), electric field (*middle*), and potential energy of a p–n junction in equilibrium

other transistors, e.g., transistors needed for logic circuits. Often, trade-offs are needed to reach the final design and process technology choices.

This is the reason dedicated processes are only developed for products that have large markets, e.g., DRAM and flash memories. By optimizing the design and processing of floating-gate transistors, for example, extremely high packing density can be achieved to create NAND flash products that are much less expensive on a per bit (or gigabyte) basis than SRAM in the logic ICs.

Similarly, for logic IC, by focusing on only a few devices, process integration and device engineers are able to optimize the processes to achieve unprecedented yield, efficiency, and total cost. This is one of the reasons that CMOS ICs are able to achieve the economy of scales and continue driving down the cost, which in turn broadens the application of digital circuits.

Integrating some or all of the earlier components, one can make a sophisticated system on a chip (SoC) that integrates the computing, storage, and communication functions on a single chip. With integration, the whole system can be smaller, cheaper, more reliable, and cost less energy. Because logic circuits, as exemplified by the MPUs, are usually the brains, most SoCs are designed around a processor and built from the base of the logic process with the necessary modifications or added process steps.

In the remainder of this section, we briefly review devices that are available from the basic CMOS process. Then in the last sections of this chapter we will review in more details the device physics of the most important devices.

2.1.1 Diodes

Diodes are two-terminal nonlinear devices and are found almost everywhere in an IC. They are probably the most numerous semiconductor devices that man has created, certainly more than MOSFETs in numbers. Some diodes are integral parts of the MOSFETs while other dedicated diodes are used in the analog functions, such as the ESD protection circuits in the bonding pads and varactors in waveform-shaping circuits.

All MOSFETs have at least two diodes, source-to-well and drain-to-well diodes. For MOSFETs built on bulk p-substrate, we also usually have the n-well to p-substrate diode, and the p-well to n-well diode. These are all the usual p–n diodes and produced during fabrication. Each bipolar transistor also has at least an emitter-to-base and a base-to-collector diode.

Parasitic diodes are also found everywhere in an IC. They can cause leakage which is one of the most important leakage mechanisms in ICs and directly contribute to the static power consumption.

Additionally, parasitic capacitances associated with the p–n junctions contribute to the active switching power consumption of ICs. Great engineering efforts have been expended on the optimization of diode design in the IC process development.

Breakdown of diodes often limits the maximum voltage that can be applied to a device. This point is very important to the operation of high-voltage circuits and the embedded nonvolatile memory circuits that we will go in depth in Chaps. 8–12.

2.1.2 Capacitors

The capacitor is another two-terminal device that is a fundamental part of the transistors. For example, the metal–oxide–semiconductor capacitor (MOSCAP) plays the most crucial function of the MOSFET. In DRAM circuits, we employ a capacitor to store the information. Actually, whether with intentional and unintentional capacitors, all information in present CMOS is related to the voltage or charge on a capacitor. Capacitors are also important in assuring stable power supplies during transient circuit functions. For example, we add decoupling capacitors to isolate the coupling between power lines of two adjacent modules.

Some capacitive structures are more parasitic in nature and could affect the speed and power consumption of circuits, so they are mostly an annoyance and should be minimized. These include the junction capacitance of source–drain junctions to the wells, parasitic capacitance from gate to source–drain not related to the channel, and well-to-substrate junction capacitance mentioned earlier.

On the other hand, junction capacitance can be used to make variable capacitors (variable capacitor diode or "varactor") that can be controlled by voltages. Varactors have found use in many analog tuning circuits, including voltage-controlled oscillators in phase-locked loops which are fundamental blocks in channel selection of communication systems. Due to their ultra-high-frequency responses, varactors are also used in many waveform-shaping circuits.

Another source of parasitic capacitances is the intermetal capacitance. The wire-to-wire capacitance (interconnect capacitance) causes wiring delays and crosstalks, which is one of the main limitations for the switching speed of logic processor. The power needed to drive those wire capacitances is a major contribution to the dynamic power consumption for an IC. The need to reduce the interconnect capacitance limits the spacing between metal wires and is a critical limitation in the scaling of ICs to smaller dimensions. The capacitance across two adjacent metals is determined by **intermetal dielectric** (**IMD**), and in recent technologies, low k dielectric with relative permittivity less than 2 has to be employed to minimize the capacitive coupling. The global interconnect concerns will be further treated in Chap. 13.

For functional circuit components, it is possible to make **metal–oxide–metal** (**MOM**) capacitors from IMD. Many processes have an add-on option for an optimized **metal–insulator–metal** (**MIM**) capacitor with additional processing steps. The capacitance per unit area is often higher than MOM capacitors, and both terminals can be at selected potentials. For a logic-only process, it is also common to build capacitors by interdigitated fingers of metal layers, including vias, where the capacitance is further enhanced from fringing fields.

Capacitors are used extensively in analog ICs as well. The important design factors are the capacitance per unit area, leakage, cost, temperature dependence, grounded terminals, linearity or voltage dependence, quality factor, and cutoff frequency. For example, MOSCAP discussed in later sections generally have a high unit capacitance but also a high voltage dependence, i.e., nonlinearity. Gate and trench capacitors often assume one terminal is at ground potentials and cannot be used in the switching capacitor for charge pumps.

As the semiconductor feature size continues to scale down and devices become more densely packed, parasitic capacitances become more and more important to the point where they become the limiting factor of floating-gate transistor scaling for flash memory. Here the capacitive coupling between floating gates in NAND flash memory becomes so large that further scaling cannot be done below about 10–15 nm for planar NAND flash chips. Fortunately, engineers found ways to pack transistors in three dimensions to continue the overall scaling trend (# of devices per chip) for NAND flash memory.

In Sect. 2.5 we discuss in more details the physics of MOSCAPs.

2.1.3 Resistors

Resistors are less commonly used in CMOS digital circuits. The parasitic resistance due to the metal interconnect is a source of wiring delay, waveform distortion, and power consumption. The desire to reduce the wiring resistance is the reason metal interconnect has moved from aluminum to copper, starting around 1997.

A resistor is often needed in power management, ESD protection, and mixed-signal circuits, including most current reference circuits. Factors important for these applications include sheet resistance, temperature dependence, voltage dependence, mismatch, parasitic capacitance, and inductance. The most commonly used resistors in ICs are polysilicon resistors built by a section of the polysilicon. There are silicided polysilicon resistors with sheet resistance on the order of 1–2 Ω/square as well as unsilicided, either p- or n-doped, polysilicon resistors with sheet resistance of about 200 Ω/square.

Temperature coefficients of polysilicon resistors can be either positive, where resistivity increases with temperature, or negative, where resistivity decreases with temperature. The sign of the temperature coefficient is useful in stabilization and compensation circuits to prevent thermal runaway or unstable thermal variations.

Another type of resistors is diffusion resistors, e.g., n-well resistors, which have sheet resistance value of about 1 kΩ/square. However, these well resistors usually have high temperature coefficients, high-voltage dependence, and serious mismatch as well as high parasitic capacitance due to the junction with the substrate, so they are less commonly used when appropriate poly resistors are available.

2.1.4 Bipolar Transistors

Bipolar junction transistors (BJT) were the first transistors that saw commercial use. They had been the workhorse of semiconductor industry before being overtaken by FETs. They are fast but generally consume more static power than MOSFETs, especially when bipolar circuits are compared to pure CMOS circuits. The main difference lies in the base current in the steady state, which is often used as controls. For FET in the steady state, the gate input current is negligible, and the complementary circuits stop most of the output current as well. However, for BJT to maintain the same state, the base input current is often sizable. When FET uses very thin gate oxide with significant gate leakage current, it will start becoming similar to BJT in terms of transconductance and static power consumption.

Two types of BJTs, npn and pnp, are still found in CMOS chips. Both types are parasitic devices that cause problems such as latchup, to the operations of CMOS chips. However, they do often find dedicated use. One particular use is in ESD protection circuits of the wiring pads. Another important circuit that uses the parasitic bipolar transistors is the bandgap reference circuit often found in CMOS ICs. We will discuss some examples in Chap. 6.

2.1.5 MOSFETs

MOSFETs are the workhorse in a majority of modern ICs. In logic CMOS process we usually use only two types of MOS transistors, n- and p-type enhancement devices, which form complementary pairs for CMOS circuits.

There are a large number of good review articles and books dedicated to the description, fabrication, and use of MOSFETs. We will not go in much detail here but in Sect. 2.4 we will review the essential physics of both n and p MOSFETs. We will highlight a few new features introduced into aggressively scaled MOSFET, especially in high-*k* gate dielectrics and FinFET, and their implications in constructing other non-logic devices. In Chap. 3 we will illustrate the basic processing steps for making both types of MOSFETs. This is also to enable the readers to understand the design opportunities in constructing the non-logic devices during the same MOSFET fabrication process by changing only the layout.

In the next sections, we will review the physics of some of these devices that are important to the subsequent chapters.

2.2 Physics of Diodes

2.2.1 The p–n Junction Diode

The p–n junction diodes and MOS capacitors (MOSCAP) are the essential building blocks for a MOSFET. In this section, we discuss the physics of p–n junctions while in the next section we discuss MOSCAP.

The p–n junction was discovered accidentally by Ohl at Bell Labs in 1939, through the observation of photocurrent current in a piece of cracked silicon crystal. A p–n junction can be formed by dopant diffusion or by ion implant. All p–n diodes in the current CMOS process are formed by implant of n (or p) type of dopants into a region of p (or n) region.

In the junction region, the mobile majority carriers, holes from the p-region, and electrons from the n-region will initially (and very quickly) diffuse into their respective regions of lower concentrations where they are minority carriers, i.e., the region of the opposite type. The carriers will recombine and reach equilibrium with no further carrier flow. The loss of an electron from the n-type region creates a positive dopant ion in the n-region, while the loss of a hole (capture of an electron) from the p-region creates a negative dopant ion in the p-region. These ions are fixed in place in the crystal lattice structure and cannot move at the usual operating temperature. Thus, they make up a layer of fixed charges on the two sides of the junction and form an electric field that will oppose further movement of the carriers. The diffusion of electrons and holes across the junction will continue until the magnitude of the electrostatic field increases to the point where the electrons and holes no longer have enough energy to overcome this electrostatic barrier. At equilibrium, when there is no external applied voltage across the junctions, there will be no current flow. A potential difference will form between the two sides. This potential difference is called the built-in or barrier potential of this p–n junction.

There will be a region around the p–n junction that has much less majority carriers. This region is called the space-charge region or the depletion region, as most of the net charges are fixed charges on the atoms that either donated or accepted an electron. The location inside the depletion region where the net charge changes sign is called the "metallurgical junction," where the electric field reaches its largest magnitude.

Figure 2.2 illustrates the carrier concentration, electric field, and potential at equilibrium for a p–n junction in a one-dimensional model. In this junction, the n and p sides are assumed to be uniformly doped with donor concentration, N_{d}, and acceptor concentration, N_{a}, respectively. The top panel in Fig. 2.2 illustrates the net charge along the direction perpendicular to the junction. The middle panel shows the internal electric field under equilibrium. The lower panel shows the electrical potential along the direction perpendicular to the junction.

The built-in potential under nondegenerate doping assumption is given by:

$$V_{\mathrm{bi}} = \frac{k_{\mathrm{B}}T}{q}\ln\left(\frac{N_{\mathrm{a}}N_{\mathrm{d}}}{n_{\mathrm{i}}^2}\right) \tag{2.1}$$

where T is the temperature measured in Kelvin, k_{B} is the Boltzmann constant, and n_{i} is the intrinsic carrier density and is ~$10^{10}/\mathrm{cm}^3$ at room temperature.

The built-in potential is usually around 1 V for the silicon p–n junction that is supported by the charge dipole through the Poisson equation.

The total width of the space charge region is given by

$$W = \sqrt{\frac{2\epsilon_{\mathrm{s}}(N_{\mathrm{a}}+N_{\mathrm{d}})V_{\mathrm{bi}}}{qN_{\mathrm{a}}N_{\mathrm{d}}}} = x_{\mathrm{n}} + x_{\mathrm{p}} \tag{2.2}$$

where ϵ_{s} is the dielectric constant of silicon. The depletion width on each side of the junction is given by

$$x_{\mathrm{p}} = W\frac{N_{\mathrm{d}}}{N_{\mathrm{a}}+N_{\mathrm{d}}}, \quad x_{\mathrm{n}} = W\frac{N_{\mathrm{a}}}{N_{\mathrm{a}}+N_{\mathrm{d}}} \tag{2.3}$$

It is clear from these equations that the width of the space charge region is dominated by the low doping side of the junction. For example, for a typical p-type source–drain doping level of $10^{20}/\mathrm{cm}^3$ in an n-well of $10^{17}/\mathrm{cm}^3$, the depletion region is mostly in the n-well region, as $x_{\mathrm{p}} = 10^{-4}$ μm is much smaller than the n-type depletion region, $x_{\mathrm{n}} = 0.12$ μm. In this case, the junction is very much a one-sided junction.

Under applied reverse-bias voltage, i.e., when V_{D} in the following equations is negative, the depletion width increases.

$$x_p = \sqrt{\frac{2\epsilon_s}{q}\frac{N_d}{N_a}\frac{1}{N_a + N_d}(\varphi_i - V_D)} \tag{2.4}$$

$$x_n = \sqrt{\frac{2\epsilon_s}{q}\frac{N_a}{N_d}\frac{1}{N_d}(\varphi_i - V_D)} \tag{2.5}$$

The peak electric field is given by

$$E_{max} = E(x = 0) = -\frac{qN_a x_p}{\epsilon_s} = -\frac{qN_d x_n}{\epsilon_s} \tag{2.6}$$

and will also increase with the magnitude of the reverse-bias voltage.

The diode junction has net charge in the space charge layer, in the reverse-bias condition, thus will have a characteristic capacitance called junction capacitance or depletion capacitance. The junction capacitance C_j can be approximated by the parallel-plate capacitor separated by $x_n + x_p$:

$$C_j = \frac{\epsilon_s A}{W} = A\sqrt{\left(\frac{\epsilon_s q}{2}\right)\left(\frac{N_a N_d}{N_a + N_d}\right)\left(\frac{1}{V_{bi}}\right)} \Big/ \sqrt{1 + \frac{V_D}{V_{bi}}} \tag{2.7}$$

where A is the cross section area of the diode.

From Eq. 2.7, the junction capacitance C_j is dominated by the lightly doped side of the junction and has a voltage dependence through V_D, which is called a variable capacitor or varactor. Notice that whenever the capacitance value depends on its bias, we call it a "nonlinear capacitance" in a varactor, even though the functional dependence on voltage is linear. When V_D changes, C_j changes almost instantaneously (on the order of picoseconds governed by the dielectric relaxation time in doped silicon), because only the majority carrier needs to adjust its position for the changes in x_n and x_p. Also, C_j is the dominant capacitance only in the reverse-bias condition (before breakdown). In the forward-bias condition, there is a much more slowly changing diffusion capacitance due to the minority carrier injection. Varactors are important components in RF channel frequency selection and waveform-shaping circuits.

As diodes are found extensively in CMOS circuits, most of the time parasitic to the active circuit elements, C_j has a big impact on the switching speed of circuits. Most parasitic junction diodes in CMOS remain reverse biased in all operations to avoid large forward-bias current and relatively slow turn-off diode transient. A lot of process development and design efforts are spent trying to reduce the parasitic capacitance associated with various junction diodes.

2.2.2 Current–Voltage Characteristics

At a given applied bias voltage, V_D, the diode current is given by

$$I = I_0\left(\exp\left(\frac{eV_D}{kT}\right) - 1\right) \tag{2.8}$$

where I_0 is given by

$$I_0 = \begin{cases} eAn_i^2\left(\dfrac{D_p}{N_d L_p} + \dfrac{D_n}{N_a L_n}\right) & \text{for diodes with long base} \\ eAn_i^2\left(\dfrac{D_p}{N_d W_n} + \dfrac{D_n}{N_a W_p}\right) & \text{for diodes with short base} \end{cases} \tag{2.9}$$

where A is the cross section area of the diode and D_n and D_p are the diffusion coefficient of the electrons and holes, respectively. L_n (L_p) is electron (hole) diffusion length in the p-region (n-region) and is related to the diffusion coefficient and the carrier recombination life time, τ_n, by

$$L_n = \sqrt{\tau_n D_n} \tag{2.10}$$

The long- and short-base cases are distinguished by the diffusion length relative to the length of the quasi-neutral region in Fig. 2.2, e.g., $W_n \gg L_n$ for long-base diodes or $W_n \ll L_n$ for the short-base diodes. The recombination lifetimes τ_n and τ_p (on the order of μs) and the diffusion lengths L_p and L_n (on the order of tens of μm) are relatively long in silicon, which is a good material quality governed by the mid-bandgap traps to give low leakage junction current, as can be seen from Eq. 2.9. Therefore, the short-base diode is the more common case in CMOS IC. Also, when the minority carrier is given enough space to relax to its equilibrium values after a few L_n, all bipolar junction effects disappear. Therefore, the base of BJT is always in a short base condition to achieve the forward-active region of current amplification.

The reverse leakage current is approximately $-I_0$ and is very small for a normal junction. This property makes p–n junctions very useful for use as isolation between devices, particularly in early IC development. See the side note on the invention of the p–n junction for isolation. A typical diode I–V curve is shown in Fig. 2.3. In the forward-biased direction, the current starts to increase exponentially when the applied bias V_D is higher than $3k_BT/e$, or about 78 meV at room temperature, and will enter a resistive region at about the built-in voltage V_{bi}. In the reverse direction, the current increases very slowly until the breakdown voltage. We will discuss the breakdown phenomena in the next section.

In addition to the exponential bias dependence, Eq. 2.9 also presents strong temperature dependence of the diode current through n_i^2, which carries an exponential dependence on temperature. From 0 to 70 °C by the commercial

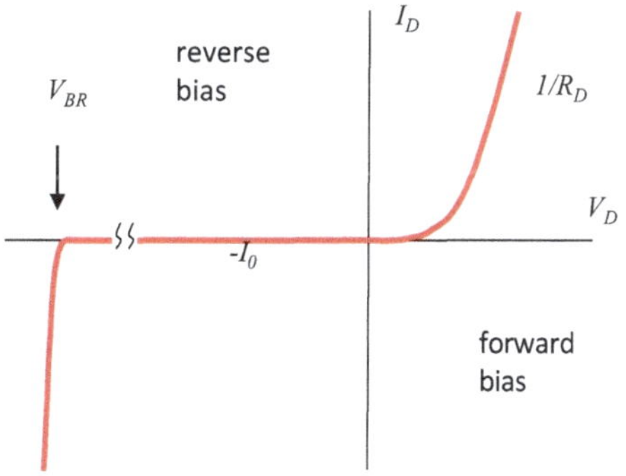

Fig. 2.3 A typical diode I–V curve. The diode I–V in the linear scale with consideration of the series resistance (RD) and junction breakdown

recommended operating conditions (ROC), n_i^2 can change from 7×10^{17} cm^{-6} to 5×10^{22} cm^{-6}, which is almost 5 orders of magnitude. Note that the industrial and military ROC can be even broader in ambient conditions. The subthreshold current in MOSFET and all operational regions of BJT have similar variations. These have two important implications: (1) We will need to carefully consider the temperature effect of the leakage current which has large variations in CMOS and current-controlled circuits in the design process; (2) the diode current has a strong positive temperature coefficient (i.e., the current increases when temperature increases), and when self-heating is significant, thermal stabilization to prevent runaway is an important design consideration, especially for non-logic devices.

2.2.3 *Junction Breakdowns*

As we saw in the previous section in Fig. 2.3 and Eq. 2.6, there is a built-in voltage V_{bi} and electric field across the depletion junction. The electric field increases with the magnitude of applied reverse-biased voltage. When the peak electric field or the depletion region at the reverse bias exceeds certain threshold, we expect the diode to breakdown electrically. If the current is limited, the breakdown is reversible and can be controlled. If current is allowed to flow freely, this breakdown can also lead to irreversible physical damage in the junction region due to thermal melts or metal migration, a scene commonly seen in ESD events.

There are three principal junction breakdown mechanisms, all under the reverse-bias condition. The I–V curve in Fig. 2.4 will be governed by whichever mechanism happens first.

1. As an example, Fig. 2.4 shows a cross section of a pFET in an n-well. The source–drain to n-well junctions can be close to the well edge, which is a junction of n-well to p-substrate. Under normal operations, the n-well may be biased at 3.3 V and the drain may be biased at ground, reverse biasing the p-type drain to n-well junction. Since n-well is much more lightly doped than the p+ drain, most of the depletion region is in the n-well. If the drain is close to the

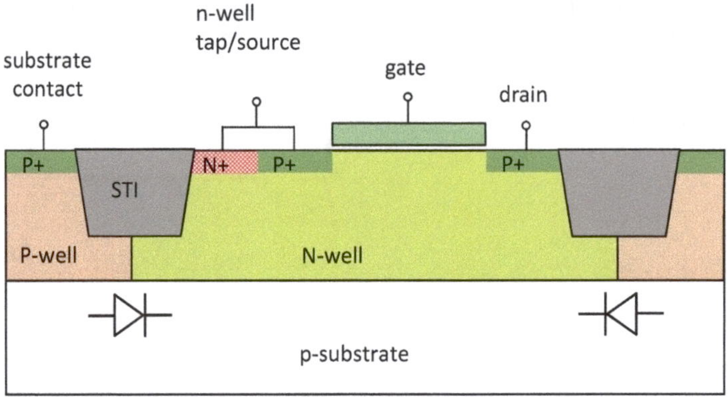

Fig. 2.4 Cross section of pFET illustrating two diodes that pose a reach through danger

n-well edge, the depletion region can reach the neighbor p-well, causing leakage between the drain and the substrate. Punchthrough to the metal contact fortunately does not usually happen in CMOS when the metal contact is always on a heavily doped region. If we construct a "Schottky" contact (to be introduced later) in silicon technology (more discussion in Sect. 7.3), care has to be taken for the punchthrough current over the reverse-bias leakage current.

Punchthrough can also happen between the two basic junctions in MOSFET and BJT. In Fig. 2.4, this will correspond to the reaching through of the source-to-n-well junction and the drain-to-n-well junction. This is often called drain-induced barrier lowering (DIBL) when the substrate doping cannot stop the extension of the source–well and drain–well junctions. In BJT, when the quasi-neural base diminishes by the extended emitter–base and base–collector junctions, punchthrough also happens, where the original "saturation" current in the active region will start increasing quickly with respect to the increase in the collector bias.

This diode punchthrough effect is one of the most important mechanisms that demands a minimum spacing between the various p–n junctions found in an IC layout. This particular restriction is one of the many **design rules** (**DR**s) in modern CMOS processes. As expected, the specific value for this particular design rule will depend on the operating voltage and the doping levels. A p-type drain that can go up to 3 V needs to be further spaced from the n-well than a drain that will only go up to 1 V.

2. If a junction is far from another junction or when the doping concentrations on both sides are higher, ***avalanche breakdown*** may occur before punchthrough can occur. This usually happens at higher voltages where electrons or holes are generated in the depletion region and are accelerated by the electric field. When the electrons or holes gain enough energy they will cause impact ionization where electron–hole pairs are created, e.g., by the breaking of a covalent bond. The created pairs of electrons and holes can be accelerated by the electric field in

opposite directions and creates more electron–hole pairs, eventually causing an avalanche of electron–hole pairs and high current across the junction.

Under reverse bias (negative V_D below), and for a one-sided n$^+$p junction, the maximum electric field in a junction can be shown to be given by

$$E_{\max} = \sqrt{\frac{2qN_a(V_{bi} - V_D)}{\epsilon_s}} \tag{2.11}$$

When the field is strong enough the junction will breakdown. This field is phenomenally expressed as a critical field, E_{crit}.

One useful note is that the breakdown voltage varies approximately inversely with the doping on the lightly doped side of a one-sided junction. This implies that for the source–drain junctions of a MOSFET in a well, the breakdown voltage is determined by the doping profile of the wells.

As we will see in Chap. 4, the breakdown voltage of an n-well adjacent to a p-well, both of approximate doping concentration of about 10^{17}/cm^3, is about 15 V. The critical field is found empirically to be related to the doping concentration, E_{crit}~$N_a^{0.2}$. Therefore from Eq. 2.11, the breakdown voltage is approximately (Hu, 2009),

$$V_{BD} = \frac{\epsilon_s E_{crit}^2}{2qN_a} \sim 15\left(10^{17}/N_a\right)^{0.6} \tag{2.12}$$

Similarly, for an n-well that is surrounded by both p-well and p-substrate, the breakdown voltage will be much lower for the n-well/p-substrate junction than the n-well/p-well source–drain junctions, as now both sides of the p–n junction can drop some voltages. As we will see in Chap. 4, this fact has a very practical application in high-voltage non-logic devices.

Another practical point to note is that the avalanche breakdown voltage for a given p–n junction increases with increasing temperature, i.e., avalanche is less likely to happen at higher temperature. The reason is that at high temperature, the electron or hole will more likely to collide with the lattice and lose some energy so that the critical carrier energy to initialize an impact ionization is harder to reach. So the worst condition for the breakdown voltage measurement is at the low temperature specification limit, and the self-heating by the reverse-biased breakdown has less opportunity to cause a thermal runaway. This is surely a silver lining of the avalanche breakdown mechanism. However, unexpected junction breakdown often injects carriers to substrate locations not considered in the design phase and can cause further complications of body effects or latchup. A rule of thumb to use the avalanche diode is to create careful device layout so that all generated electron–hole pairs by impact ionization are collected by the intended terminals, often substrate taps.

The critical field for impact ionization is roughly proportional to the bandgap, and silicon has a relatively low E_{crit} in comparison to GaN. This sets a fundamental limit for the length of the drift region to sustain the high voltage in silicon. The higher the voltage needs to be accommodated, the longer the drift region, the slower the carrier transit time, and the lower the cutoff frequency. Silicon devices can be used for high voltages (over a few thousand volts in power switches), as long as there is sufficient length to make the electric field lower than E_{crit} everywhere in the device region. However for RF power amplifiers, GaN can allow both high voltage and high frequency together, much more than what can be achieved in silicon. This is one of the remaining niche areas where placing a GaN transistor on a silicon substrate will have a strong advantage: allowing much higher output power to drive RF antennas. The topic of RF output power will depend on the quality factor of the passive resonators and the associated circuit architecture and is not treated in this book.

Note that although we parametrize the avalanche breakdown by E_{crit} or N_a for simplicity here, impact ionization is a "nonlocal" effect initiated by "lucky electrons" travelled some distance without phonon scattering to collect sufficient energy. More detailed treatment is required to predict the avalanche breakdown or similarly hot carrier injection later in the flash memory operations. The original formalism was started by W. Shockley and had been discussed in many more specific textbooks on hot carriers. In comparison with phonon scattering, impact ionization is a more effective way to reduce the energy in the carrier system by generating electron–hole pairs without generating heat. The effect is entirely reversible without any permanent damage to the material unless high current generates too much heat, or too many generated carriers cause other effects such as migration.

3. ***Zener breakdown***—When doping levels on both sides of a p–n junction are high, band-to-band tunneling can occur before impact ionization. Substantial tunneling current signifies a breakdown.

   Zener breakdown happens when the energy band on the n-type side is pulled down heavily by the large reverse bias, and the valence band on the p-type side is now overlapping with the conduction band on the n-type side. The triangle-shaped potential barrier between these two sides has a height of V_{bi} and the width depending on the electric field. When the triangular barrier becomes very thin, band-to-band tunneling efficiency becomes high, and Zener breakdown happens.

   For Zener breakdown, the breakdown voltage decreases with increasing temperature since the energy bandgap, E_g, usually decreases. The difference in temperature response is one of the experimental clues to pin down the breakdown mechanism.

In the following, we go into a bit more details how the junction breakdown impacts the process and device layout design in highly scaled ICs. Figure 2.5 shows the layout of part of a CMOS inverter that has one p-MOSFET and one n-MOSFET.

The cross section along the AA′ line is shown in Fig. 2.6, where there are two diodes: a N+/p-well diode that is at a distance of d_1 to another n-doped region and a P+/n-well diode that is distance d_2 away from a p-doped region. The N+/P+

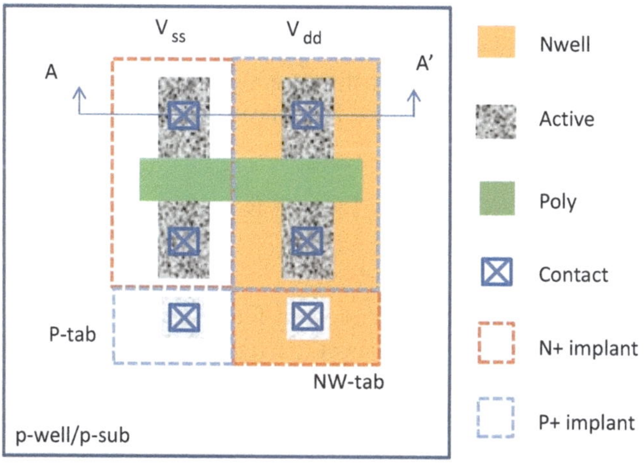

Fig. 2.5 Layout of a CMOS inverter to illustrate the important parameters for junction breakdown

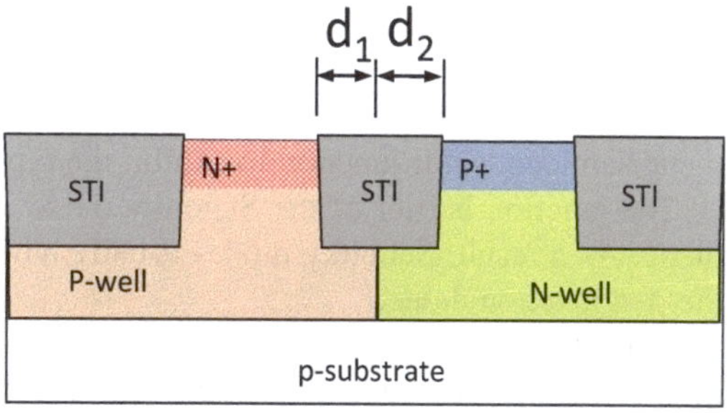

Fig. 2.6 Cross section of an inverter to illustrate diode punchthrough parameters

regions are often separated by the shallow trench isolation (STI), which is filled with SiO_2. Surely a deep trench isolation will have different breakdown behavior.

The minimum distances d_1, n+ active area to p-well edge, and d_2, p+ active area to n-well edge, are usually given by the process developers, e.g., foundries, in a design rule set. Note that d_1 and d_2 are as follows:

1. Dependent on the doping levels
2. Usually given for a certain voltage spec.

These design rules are different for, e.g., core pFET in the core n-well that only goes to 1.2 V, while a pFET in the I/O circuit in a differently doped n-well can go up to 3.6 V. Even though these are both one-sided junction as the n+ (and p+) source–drain region has much higher doping (10^{20}/cm^3) than the well region (10^{17}–10^{18}/cm^3), the design rule for d_1 and d_2 for these two situations will be different from the voltage dependence in the punchthrough condition.

As we will discuss in Chap. 6, often we need to operate a device at a voltage level beyond that is specified by the foundry. For example, we may need to take a

3.3 V device to 10 V in a high-voltage switch circuit. This is possible without any process modifications by exploiting the delay of the breakdown from layout techniques, as we will show in Sec. 4.3. In this case, the design rules need to be modified and set at a value that can be obtained from measured data.

2.2.4 Schottky Diodes

Another type of diode that can be occasionally found in an IC is the Schottky diode, which is formed when a metal is adjacent to a lightly doped semiconductor region. When the metal is in contact with a highly doped region, as are found in metal to source and drain, the contact is mostly ohmic due to strong band-to-band tunneling.

At equilibrium, in a metal–semiconductor junction, the Fermi levels line up and the energy bands in metal and in semiconductor is shown in Fig. 2.7. There is an energy barrier that blocks the moving of carriers between the metal and the semiconductor region. The energy barrier is given by

$$\text{Barrier height} = \text{work function of metal} - \text{Si electron affinity.}$$

This energy barrier resembles the barrier in the p–n junction and causes the metal/semiconductor junction to exhibit the typical diode behavior.

The junction barrier of the Schottky diode is usually less than that of the p–n diode. As a result, Schottky diodes usually have lower forward voltage drop than the p–n junction diode.

When the Schottky diode is forward biased, most often the thermionic emission of the majority carriers dominates over the diffusion of the injected minority carriers, and therefore, there is NO diffusion capacitance or minority carrier to be

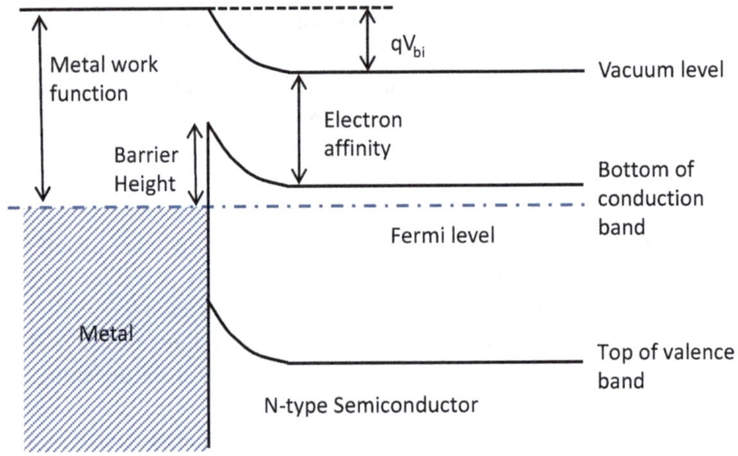

Fig. 2.7 Metal–semiconductor junction

removed during the turn-off process. So the Schottky diode is more advantageous for high-speed and high-current applications. In Sect. 7.3 we discuss some examples of making Schottky diodes in logic CMOS process.

2.3 MOS Capacitors and Oxide Device Physics

2.3.1 MOS Capacitors

The metal–oxide–semiconductor capacitor or MOSCAP is another fundamental building block for MOSFET-based devices and circuits.

Until very recently, the "metal" in MOS is actually highly doped polysilicon. For sub-40 nm processes polysilicon has been replaced by real metal or metal silicides, or at least a significant part of it. While the oxide commonly refers to silicon oxide, for the most advanced processes, it has been replaced by oxynitride (SiON) for prevention of dopant migration or high dielectric-constant materials such as $HfSiO_4$ or HfO_2.

We will still use the conventional metal–SiO_2–silicon MOS to illustrate the physics of MOSCAP as this system has been thoroughly studied and the heavily doped polysilicon behaves similarly to metals.

The electronic band structure of the metal–oxide–semiconductor structure at equilibrium is illustrated in Fig. 2.8. As the current flow between the two sides can be most often ignored, the Fermi levels on the metal and semiconductor side are always flat and will align at equilibrium. As a result of the difference in electron

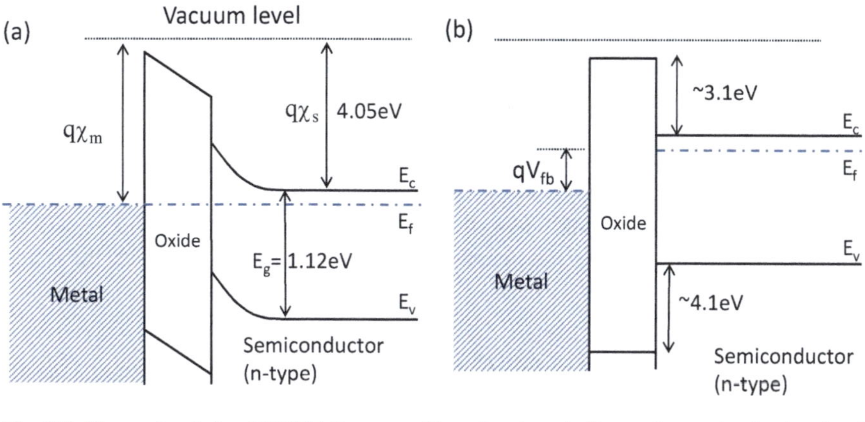

Fig. 2.8 Energy bands in a MOSCAP system. The subscript c indicates the conduction band and subscript v indicates the valence band. When the metal, oxide, and semiconductor are in contact and in electrical equilibrium, the Fermi levels, E_f should align. The electron affinities in metal and semiconductors are different, which cause the energy bands to bend at equilibrium (**a**). With applied bias of the flatband voltage V_{fb}, the energy bands will be flat (b). The misalignment of the conduction bands of silicon and SiO_2 is about 3.1 eV, while their valence bands are offset about 4.1 eV. These will form the barrier for electron and hole injection, respectively

Table 2.1 Examples of the flatband voltage for n and pMOS assuming a substrate doping of 10^{17} cm^{-3}

	p+ poly gate	n+ poly gate
nMOS	0.14	−0.98
pMOS	0.98	−0.14

affinity of these dissimilar materials, the energy bands will bend even when there is no bias applied.

A flat band condition is illustrated in Fig. 2.8b, which can be used to extract a flatband voltage V_{fb}, defined as the applied voltage when the band bending disappears and is given by

$$V_{fb} = \chi_m - \psi_s \tag{2.13}$$

where ψ_s is given by

$$\psi_s = \chi_s + (E_c - E_f) \tag{2.14}$$

χ_s is the electron affinity of silicon. At the flat band condition, the gate leakage current, although already small in most operating regions, will reach a minimum. Table 2.1 listed the flatband voltage for MOSCAP for gate that is either n+ or p+ type of polysilicon, assuming a substrate doping level of $10^{17}/cm^3$.

When a MOS capacitor is under bias, three bias regions can be distinguished by the nature of the charge distribution in the semiconductor. We use the case of the p-substrate as an example (this is the called N-MOSCAP, as defined by the source–drain doping in the MOSFET case). A negative bias on the gate electrode will cause the holes in the substrate to be attracted to the surface under the gate, this is the ***accumulation*** condition and is the easiest to analyze.

The voltage drop across the oxide is given by

$$V_{ox} = V_{GB} - V_{fb} \tag{2.15}$$

where V_{GB} is the gate to bulk bias voltage.

Under this condition, the total capacitance of the system is given by the oxide capacitance C_{ox},

$$C_{ox} = \frac{\varepsilon_{ox} A}{t_{ox}} \tag{2.16}$$

where A is the area of the capacitor, ε_{ox} is the dielectric constant, and t_{ox} is the thickness of the silicon oxide.

Under moderately positive gate-bias condition passing V_{fb}, holes in the semiconductor will be driven away from the gate, the surface underneath the gate will be depleted of holes, leaving the near-surface region to be in a ***depletion*** state. The voltage drop across the oxide, V_{ox}, is related to the width of depletion region, W_{dep}, by the following relation:

$$V_{\mathrm{ox}} = -\frac{Q_{\mathrm{del}}}{C_{\mathrm{ox}}} = \frac{qN_{\mathrm{a}}W_{\mathrm{dep}}}{C_{\mathrm{ox}}} = \frac{\sqrt{2qN_{\mathrm{a}}\varepsilon_{\mathrm{Si}}\phi_{\mathrm{Si}}}}{C_{\mathrm{ox}}} = V_{\mathrm{GB}} - V_{\mathrm{fb}} - V_{\mathrm{B}} \tag{2.17}$$

where V_{B} is the voltage applied to the bulk of the silicon.

In this case, the total capacitance is a series combination of the oxide capacitance and the capacitance of the depleted Si region, i.e.,

$$C = \frac{C_{\mathrm{ox}}C_{\mathrm{dep}}}{C_{\mathrm{ox}} + C_{\mathrm{dep}}} \tag{2.18}$$

Under even more positive bias, the semiconductor surface will have strong attraction of electrons due to the positive voltage on the gate, when electrons created in the depletion region migrate to the oxide–semiconductor interface. The surface will begin to invert its carrier type and form an inversion layer of n-type carriers, i.e., electrons. Note that this is the p-type substrate, and electrons are naturally the minority carriers. Generation of significant amount of electrons to compensate the gate charge will be a very slow process. When there is insufficient time for minority carrier generation to populate the inversion layer, deep depletion of the substrate will continue to happen until the maximum field hits E_{crit} of impact ionization, after which the electrons can be efficiently generated. Generation of minority carriers is only a limit for MOSCAP, not MOSFET where the minority carrier is injected from the forward-biased source–substrate junction.

The gate bias that separates the inversion and the depletion region is defined as the threshold voltage for this MOS system and given by

$$V_{\mathrm{t}} = V_{\mathrm{fb}} + 2\phi_{\mathrm{B}} + \frac{\sqrt{4qN_{\mathrm{a}}\varepsilon_{\mathrm{s}}\phi_{\mathrm{B}}}}{C_{\mathrm{ox}}} \tag{2.19}$$

where $\phi_{\mathrm{B}} = kT\log\left(\frac{N_{\mathrm{a}}}{n_{\mathrm{i}}}\right)$. After the formation of the inversion layer, the differential quasi-static capacitance is again C_{ox} (Taur & Ning, 1998). The factor of 2 in front of ϕ_{B} results from the "inversion" condition: the surface minority carrier is as high as the original majority carrier when the bias at the metal contact reaches V_{t}.

From the earlier description, the C–V curve will look like that in Fig. 2.9 for the p+ gate/p-well MOSCAP structure. The V-shaped C–V curve only works under quasi-static condition, where all mobile carriers are assumed to be able to respond to changes in bias condition instantaneously.

In reality, when the capacitance is measured at high frequency (say 1 MHz), the inversion charge will not be able to follow the change in the AC bias. This is because the inversion charge is generated in the depletion region and the starting minority carrier has very small concentration. In this case, the capacitance in the inversion region will be like those from the depletion, e.g., at the minimum C_{min}. Capacitance measured by intermediate frequencies will be between C_{ox} and C_{min}, where part of the responding charge will be from the inversion layer, and part of the charge be from the extension of the depletion region.

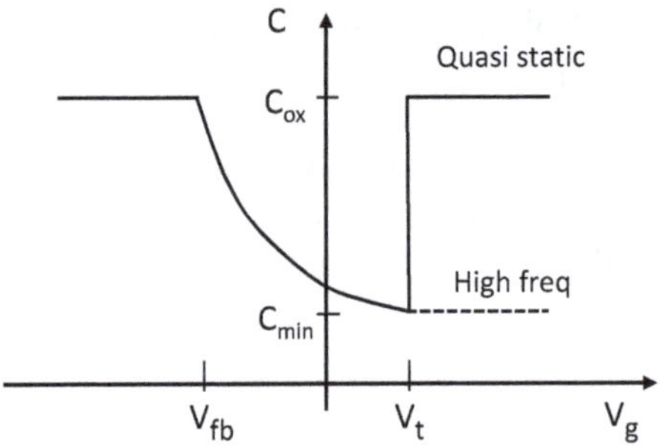

Fig. 2.9 C–V curves for an n-MOSCAP structure. The solid line is the C–V curve measured under the quasi-static condition. At higher frequencies, the capacitance in the inversion condition will remain at the minimum $C_{\min}$ since the mobile inversion charge cannot respond fast enough to the AC perturbation. Deep depletion is not shown, when the mobile inversion charge cannot respond to the Vg sweep

Fig. 2.10 MOSCAP with fixed charges

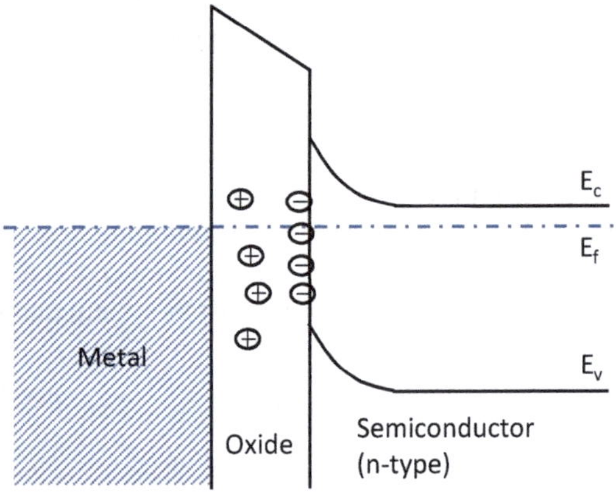

Due to the variation of the capacitance in different bias conditions, besides being the fundamental component for the MOSFET, MOS capacitors have been used as a voltage-controlled variable capacitor (varactor) similar to the reverse-biased diodes. The use of MOS capacitor coupled with the floating gate as a novel variable capacitor is discussed in Sec. 5.3.

If there is fixed charge between the gate and the substrate, for example, charge trapped in the oxide as shown in Fig. 2.10, the flatband voltage will be shifted by

$$V_{\mathrm{fb}} = V_{\mathrm{fb0}} - Q_{\mathrm{ox}}/C_{\mathrm{ox}} \tag{2.20}$$

where Q_{ox} is the effective fixed charge density, equivalent to the charge evaluated at the interface.

This flatband voltage shift translates into a change in the threshold voltage for MOS transistors. We will discuss in Chaps. 8–10 the use of this charge trapping phenomenon as a mechanism for storing information. A whole class of nonvolatile memory, including the most advanced 3D NAND flash memory, is based on charge trapping.

2.3.2 *Gate Tunneling Current*

In the classical MOS description, the oxide is usually assumed to be an insulator with infinite resistance and impervious to electron conduction. As the gate oxide gets thinner from the continued scaling, quantum mechanical tunneling current across the oxide becomes significant and can no longer be ignored. In particular, direct tunneling causes significant leakage through the gate oxide even at low-bias voltage. It is for this reason silicon oxide thickness scaling stopped scaling at about 1.2 nm and silicon oxide has been replaced by high dielectric constant (high-*k*) materials as the gate dielectrics starting at the 40 nm node.

Tunneling is a pure quantum mechanical phenomenon. Figure 2.11 illustrates the three dominant tunneling mechanisms for a MOSCAP system.

Fowler–Nordheim (**FN**) tunneling is a special case of tunneling. The term specifically means tunneling through a triangle barrier that is caused by the application of an electric field across the insulator. FN tunneling only happens after eV_{ox} is larger than the silicon-oxide conduction band barrier at 3.1 eV. Notice that the tunneling distance will be smaller than the thickness of the oxide, as the electrons are injected into the conduction band of the oxide.

Direct tunneling is defined for the situation when the applied voltage is lower and electrons go through a trapezoidal barrier and inject into the conduction band of the anode.

Band-to-band tunneling (**B2BT**) for electrons dominates when the silicon interface is in accumulation of holes, because the conduction band electrons have very low concentration at the interface and the valence-band hole has a larger barrier (around 5 eV), the valence-band electron has large supply and a smaller barrier (around 4.2 eV). BTBT is the main mechanism in MOSFET **gate-induced drain leakage** (**GIDL**). BTBT had found some use in programming nonvolatile memories.

FN tunneling is used in most of the floating-gate-based nonvolatile memories which will be discussed in more details in Chap. 8. FN tunneling is well studied and the tunneling current as a function of the electric field in the oxide is given by

$$I_{FN} = AE_{ox}^2 e^{-B/E_{ox}} \tag{2.21}$$

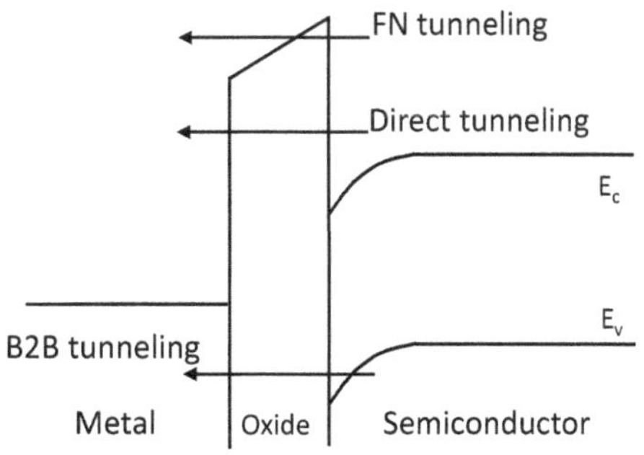

Fig. 2.11 Tunneling currents from a MOS structure

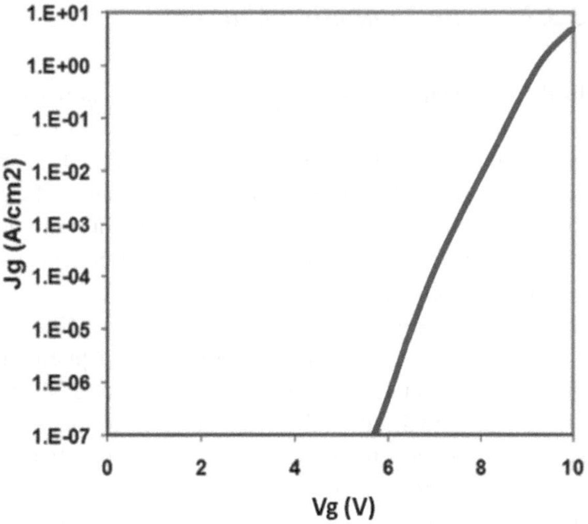

Fig. 2.12 Gate current of a 7 nm thick oxide, from a 3.3 V pMOSFET transistor device with 7 nm gate oxide

where E_{ox} is electric field at the injection surface and is related to the voltage across the oxide and the oxide thickness by, $E_{ox} \sim V_{ox}/t_{ox}$ for a uniform oxide. A and B are constants that depend on electron effective mass and the barrier height. As the conduction band barrier is very high at 3.1 eV, I_{FN} is often very small which is the main reason of slow programming and erase in today's flash memory. The field sensitivity of I_{FN} derives from E_{ox} making the triangular barrier thinner, but the inverse slope (voltage change to cause a decade of current change) is often much larger than the turn-on characteristics of p–n junctions at 60 mV/decade at the room temperature.

Figure 2.12 shows the tunneling current density for a 7 nm oxide available in most foundries as the gate oxide of 3.3 V I/O transistors. The inverse slope here is about 600 mV/decade. Another good number to remember is that E_{ox} needs to reach 0.9 V/nm before any appreciable I_{FN}. For 7 nm oxide here, it would be around 6.3 V.

It should be emphasized that the earlier discussion, in fact in almost all literature discussions, the density of available carriers or empty states to receive the injected carriers are not considered. This is a factor that can have huge effect on the absolute current and be a factor that can be controlled by the selection of the gate work function and the associated bias. For example, metals of large work functions are often used in the control gate of Flash memory to prevent electron injection from the control gate to the floating gate.

2.3.3 *Oxide Reliability*

The reliability of gate oxide plays a very important role in the reliability of virtually all MOS-based ICs and, as a result, has been extensively studied. As described in the last section, under bias, there are various mechanisms whereby carriers, electrons or holes, can go through the oxide. Each carrier with enough energy may cause damage to the oxide or change of the interface charge conditions. When a large number of carriers with high enough energy have gone through the oxide, the damages may accumulate and cause the oxide to break down.

Various empirical and analytical models, including percolation models, have been proposed to explain the oxide breakdown. The wear out and breakdown mechanism has been qualitatively understood to go through the following sequence (Yeo, Lu, & Hu, 2001).

With the passing of carriers through oxide under bias, the carriers will gain energy and interact with the lattice through electron–phonon interactions. Some of them may gain enough energy to knock off lower energy interfacial or interstitial hydrogen atoms, which are present during the oxide growth process, and create traps for more charged carriers. Some of these carriers may also lose all of the excess energy and get trapped by defects in the oxide. If short lifetime electron traps are formed, they will induce trap-assisted tunneling to cause additional leakage especially for traps generated close to the center of the oxide, even at low electric field. This mechanism causes low resistance leakage path which in turn causes more carriers to be injected into the oxide. The gate current increases with oxide trap generation during the cycles of carrier injection, which is called **stress-induced leakage current** (**SILC**). This will shorten the retention time of flash memory cell and cause either a stuck-at-erase fault or a tail bit with shorter-than-expected retention time. Charge trapping and trap generation in oxide are critical for flash reliability and are often tackled with thick tunnel oxide and **error correction codes** (**ECC**).

From another point of view, when the electric field is sufficiently high, Fowler–Nordheim tunneling or hot carrier injection of electrons occurs. Many electrons will get injected into the conduction band of the oxide. At the anode they fall into the conduction band of Si and lose the excess energy by creating electron–hole pairs. The holes created at the anode may get injected back into the oxide to cause more damage or trapped charge.

Eventually when the number of traps created exceeds a critical density, a conductive path may form to cause a positive feedback effect, more electrons go through this conductive path causing local heating and eventual rapture—a hard breakdown—of the oxide. Oxide breakdown can be caused by electrode material migration to create strong percolation path or severe drift of impurities or oxygen atoms. On the positive side, this mechanism enables a nonvolatile memory element based on oxide soft/hard breakdown in **one-time-programmable** (**OTP**) read-only memory (ROM) and is the similar mechanism for the new resistive memory (ReRAM).

Our best understanding of oxide breakdown is that it is a statistical event and occurs randomly over the area of the oxide. Weibull distribution has been found to be the best model to describe the breakdown. The oxide reliability can usually be measured either by **time-dependent dielectric breakdown** (**TDDB**) or **charge to breakdown (Qbd)**. TDDB is usually measured using constant voltage stress (CVS) while *Qbd* is usually measured using constant current stress (CCS) condition.

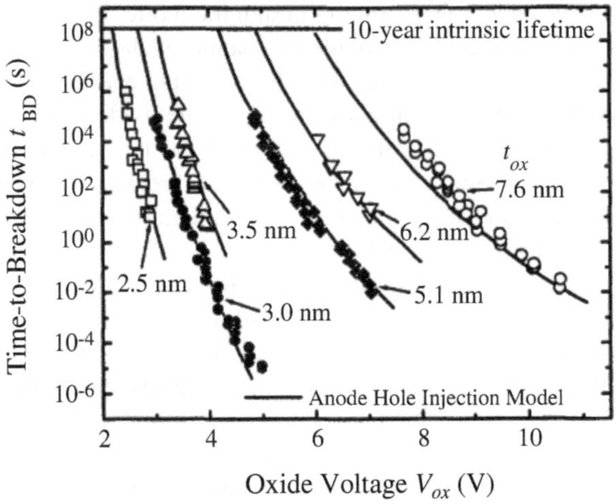

Fig. 2.13 Oxide TDDB for several oxide thicknesses (Yeo et al., 2001)

Qualitatively, time to breakdown (*Tbd*) or charge to breakdown (*Qbd*) of a given oxide decreases exponentially with increasing electric field or applied voltage. In Fig. 2.13 the TDDB vs. gate voltage for a few oxides of different thickness is plotted (Yeo et al., 2001). We can see than a 6.2 nm oxide can sustain 8 V for over 1 s, at 10 V for about 10 ms, and breaks down at 1 ms or less of bias at 11 V. The oxide is designed for operation at 3.3 V for 10 years.

The voltage dependence of Tbd or Qbd for a given oxide is strong and it is still being debated whether the dependence is $\exp(-V/V_0)$ or $\exp(V_0/V)$ dependence, i.e., described by the following two models:

$$\text{Tbd} \propto \exp(-E/E_{\text{crit}}) = \exp(-V/V_0) \propto 1/\exp(V)$$

and

$$\text{Tbd} \propto \exp(E_{\text{crit}}/E) = \exp(V_0/V).$$

The E dependence can be explained by a thermochemical model based on the interaction of molecular dipoles with the electric field. On the other hand, the FN tunneling and the associated charge trapping mechanism form the foundation to the $1/E$ model (Yeo et al., 2001). In practice, a power law V^{-n} model is often used, where n is a positive constant.

These two models as well as a power law model are illustrated in Fig. 2.14. Over the limited region that can be tested, it is found that both V and $1/V$ models describe the data reasonably well. But as shown in Fig. 2.14 by extrapolation, these two models predict very different oxide lifetimes at lower, operationally more important voltage region.

For the Weibull distribution based on percolation path formation, the failure probability of an oxide at time t is given by

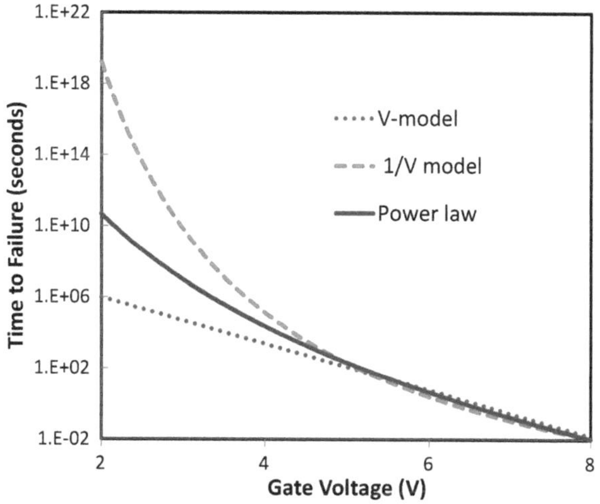

Fig. 2.14 Voltage-dependent models for the time to breakdown of an oxide

$$F(t) = 1 - \exp\left(-\left(\frac{t}{\alpha}\right)^{\beta} A_{ox}\right) \tag{2.22}$$

where A_{ox} is the area of the oxide, α is the characteristic breakdown time at 63 %, and β is the Weibull shape factor.

For an oxide with area A_{ox}, the time to breakdown can be predicted from the Weibull model by

$$T_{bd} = kA_{ox}^{-1/\beta} \tag{2.23}$$

or

$$\frac{T_{bd1}}{T_{bd2}} = \left(\frac{A_2}{A_1}\right)^{1/\beta} \tag{2.24}$$

So the smaller the area, the longer it takes for an oxide to break under identical bias condition due to the smaller probability of percolation path formation.

Putting these influences together, the oxide reliability may be described by a failure model as follows:

$$\text{Tbd} = kV^{-n}e^{E_a/kT}\,A_{ox}^{-1/\beta} \tag{2.25}$$

Foundries usually provide lifetime prediction models for their oxides.

For typical ICs, to guarantee 10 year or more lifetime for a whole chip, gate oxide thickness of around 3 nm is usually chosen for MOSFET under 1.8 V operations, 5 nm for MOSFETs that will be operating with 2.5 V, and around

7 nm for 3.3 V MOSFETs. These oxides are fixed for a given process and, as shown in Chap. 3, are quite difficult to change.

Given this physical constraint one interesting and practical question is how much we can use a given oxide outside the reliability specification given by the foundry. For example, for a 7 nm oxide that is rated for 3.3 V operations, can we take the oxide to 7, 8 V, or even up to 10 V? And if yes, under what condition?

Indeed, as evident from Fig. 2.13, we may indeed take a 7 nm oxide to over 10 V, albeit only for a shorter time, e.g., for 1 s or less. Fortunately, there are other factors that may prolong the operating lifetime of an oxide. As suggested by Eq. 2.24, for an oxide with smaller total oxide area, the time to breakdown increases. In any other words, we may subject a small number of devices to the high-voltage condition for a longer time.

The key take-home message from the earlier discussion is that the reliability of gate oxides depends on many factors:

1. The gate oxide thickness is fixed by the foundry. Taking into account the oxide thickness variations induced by the process variations, some margins need to be built into our lifetime assumption.
2. Nominal voltage specifications are for general, fairly large, ICs and normal operating conditions. It is possible to use an oxide under out-of-spec conditions, with limitations on the total area and stress time for this particular oxide.

In later chapters, particularly in Chaps. 8–10, we will discuss many examples of using an oxide that is rated for 3.3 V operations to higher voltages, up to 10 V.

Finally, we note in general the oxide breakdown is a bad thing to be avoided either by design or by process technology improvement. However, the breakdown of oxide can sometimes be used beneficially. As will be discussed in more details in Chap. 9, oxide can be used as antifuse where low, and uniform, breakdown voltage are desired. Use of this kind of antifuse as one-time programmable memory has found extensive use commercially.

2.4 MOSFET Physics

Figure 2.15 illustrates the basic structure of a planar MOSFET, which is a four-terminal device comprised of gate, source, drain, and bulk (or well) terminals. Gate of a MOSFET is insulated to the substrate by a gate insulator, i.e., the MOSCAP

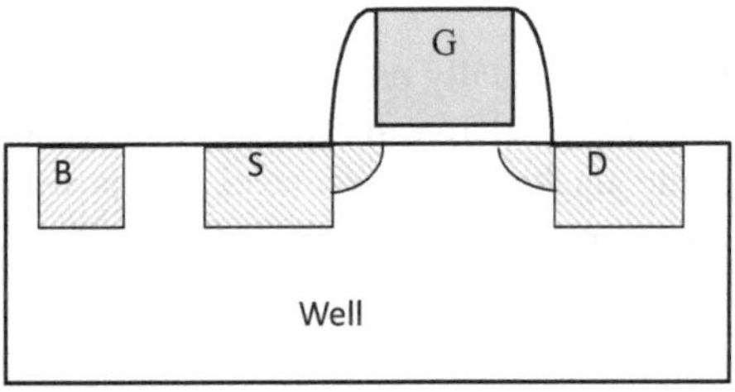

Fig. 2.15 Cross section of a typical MOSFET illustrating its four terminals: gate (G), body or well (B), source (S), and drain (D)

discussed in the previous section, and can modulate the channel through capacitive coupling.

MOSFET can be an n-channel or a p-channel device, depending on whether the carriers are electrons or holes injected from the source contact. N-channel devices are defined with n-type source and drain, and typically built on the p-type substrate. When a positive voltage is applied to the gate of an n-MOSFET that exceeds the threshold voltage, the p-doped channel region under the gate can be inverted to form an "inversion" layer. The electrons in the inversion region form a conducting channel from the source to drain.

2.4.1 Basic Long-Channel MOSFET Equations

When the channel condition is mostly controlled by the gate-source MOSCAP in the vertical direction instead of the drain-source back-to-back diodes (i.e., the gate oxide thickness is much smaller than the channel length, normalized by the respective dielectric constant), the MOSFET operation is in the "long-channel" approximation. For the long-channel MOSFET, a key electrical parameter is the threshold voltage V_t, which determines if a MOSFET is in a conducting or nonconducting state, or equivalently, if the MOSCAP has entered the inversion region.

The threshold voltage is hence the same as that of the MOSCAP, discussed in Sect. 2.3, and is given by

$$V_t = V_{fb} + 2\phi_B + \frac{\sqrt{4qN_a\varepsilon_s\phi_B}}{C_{ox}} \tag{2.26}$$

The threshold voltage can be varied electrically by the body effect, when the source is biased at a voltage V_{sb} to the body or well,

$$\begin{aligned} V_t &= V_{t0} + \frac{\sqrt{qN_a 2\varepsilon_s}}{C_{oxe}}\left(\sqrt{2\phi_B + V_{sb}} - \sqrt{2\phi_B}\right) \\ &\equiv V_{t0} + \gamma\left(\sqrt{2\phi_B + V_{sb}} - \sqrt{2\phi_B}\right) \end{aligned} \tag{2.27}$$

When $V_{sb} = 0$, $V_t = V_{t0}$, which is the threshold voltage given by Eq. 2.25.

The body effect, measured by the body coefficient γ in Eq. 2.27 and also called the backgate effect, is important in many practical circuits. For example, the charging of the capacitor shown in Fig. 2.16 is the commonly seen DRAM cell.

In the charging of the capacitor, if $V_d = V_g = V_{dd}$, then source node voltage V_s can be charged to only $V_s = V_{dd} - V_t$ before the channel is shut off, where V_t is the threshold voltage with the back gate bias of V_s. The threshold voltage loss when the nMOSFET is used to pass a high voltage is aggregated by the body effect. One solution is to boost V_g above V_{dd} for a short period of time, which is called the

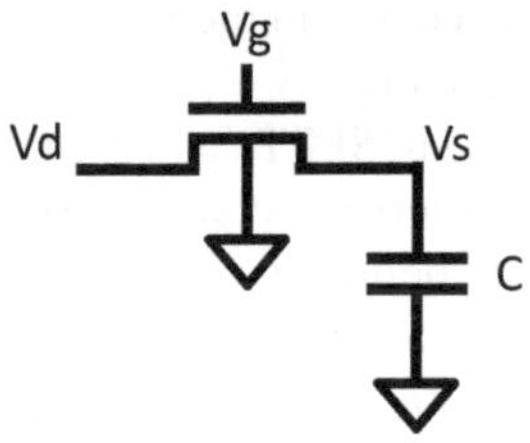

Fig. 2.16 Back gate effect on the charging of capacitor

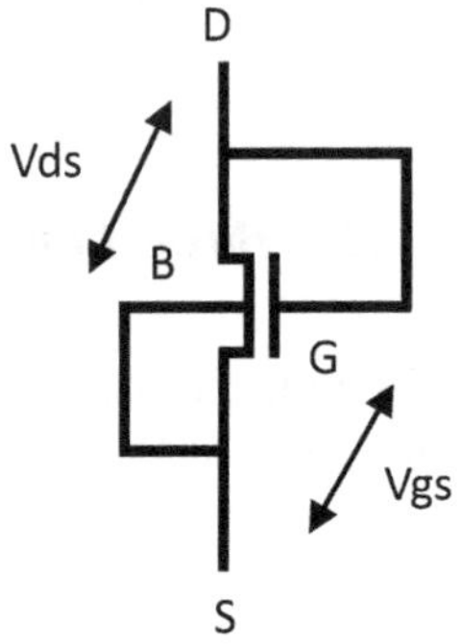

Fig. 2.17 Diode-connected MOSFET configuration is commonly used in current mirrors and current source circuits

"word line" boosting. Notice that it is not useful to boost V_d, as the channel with still shut off at $V_g - V_t$.

When the gate voltage is below V_t, the MOSFET is in the cutoff region and no current will flow between source and drain except the subthreshold leakage current. The subthreshold leakage current is important to the static power consumption for an IC.

The current between the source and drain in the triode region (also called the linear region, due to I_d relation with V_{gs}), defined by $V_{gs} > V_t$ and $V_{ds} < V_{gs} - V_t$, is given by

$$I_{ds} = \frac{W}{L}\mu_s C_{ox}\left(V_{gs} - V_t - \frac{1}{2}V_{ds}\right)V_{ds} \tag{2.28}$$

where W is the gate width, L is the channel length defined by the source–drain metallurgical junctions, μ_s is the universal channel mobility in cm^2/Vs, and C_{ox} is the gate capacitance of the unit area in F/cm^2. In the saturation region (i.e., I_d saturates to, or no longer change with, V_{ds}), $V_{ds} > V_{dsat} = V_{gs} - V_t$, the current is given by

$$I_{ds} = \frac{W}{2L}\mu_s C_{ox}\left(V_{gs} - V_t\right)^2 \tag{2.29}$$

Diode-connected MOSFET, shown in Fig. 2.17, is a special and useful configuration of MOSFET. In this configuration, $V_{ds} = V_{gs}$, where V_{ds} normally is always more than $V_{dsat} = V_{gs} - V_t$. Therefore, diode-connected MOSFET is always in the saturation region. The bulk/well is connected to source, and the four-terminal MOSFET is turned into a two-terminal diode.

We can rewrite Eq. 2.29 to obtain

$$V_{gs} = V_{ds} = V_t + \sqrt{\frac{2I_{ds}}{\frac{W}{L}C_{ox}\mu_s}} \tag{2.30}$$

When the current is small, for example, when supplied by a small current reference, the second term maybe neglected and we have

$$V_{gs} = V_{ds} = V_t \tag{2.31}$$

Thus, the voltage is fixed to the threshold voltage of the MOSFET, insensitive to the bias current. This is a very useful relationship and the diode connected FETs are used often in voltage reference or current source designs.

2.4.2 Capacitances of MOSFET

Capacitances associated with a MOSFET are important to the speed and the energy consumption. We will see in later chapters that these capacitances are also important to the operations of floating-gate devices and memories. Hence it is important to review the MOSFET capacitance in the various operation regions. Figure 2.18 illustrates the capacitances that we are most concerned about—the capacitances involving the gate of the MOSFETs. Other capacitances, such as the capacitance of the source and drain junctions with the well, are important as well, specially related to the speed of the MOSFETs. We leave the coverage of those capacitances to other textbooks on the MOSFETs.

The gate capacitances contain a parasitic component and an active component. For example, the gate-to-source capacitance, C_{gs}, contains a parasitic component that is the overlap of the source to the gate region, C_{gso}, which is usually small but

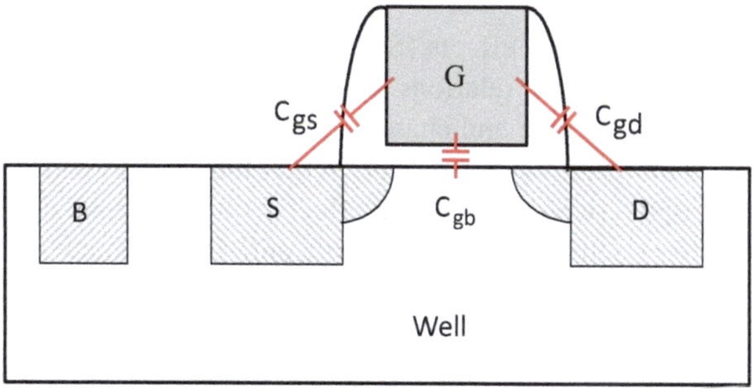

Fig. 2.18 Capacitances of a MOSFET

not negligible and independent of the operating condition of the MOSFET. The active components are usually dependent on the operating regions of the MOSFET.

1. In the cutoff region, $V_{gs} < V_t$, the channel is in the accumulation and the gate-to-bulk capacitance C_{gb} is the gate capacitance WLC_{ox}. The gate-to-source and drain capacitances are only the parasitic overlap capacitance, i.e.,

$$\begin{aligned} C_{gb} &= WLC_{ox} \\ C_{gd} &= C_{gdo} \\ C_{gs} &= C_{gso} \end{aligned}$$

2. In the triode or linear region, the channel is inverted. The gate-to-well capacitance is approximately zero as the channel shields the well from the gate. The gate oxide capacitance is equally distributed to the source and drain.

$$\begin{aligned} C_{gb} &\sim 0 \\ C_{gd} &= WL/2\ \ C_{ox} + C_{gdo} \\ C_{gs} &= WL/2\ \ C_{ox} + C_{gso} \end{aligned}$$

3. In the saturation region, again the gate is shielded from the well. Similarly the drain voltage does not affect much the charge in the channel. The gate-to-drain capacitance is only the overlap capacitance. The source capacitance varies and can be approximated by $2C_{ox}/3$.

$$\begin{aligned} C_{gb} &\sim 0 \\ C_{gd} &= C_{gdo} \\ C_{gs} &= 2/3WL\ \ C_{ox} + C_{gso}. \end{aligned}$$

2.4.3 Nonideal Effects

1. Short-channel effects (SCE)

   The earlier relations for MOSFET were derived for long-channel devices, where the gate and channel directions are decoupled for the charge contribution in the electrostatic model. When the depletion depth is similar to the channel length, two-dimensional effects, often termed collectively as **short channel effect** (SCE), become significant. In particular, the structure of source and drain will influence the electrostatic behavior.

   Due to charge sharing by the source and drain, the depletion of the source and drain junctions affects the onset of the depletion in the channel region.

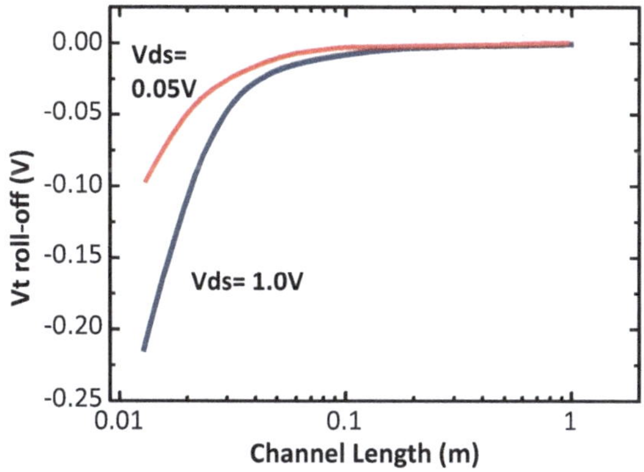

Fig. 2.19 V_t roll-off due to short channel effect. The V_{ds} dependence can be attributed to the so-called DIBL effect

Therefore, the gate loses some control of the channel charge to account for the static charge in the source–drain depletion region. The result is the dependence of threshold voltage on the channel length. An example of this dependence, called V_t**roll-off**, is shown in Fig. 2.19. In particular, V_t roll-off is more prominent at higher V_d, which is caused by **drain-induced barrier lowering** (DIBL). DIBL is a critical SCE in the submicron MOSFET. In addition to the charge view earlier, we can "visualize" DIBL from the electric field and potential points of view as well. Notice that the charge, field, and potential are three viewpoints of the Poisson equation solution and are not additive.

We will keep using nMOSFET for the following illustration. As the electric field originates from a positive charge and end at a negative charge, the positive charge at the drain/gate sides caused by the positive drain/gate bias has to end at a negative charge. When the channel electron serves mostly to terminate the gate positive charge and the electric field originated from the drain mainly terminates at the charge within the drain depletion region, we retrieve the long-channel condition. When the drain field terminates at the channel charge, it will cause a finite output resistance, i.e., V_{ds} will affect I_d at the saturation region. When the drain field reaches all the way to the source junction, V_{ds} will additionally modify V_t, lowering the source injection barrier. The drain field penetration is a useful picture especially for SOI (silicon-on-insulator), isolation oxide in the width direction and high-*k* gate dielectrics. Those insulators can stop parasitic current but cannot stop parasitic field penetration. The higher the dielectric constant, the more severe the field penetration, as long as no net charge is involved. For example, fully depleted SOI (FDSOI) with a thick buried oxide (BOX) will have severe field penetration through BOX, and hence large DIBL. When high-k gate dielectric is used, field penetration from drain to source through the high-*k* region will be more than the case with SiO_2 gate oxide, which degrades the output resistance in the saturation region and lower V_t.

We can also observe DIBL from the potential point of view. As the Laplacian (second derivative or curvature) of the potential is the local net charge, before the channel is formed in the subthreshold region, the only negative charge is from the channel p-type coping, i.e., the potential curvature at subthreshold has a maximum. When we increase the potential at the drain, the potential curvature at the source will increase as well. However, when maximum curvature is reached, the drain potential will start lowering the source barrier, which is a direct view of DIBL in the subthreshold region.

The V_t roll-off is most often undesirable because circuit designers like to have V_t to be constant in their designs independent of V_{ds}. One of the major challenges during process development is to reduce the V_t roll-off through thinner (or equivalently higher-*k*) gate oxide and higher doping at the source–drain junctions, which is often tuned by a tilted source–drain p-type implant called TIPS (tilted implants for punchthrough stopper) or halo/super-halo (a ring of high p-type doping around the n-type source–drain doping).

It is also possible to reduce SCE and V_t roll-off by adding source–drain extensions that are shallow—the **light doped drain** (**LDD**). The shallow junctions reduce the effect of source–drain without substantially increasing the series resistance. Due to the lighter doping than the source–drain, LDD also reduces the electric field in the drain region so that the oxide reliability improves.

Even though SCE is a nuisance and many efforts have been applied to minimize it, we can use SCE to obtain novel MOSFETs with nonstandard V_t. Some examples of this "turning lemon into lemonade" practice will be discussed in Chap. 4.

2. Narrow width effects (NWE)

Similar to the short-channel effect but in the width direction of a MOSFET is the narrow width effect (NWE), where the width of the MOSFET has an effect on the threshold voltage, as illustrated in Fig. 2.20.

For older (>0.25 μm) processes with LOCOS (local oxidation of silicon) isolation, the area under the bird's beak close to the channel has thicker oxide

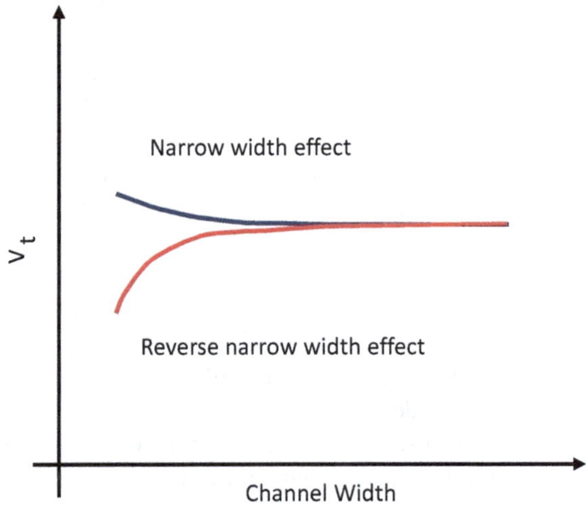

Fig. 2.20 Narrow width and reverse narrow width effect

and needs higher gate voltage to be inverted. In this case V_t increases with decreased width. This is referred as the normal narrow channel effect.

For newer transistor with shallow trench isolation, the effect is reversed. The narrow width effect is due to the effect of the end caps which create a fringe field. Basically, due to this fringe electric field enhancement at the edge, it is easier to form a conducting channel (by inversion) at a lower voltage. The threshold voltage is effectively reduced for a narrower transistor width. This is termed the "**reverse narrow channel effect**" or "**inverse narrow width effect**". In the severe case, V_t of the side channel is very different from that of the main channel region, a small hump can be developed in the subthreshold region.

2.4.4 Hot Carrier Reliability of MOSFET

Gate oxide reliability discussed in Sect. 2.3.3 is one aspect of the overall MOSFET reliability that is important to build ICs that can last many years. Another important aspect of MOSFET reliability is the so-called hot carrier effect, which does not exist for MOSCAP. This effect causes the shift of MOSFET V_t over time. Although not as destructive as the oxide breakdown, hot carrier effects do affect the performance of the MOSFETs over time—the aging of the IC—and hence the long-term circuit behavior, including the functionality of the IC. Therefore, hot carrier reliability has been the focus of intense studies.

There are several kinds of hot carrier effects, as classified by the type of carriers and the mechanisms.

1. Channel hot carriers (CHC)
   These include the **channel hot electron** (CHE) for nMOSFET and **channel hot hole** for pMOSFET. Electrons in the channel gain energy from the electrical field along the source-to-drain direction. The hot electrons can undergo elastic or near elastic collisions and be redirected to the gate, rather than to the drain, while retaining most of the collected energy. Near the drain, some of the electrons could have gained enough energy to be considered hot, e.g., having energy much higher than the ambient lattice temperature. Thus, some electrons may be lucky, both gained enough energy and also undergo collision to be directed to the gate, enough to overcome the energy barrier to go into gate oxide and be collected by the gate, as illustrated by Fig. 2.21. The injection of these energetic electrons into the gate oxide can cause damage to the oxide. For example, traps may be created by these hot electrons which, when subsequently occupied by electrons or holes, can cause a V_t shift. Figure 2.22 shows the injected gate current vs. gate voltage for nMOSFET.

   The injection efficiency in the CHE regime is usually low and usually below 10^{-5}. This effect peaks around $V_{gs} = V_{ds}$ for nMOSFET. As shown in Fig. 2.22 where the peak gate current corresponds to $V_g \sim V_d$.

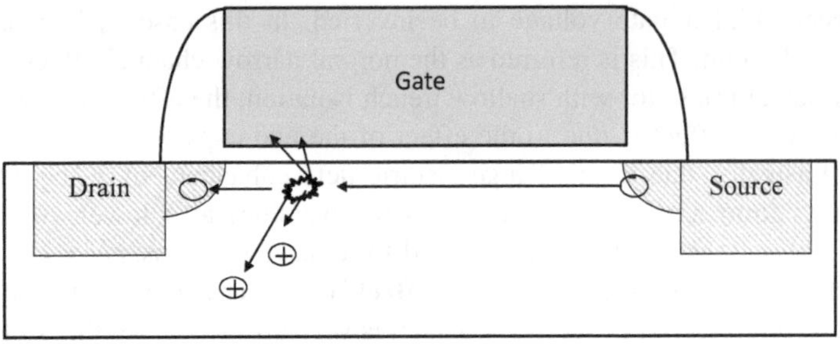

Fig. 2.21 Channel hot electron generation

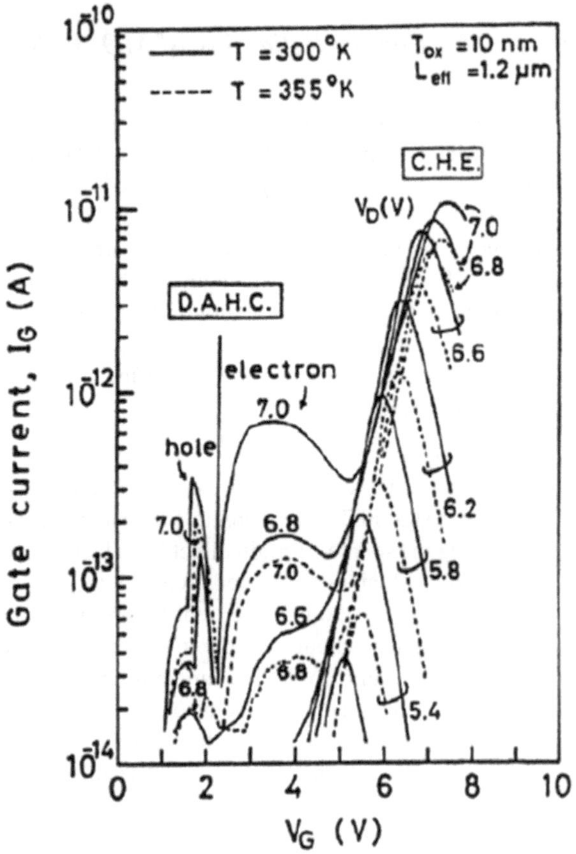

Fig. 2.22 Gate current due to CHI for nMOSFET, from (Takeda, Yang, & Hamada, 1996)

2. Drain avalanche hot carrier effect
   The second mechanism is impact ionization related where channel hot carriers gain enough energy to cause a collision where one of the two things happens: (1) the break of a Si–H bond in the interface region; or (2) the creation of an electron–hole pair. The second effect is called DAHC, **drain avalanche hot carrier** effect.

Due to the bias conditions, the generated electron will be collected in the drain with a high bias. The holes may be collected by the well and contributed to a substrate current. If there is no substrate contact as in SOI, these generated holes will accumulate at the BOX interface and cause a virtual body effect. As holes have positive charge which will lower the source injection barrier, DAHC will cause a slight increase in I_d. This effect is prominent when V_{ds} is high and V_{bs} is low.

3. Substrate hot electron (SHE) effect
   The substrate hot-electron (SHE) injection is a result of a high positive back bias, i.e., for an nMOSFET, V_b is highly negative and V_{sb} is positive. This leads to carriers in the substrate being driven to the gate oxide/substrate interface which gain further kinetic energy in the surface depletion region and eventually get injected into the oxide.

For pMOSFET, the gate current peak around $V_{sg} \sim V_{sd}/3$ to $V_{sd}/2V_{sd}/2$ and has been modeled by an impact-ionized hot electron injection (IHEI) model (Diorio, 1997). For example, in Fig. 2.23 the gate current due to hot carrier for a 3.3 V pMOSFET is illustrated. The source and well are biased at 3 V, giving $V_{sd} = 5$ and 5.5 V. The peak gate current is at $V_g \sim 1.7 - 1.8$, or $V_{sg} = 1.3$ V or about ¼ of V_{sd}.

One effect of the HCI is the damage to the gate oxide by the hot carriers, e.g., carrier injection into the gate oxide can lead to hot carrier degradation effects such as threshold voltage changes due to occupied traps in the oxide. Hot carriers can also generate traps at the silicon–oxide interface leading to subthreshold swing deterioration and stress-induced drain leakage. In general, these degradation effects set a limit to the lifetime of a transistor and therefore they have to be controlled as well as possible.

One reason for the invention of the LDD is the reduction of the electric field in the channel region. This reduces the hot carrier effect that can impact the reliability of the MOSFET.

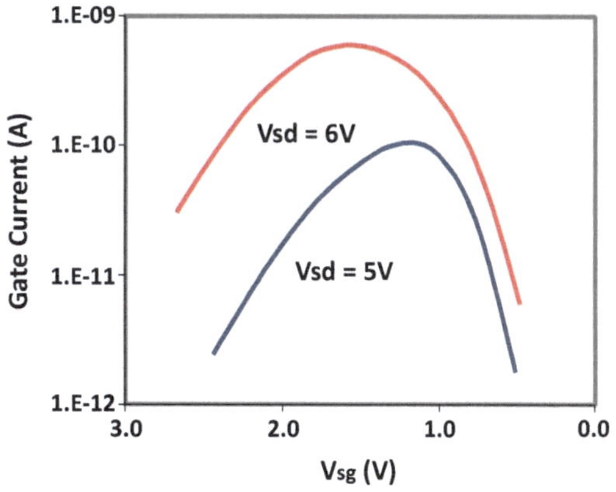

Fig. 2.23 Gate currents due to hot carrier injection for a pMOSFET vs. gate to source bias, at two V_{sd} (at −5 V, and −6 V) conditions

On the other hand, HCI has been used advantageously as the programming mechanism in a number of **nonvolatile memories** (**NVM**s). In fact, the first floating-gate memory, an electrically programmable read-only memory (EPROM), is a so-called FAMOS device, which stands for floating-gate avalanche MOS.

Currently, HCI is the programming method for a majority of the NOR flash memory ICs. The advantage of using HCI is that the programming speed can be very high with lower voltages than the case in the FN current. The drawback is the high drain current, most of which is wasted as the hot carrier injection efficiency is very low.

Most of the early HCI-based NVMs are nMOSFET based. PMOSFET-based NVM using HCI programming is also interesting and has found niche applications. Some examples of using pMOSFET for embedded NVMs in logic CMOS process are shown in Chaps. 9 and 10. The preference of using pMOSFET for these applications can be attributed to the HCI characteristics discussed earlier which enables the programming of pMOSFET floating gate in single-poly logic CMOS processes.

For nMOSFET the injection efficiency is low because of the direction of the field. Of particular interest is that the gate current is maximum near $V_g \sim V_d$, as seen in Fig. 2.22. To arrive at this optimal injection condition, a control gate is then necessary to pull up the floating gate through capacitive coupling. A quantitative discussion of the concept of capacitive coupling is presented in Sect. 5.2.

This is in contrast to pMOSFET where HCI can occur under self-biased condition. Figure 2.23 shows that the peak injection occurs at about $V_g \sim 1/3$ to 1/2 of V_{sd}. Therefore, there is no need for a control gate and suitable for single poly NVM in logic CMOS process. This fact has been used to make one-time programmable memory using single-poly processes, discussed later in Chap. 9.

The difference between nMOSFET and pMOSFET can be understood from the voltages of the terminals in the respective devices. For nMOSFET, the highest voltage is at the drain. The hot electrons are also near the drain and will be attracted to the drain. The gate can attract significant share of these hot electrons only when the gate is biased at a voltage similar to or higher than the drain voltage.

On the other hand, for pMOSFET, the highest voltage is at the drain while the hot holes are near the source. The electron–hole pairs generated by impact ionization of the hot holes are initially near the source. The secondary holes are mostly attracted to and collected by the source while the electrons are repelled. These secondary electrons are then collected either by the n-well or positively biased gate.

One peculiar fact due to the alignment of the band structure in the Si/SiO_2 system, illustrated in Fig. 2.8, is that the energy barrier for an electron to go from Si conduction band to the SiO_2 conduction band is about 3.1 eV. On the other hand, the barrier for a hole to cross from the Si to the SiO_2 is about one eV higher. As a result, the probability for hole injection into the oxide and to the gate is much lower than that of the electron efficiency.

2.5 CMOS Circuits and Power Consumption

As discussed in Sect. 2.1, one of the principal barriers to the continued scaling is the power consumption and the consequent need for cooling of integrated circuits. For example, a practical power limit is about 100 W/cm^2 for air-cool packages. In this section, we will illustrate, using the simplest CMOS circuit, the basic power consumption concepts in CMOS ICs.

The simplest CMOS circuit is a CMOS inventor, which is shown in Fig. 2.24. Other simple gates, including NAND and NOR gates, transmission gate, etc., work on the similar principle and are discussed extensively in CMOS circuit books. The inverter circuit illustrates the unique power characteristics of CMOS circuits, i.e., there is no DC current in the circuit in the two stable states, especially between the supply and ground, and hence no power is consumed in the steady-state situation. This is because there is no conduction path between V_{dd} and the ground except for the subthreshold leakage, due to either the pMOSFET or the nMOSFET is off in the steady state. Only when the input is being switched, from low to high or high to low, there is a transient current. Since this switching occurs very fast, very little energy is consumed during the switching process.

The energy efficiency of the CMOS circuits is the main reason CMOS circuits became the building blocks of virtually the whole semiconductor industry. Like in the case of the inverter, CMOS circuits only consume significant power during switching. As a result, the dynamic power consumption of CMOS-based electronic systems is generally given by

$$P_{\mathrm{dynamic}} = \frac{1}{2} f C V^2 = I_{\mathrm{dd}} V_{\mathrm{dd}} \tag{2.32}$$

where f is the switch frequency, C is the total capacitance of the part of the circuit that are switching, and V is the voltage swing of the signal, and I_{dd} is the average current. Typically for full rail systems, the voltage swing is the supply voltage, V_{dd}. Thus, it is desirable to keep V_{dd} as low as possible to minimize the dynamic power consumption.

The earlier statement is only correct when leakage currents, including gate to substrate, source–drain leakage, etc., can be neglected. For advanced process with ultrathin gate oxide and very short channel, various leakage mechanisms can cause

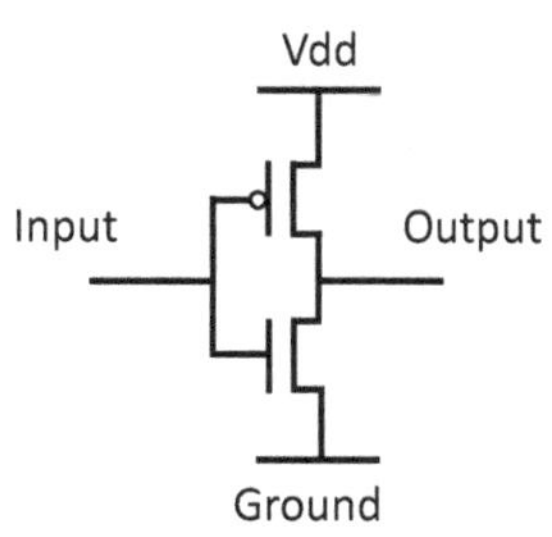

Fig. 2.24 CMOS inverter circuit

substantial DC leakage current. For example, for the inverter when the input is high, the gate of the nMOSFET is at V_{dd}, while the output is at ground. There may be substantial leakage from the gate to the p-well, due to the direct tunneling current when the gate oxide is very thin.

The leakage power can be written as

$$P_{static} = I_q V_{dd} \tag{2.33}$$

where I_q is the leakage current. The leakage is especially important for battery-powered mobile devices as most of the time the devices are in the standby mode.

Reducing supply voltage V_{dd} will reduce the total power consumption of a chip. However, the speed of a CMOS circuit, as represented by CV_{dd}/I_{dsat}, will also be reduced. Thus, the design of an IC always involves the trade-off between the power consumption and its speed, among other important considerations such as cost and reliability.

The core functional blocks are usually powered by the core supply that is as low as possible, with voltage of 1 V or less are common for the advanced processes, subject to the speed requirement. The core voltage is determined by the $I_{on} - I_{off}$ characteristics of the core transistors, with the threshold voltage being the main determination factor. Speed of transistors is directly proportional to I_{on}, which in turn is proportional to the voltage over drive $(V_{dd} - V_t)^2$. The leakage current is exponentially dependent on V_t through the subthreshold slope (mV/decade of current). Thus, the core V_{dd} and V_t are chosen by the trade-off between the speed of the circuit and the power consumption—both dynamic power and the leakage power considerations.

On the other hand, the **input and output (I/O)** circuits are usually powered by a second supply. I/O voltage is typically set by the interface requirement with other chips, often determined by legacy circuits. This voltage is usually higher than the core supply voltage in order to drive the large capacitance of the long bond wires and PCB traces. Common I/O voltage is 1.8, 2.5, or 3.3 V.

As a result of this requirement, a CMOS process will usually provide transistors for both core voltage and a separate set of transistors optimized for higher I/O voltage.

Sometimes to reduce cost and for simplicity of use, one supply voltage, usually the high I/O voltage, is used. Internal on–chip down-converter circuit is then used to generate the core voltage. This usually only works well if the core circuits do not consume too much power and the current consumption does not fluctuate very much (i.e., the IR drop on power line or the switching noise is low). Otherwise the noise generated from the fluctuation and the conversion loss may be too great to justify the benefit of the single supply.

On a system level, often different blocks of the IC may be run at different voltages at different time, using the so-called dynamic voltage frequency scaling (DVFS). Many circuit techniques for reducing power consumption are detailed in the book CMOS VLSI Design (Weste & Harris, 2011).

2.6 Non-logic CMOS Devices Discussed in This Book

All the devices discussed in the previous sections in this chapter can be obtained from the basic CMOS processes which we will describe in the next chapter. For cost reasons, most often only the basic versions of these devices are available in the standard process offerings. Variations of these devices are available but often with additional process steps or modifications and come with a cost added to the basic process. The rest of this book teaches the art of designing some of the nonstandard devices using the standard processes.

The devices we will discuss include

- MOSFETs with nonstandard threshold voltage and MOSFETs that can tolerate much higher voltage than those offered by the foundries (Chap. 4);
- Floating-gate device transistor and capacitors (Chap. 5);
- Parasitic bipolar transistors and punchthrough devices (Chap. 6);
- Poly p–i–n diodes and Schottky diodes (Chap. 7).

The rest of the book (Chaps. 8–13) discusses applications of these nonstandard devices.

References

Diorio, C. (1997). Neurally inspired silicon learning: From synapse transistors to learning arrays. Caltech, Ph.D. thesis.

Hu, C. (2009). *Modern semiconductor devices for integrated circuits*. Upper Saddle River: Prentice Hall.

Takeda, E., Yang, C., & Hamada, A. (1996). *Hot-carrier effects in MOS devices*. San Diego: Academic.

Taur, Y., & Ning, T. H. (1998). *Fundamentals of modern VLSI devices*. Cambridge: Cambridge University Press.

Weste, N., & Harris, D. (2011). *CMOS VLSI design: a circuits and systems perspective*. Boston: Addison-Wesley.

Yeo, Y.-C., Lu, Q., & Hu, C. (2001). MOSFET Gate oxide reliability: anode hole injection model and its applications. *International Journal of High Speed Electronics and Systems, 11*, 849–886.

Chapter 3
Overview of Logic CMOS Processes

Abstract In this chapter, we review the process steps of a generic, planar logic CMOS process. The fabrication of I/O devices, which has a different gate oxide from that of the core devices, is shown in parallel with the core devices. We also discussed the fabrication of native devices and potential use of spacer for charge storage—two features that will see use in later chapters. The basics of mask design, process monitoring, and wafer fabrication economics are then reviewed to highlight the best practices discussed in Chap. 2, i.e., the advantages and trade-offs to use the simplest and basic CMOS process.

Keywords Semiconductor processing • Design rules • Mask • Photoresist • Implantation • Photolithography • Silicide • Spacer • Charge trapping • Native transistor • I/O transistor • Process control monitor • Yield • Layout • LOCOS • STI • Foundry

3.1 Generic Logic CMOS Processes

3.1.1 Introduction

As mentioned in the previous chapter, because of the energy consumption constraints, virtually all digital ICs today are CMOS based. The basic building blocks of CMOS circuits are p-type and n-type MOSFETs. In addition, I/O transistors are also required to interface with other chips or the physical world, often at a voltage that is different from the core supply voltage. So the most basic CMOS process that a foundry usually offers include a pair of core MOSFETs and a pair of MOSFETs for I/O. This is the generic CMOS process that we will review in this chapter.

What we refer to as a generic logic process is the *de facto* standard logic CMOS process offered by most of the semiconductor foundries. Virtually all general-purpose foundries follow the International Technology Roadmap for Semiconductors (ITRS) (www.itrs.net) closely, which, for each process technology node, specifies the basic device performance parameters. Detailed process steps from each foundry may differ, but market forces in both equipment and electronic products usually make the generic logic process from each foundry very similar. Large fabless semiconductor companies typically require multiple manufacturing

Y. Ma, E. Kan, *Non-logic Devices in Logic Processes*,
DOI 10.1007/978-3-319-48339-9_3

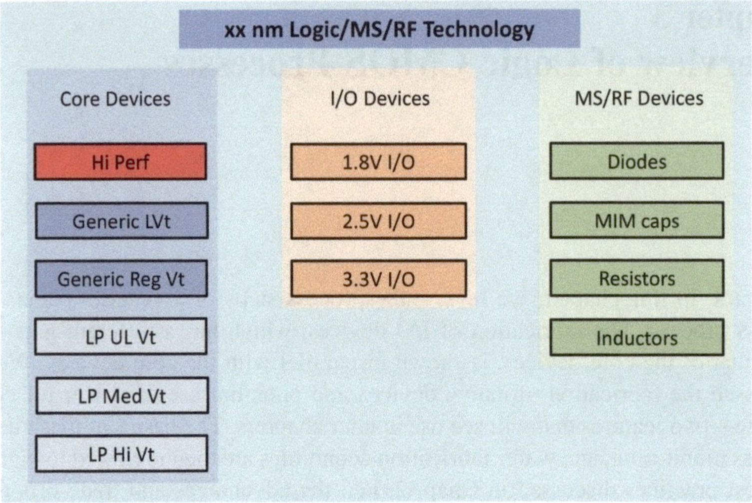

Fig. 3.1 Device palette from an example xx nm technology

partners for their ICs—to avoid supply disruptions and to instill competitions among their suppliers in order to get the best price. As a result, the smaller foundries are motivated to make their processes similar to those from the leading foundries in order to win some of the second sourcing business.

For example, since TSMC (Taiwan Semiconductor Manufacturing Co.) dominates the foundry business at the 28 nm process node, other foundries usually try to match the process and device parameters of the corresponding TSMC process. As a result of this kind of business dynamics, we usually have a *de facto* standard logic CMOS process at each technology node.

Figure 3.1 shows the basic devices that may be available in a nanometer-scale process, e.g., 28 nm technology from a foundry. We start from the core devices listed on the left column. Several process variations are given, each optimized for high performance, generic, and low-power applications, respectively. Each process option has several transistors with different threshold voltages. For example, the low-power process option has transistors with ultra-low V_t, regular V_t, and high V_t available. Usually the basic process variation has only one V_t choice. The additional V_t's are available only with additional processing steps, usually involving extra masks, which as mentioned in the previous chapter may incur additional cost.

In the middle column, three options are available for the I/O devices at 1.8, 2.5, or 3.3 V operations. Usually only one of these options can be chosen. This is because reliability of the gate oxide at different operating voltage requires different oxide thickness. As we will see in the next section, usually one CMOS process can only support two gate oxide thicknesses, one for the core device and one for the I/O devices. Supporting too many different gate oxides is logistically complex and expensive to do, and usually not necessary.

At the right are devices for mixed-signal and RF operations, including diodes, metal–insulator–metal (MIM) capacitors, different resistor options, and inductors. Often extra masks are necessary for some of these device options. Therefore, the use of these devices usually comes with an increase in the manufacturing cost of each chip.

In the remainder of this chapter, we will discuss the process for manufacturing the basic core and I/O MOSFETs. In Chaps. 4–7, we will discuss the methods of designing and making other devices using the generic logic CMOS process.

3.1.2 Process Flow and Mask Design

In this section, we describe a generic *planar*[1] logic CMOS process with one polysilicon layer and several metal interconnect layers. This generic CMOS process is capable of providing four types of MOSFETs, core and IO transistors of n- and p-type. A number of other devices, such as diodes, poly resistors, and variable capacitors (varactor), will also be available with these MOSFETs without any additional processing steps. There are many excellent books or book chapters showing the basic CMOS processing steps, e.g. (Jessing, 2007) completed with SEM cross sections. In the following, we will review only the basic steps for making all four types of devices.

The semiconductor manufacturing process is like a traditional machine manufacturing process—the essence is to transfer the circuit design pattern that exists in the digital media (like the blueprints on paper) to the substrate which is, in our case, the silicon wafers. The key is to reproduce, within the manufacturing tolerance, the design onto the physical substrate with high fidelity and low cost. The manufactured device needs also to be durable, i.e., with high reliability.

Semiconductor fabrication consists of a series of materials addition (through deposition, growth, or ion implantation) and subtraction (etching and polishing) steps to produce the designed circuit pattern. These series of steps are translated from circuit layout patterns which are divided into layers. Each layer can be patterned on photomasks which is used to transfer, through photolithography process, the pattern on to the silicon substrate.

Our four types of FETs need about 20 masks in total. The number of mask layers is usually the most important metric for a given process as the processing cost is directly related to the number of masking layers. Within a given number of mask layers, another important quantity is the number of critical masking layers, which are needed to pattern the smallest features of the circuits. The masks for patterning the polysilicon gate and for patterning the contacts as well as the first metal layer are

[1] As opposed to the finFET process below ~20 nm process node, which makes 3D FETs. Chapter 13 contains a brief description of the process.

Fig. 3.2 Cross section of the wafer after nitride deposition

usually the most critical mask layers with the smallest size and pitch. The mask layout and generation will be covered in Sec. 3.2.

We start the wafer processing with a blank or raw silicon substrate. Most commonly used substrate in CMOS processing is p-type silicon wafer with about $10^{15}/cm^3$ boron doping, which translates to a substrate sheet resistance of about 10 Ω/square. Two kinds of starting wafers: prime and epi, where prime wafers are cheaper with more defects and epi-wafers have an additional 8-μm epitaxial silicon layer grown on top of the prime wafer. The epitaxial layer is grown at lower temperature than the prime wafer made by the Czochralski cylinder-pulling method, and hence has lower defect density and a better control of the doping profile. The epi-wafer also has better latchup resistance.

Integrated circuits are usually built on bulk silicon substrate.[2] Since the bulk of the wafer is semiconducting, there is a need to isolate devices from each other. Though passive and lacking any active electrical function, isolation technology nevertheless plays a critical role in overall IC scalability. Before the 1980s, the 0.25 μm and larger process technologies used **LOCal Oxidation of Silicon** (**LOCOS**) technology for isolation. Most modern processes, at 0.18 μm and below, have used **shallow trench isolation** (**STI**) or deep trench technology which enables denser packing of devices without better electrical isolation, particularly in view of field penetration.

The first major step of our generic CMOS process is to make the trenches for the STI. We start by a ~10 nm thermally grown silicon oxide on the surface of the silicon wafer, called a pad layer. Next ~100 nm silicon nitride is deposited on top of the pad oxide to use as a hard mask, as illustrated in Fig. 3.2. The pad layer is a buffer layer to promote the adhesion of the nitride on the silicon surface. Nitride films usually have much higher internal stress as well as have a very different thermal expansion coefficient than silicon. As a result, the pad oxide layer is needed to help with the adhesion and promote the stability of nitride film on the silicon substrate.

Next photoresist (PR) is coated and is patterned with the active area (AA) photomask, which is also often called diffusion, OD, and thinox layer. The active area is where the transistors are to be built on. The rest of the area is used for isolation. Figure 3.3 shows a layout of the active area.

[2] Silicon-on-insulator (SOI) wafers have been used in some applications.

Fig. 3.3 Layout of the active area, which is indicated by the *dark rectangles*. The rest of the areas will be etched and filled with oxide for isolation

Note that the actual pattern on the masks will depend on the PR used in the patterning processes. There are two types of PR:

1. For *positive resist*, the part of the PR that is exposed to light becomes soluble to the PR developer by breaking photosensitive linkage of resin molecules. The portion of the PR that is unexposed remains insoluble. During development, the pattern is sharper than the aerial image by photosensitive proton generation, which creates a threshold effect called chemical amplification. A direct consequence for chemical amplification is the line-edge roughness, which remains one of the major factors in process variation. Because the mask is used to block the exposing light, the mask for positive PR contains an exact copy of the pattern of PR which is to remain on the wafer.
2. For *negative resist* the portion of the PR that is exposed to light becomes insoluble to the PR developer. The unexposed portion of the PR is dissolved. Masks used for negative PR, therefore, contain the inverse (or complement) of the pattern to be transferred.

Most IC processes use positive resists because they offer better process controllability for small geometry features. The mask contains an amplified pattern image, often from 4:1 to 10:1, which will be proportionally reduced to the actual PR pattern on the wafer.

The AA patterning steps include spin coating of PR (with thickness of around 100–1000 nm), and the PR exposure inside the photolithography tool called a "stepper" or "scanner," as repetitive cm-level patterns are tessellated on the whole wafer step by step. For an example of 10:1 amplification, a 1 cm^2 area on the wafer will need a mask of 100 cm^2 in size. This is followed by the development of the exposed PR. The result will look like the cross section shown in Fig. 3.4.

Next the patterned wafers are transferred to multichamber etchers for a multistep etching process using the patterned PR as the etch mask, which may include:

Fig. 3.4 Cross section of the wafer including photoresist after exposure and development

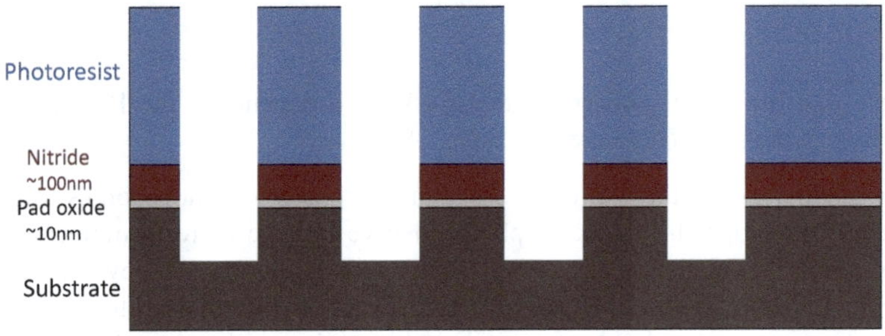

Fig. 3.5 Cross section after etching of shallow trenches for STI

1. Dry nitride etch;
2. Dry pad oxide etch;
3. Dry etch of deep (~ 300 nm) trench into the silicon substrate.

These etch steps form the trench for the shallow trench isolation. The cross section now looks like what is shown in Fig. 3.5. Dry etch is assisted by directional electron or ion bombardment of the surface, and hence can create a straight side wall instead of undercut in an isotropic wet etch.

The PR is then removed, wafer cleaned, and a thin layer of oxide is grown through thermal oxidation. This oxide layer will help with the adhesion of the deposited oxide in the next step.

Next a thick layer of oxide is deposited to fill the trench. As the oxide will be deposited everywhere, both in the trench and on the active area, after deposition the wafer surface will not be planar. The wafer can be planarized using a chemical mechanical polish (CMP) process.

In a common practice, another mask that is the complementary as the AA mask is used to pattern and remove the tall and excess oxide on top of the active area to achieve a more planarized surface. This practice increases the uniformity that can be achieved from the CMP process and has less contamination from the CMP slurry which contains metal particles (Fig. 3.6).

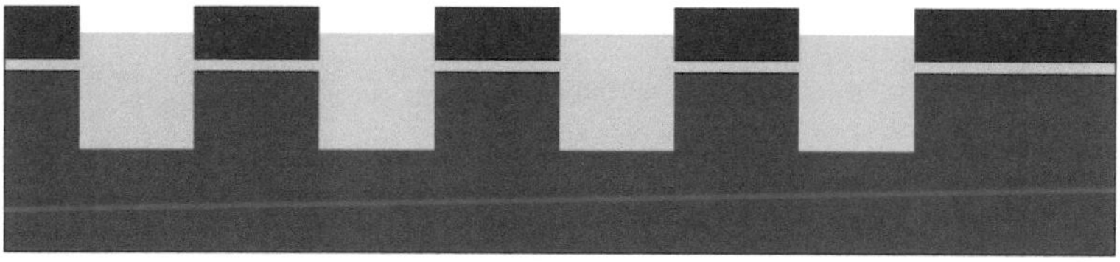

Fig. 3.6 Cross section after filling and CMP of trench isolation

Fig. 3.7 Cross section after the formation of trench isolated active regions

In retrospect, we can see the importance of the nitride hard mask. As selective areas will need to grow additional pad thermal oxides and deposited oxide, which are high-temperature processes above what PR can tolerate, it is necessary to transfer the PR pattern to a "hard" material for masking in the high-temperature process. In the CMP option, the nitride hard mask can also serve naturally as the CMP stop. It is difficult to use an estimated rate to accurately stop the etch or CMP process. An etch or CMP stop layer (i.e., the material will not be etched or CMP away) together with end-point detection will help greatly the layer removal control.

In the next step, the nitride is removed, usually by wet etch, which has a higher etch selectivity than dry etch. This step leaves a nearly flat surface to build devices on, as shown in Fig. 3.7, where the active areas are laterally separated from each other by trenches that have been filled with oxide.

Next, we regrow a sacrificial pad oxide before the ion implant step. The wafers are then patterned with n-well implant mask (NW), which is illustrated in Fig. 3.8.

Note that the NW mask is drawn similar to AA. However, during ion implant we actually want to implant in the regions that were drawn, i.e., we want to block the ion implant from the regions that are outside the drawn n-well regions. PR can be directly used as the implant mask without resorting to tranfer to a hard mask, as ion implantation is a low-temperature process, where the temperature is chosen from the consideration of the amorphotization threshold. The implant damage will be annealed in later steps. So during the actual mask making step, the n-well layer is specified as having a dark tone, while the AA layer was to have to the clear tone. So a pattern that is inverted of the drawn n-well pattern is etched on the mask. After exposure, the photoresist is removed in the areas where n-wells are drawn, in contrast to the AA mask where the PR remains in the areas where the AA is drawn. Figure 3.9 shows the cross section of the PR after the n-well exposure. Note the contrast to Fig. 3.4, where the PR covers the region of AA mask. On

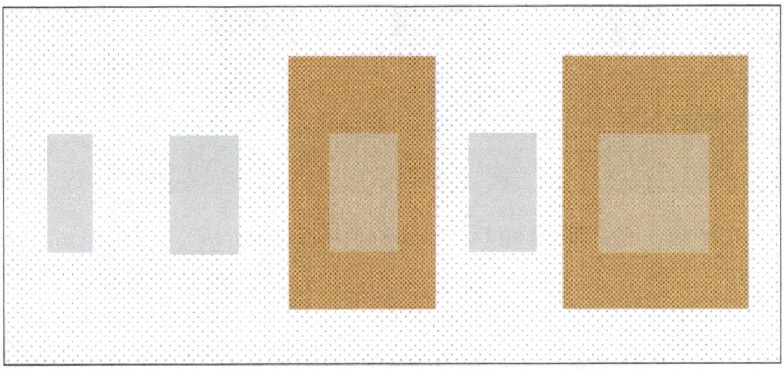

Fig. 3.8 Layout of n-well mask (*yellow squares*), shown together with the active regions

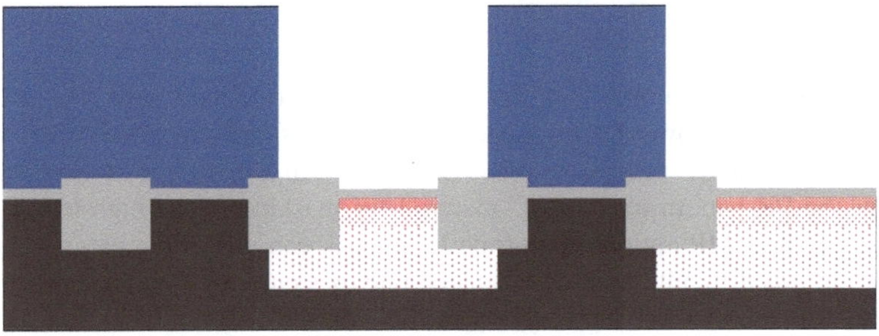

Fig. 3.9 Cross section after n-well pattern and ion implant

Fig. 3.9, where the n-wells are drawn, the PR has been removed. Both have used positive PR.

With the PR pattern, the wafer is sent through an ion implanter. Three implants are performed at this step, including the n-well implant, an anti-punchthrough implant, and a threshold voltage adjustment implant that is used to fine-tune the threshold voltage of the pMOSFET. Implant angles between the source and the wafer are additional control factors for eventual doping profiles.

The earlier process for n-well is repeated with a p-well mask (PW) to form the p-wells. Cross section after the p-well patterning is shown in Fig. 3.10.

To ease the layout effort, the p-well mask is usually generated automatically using logic operations, as a complement of NW. As a result, all areas that do not have the NW implants are implanted with PW implants. Similar to n-well implants, p-well pattern also has triple implants for anti-punchthrough protection and for the threshold voltage adjustment of nMOSFET.

A special case during p-well pattern generation is the formation of the so-called native n-MOSFET. ***For native nMOSFET, all three implants are skipped during p-well formation***. The threshold voltage of the native nMOSFET is determined by

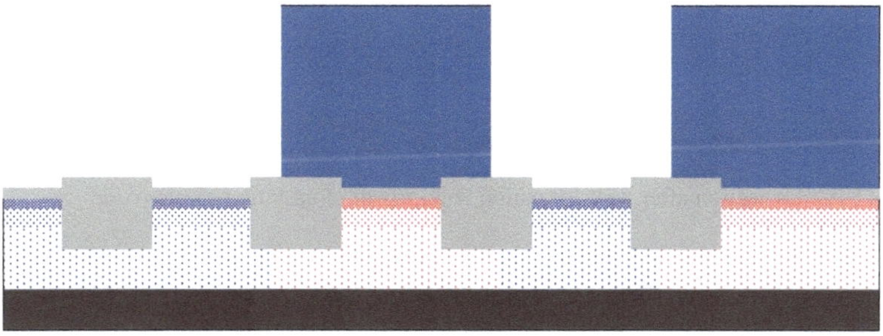

Fig. 3.10 Cross section after p-well pattern and implant

the doping concentration of the substrate and, because of the low doping, is usually close to 0 V.

To make a region of the native nMOSFET that is devoid of the usual p-well implants, we can define a dummy layer that marks the areas for the native nMOSFET. The dummy layer, which we will call NAT, is not made into a real mask. It is only used during the logic operation to generate the PW mask. During the PW mask generation, the area covered by the dummy layer will be blocked. This can be accomplished with a logic operation by defining

$$\text{PW} = \text{NOT}(\text{NW} \ \text{ OR } \ \text{NAT})$$

After the wells are formed, we begin to grow the gate oxides.

Gate dielectric is one of the most crucial components of the modern MOSFET. For the longest time, SiO_2 is used as the gate dielectric in MOSFETs. SiO_2 has a reasonably high relative dielectric constant around 3.9, can be grown natively on Si surface, and has other properties well suited for MOSFET with excellent manufacturability. In fact, SiO_2 may well be the reason that silicon, rather than germanium, became the dominant semiconductor substrate in the history, especially considering the fact that the first transistor invented at Bell Labs is a Ge point-contact transistor. The native germanium oxide is water soluble, thus not suited for mass manufacturing which uses a lot of wet chemistry. One of the main reasons Ge did not become the substrate of choice for ICs is the lack of a high-quality germanium oxide that can be used to form the MOS structure.

The other dominant advantages of the silicon wafers include the high Young's modulus to make flat and large-area wafers and the high material purity with suitable impurity getters. It would be nearly impossible to replace silicon wafers as long as the present lithography tools are used, even though we can now grow good-quality dielectrics on other semiconductor wafers by deposition. If we really need another semiconductor such as GaAs or GaN for special-purpose devices, the current trend is to grow these materials on selective islands on the silicon substrate.

Traditionally SiO_2 gate dielectric is grown using thermal oxidation as this method produces oxide and interface of highest quality. During the oxidation

Fig. 3.11 Cross section after well formation and growth of an intermediate oxide

process, oxygen atoms diffuse through existing SiO_2 film on the surface to the Si–SiO_2 interface and react with the Si atoms at the interface. This model results in an oxide growth rate that decreases as oxide film thickness gets thicker (Deal & Grove, 1965).

As mentioned in the beginning of this chapter, typical logic CMOS processes usually offer two gate oxide thicknesses: one thin oxide for core FETs that are rated at lower voltages and a second, thick oxide for I/O FETs that need to operate at the higher I/O voltage. An example is the 0.18 μm process, the core FET has oxide at 3 nm while the 3.3 V I/O FET has oxide at 7 nm. For a 90 nm process, the core FET may have oxide at 1.2 nm with operating voltage of 0.9 V while the I/O FET may have oxide at 5 nm for 2.5 V operation.

To produce these two oxides, a two-step process has been deployed—we first grow an intermediate oxide that will form the eventual thicker oxide, and then grow the thin oxide. For the partial oxide growth, we first remove the photoresist from the wafer in the previous step. Then the pad oxide is removed and cleaned. A high-quality oxide is then grown that has the thickness of immediate between the thin and the thick oxide. For example, a 5 nm oxide may be grown in the earlier 0.18 μm process example (Fig. 3.11).

A thick oxide mask (AA2) is provided to cover the thick oxide region. After photo process the PR will cover the thick oxide region. With this protective layer, the immediate oxide is wet etched in the thin oxide region and preserved in the thick oxide region.

The PR is then removed and the wafer cleaned. The whole wafer is then put into the growing furnace and oxidized again. The growth time is set to grow a thin oxide in the region specified for the core devices, 3 nm oxide in the 0.18 μm example. During the same growth process, the growth will also occur in the regions with the intermediate oxide, albeit at slower growth rate. This intermediate oxide eventually grows to have the targeted oxide thickness; as illustrated in Fig. 3.12 in the earlier example, this intermediate oxide of 5 nm will grow into a 7 nm final oxide thickness.

Figure 3.13 shows a layout example for the thick oxide area mask (AA2), which defines the area for the I/O FET.

Note that for advanced process nodes that require gate oxide thickness of 1.2 nm or thinner, SiO_2 becomes too leaky due to the direct tunneling discussed in Sect. 2.3.2. As a result, SiO_2 has been hybridized with high dielectric constant (high-k) gate dielectrics such as HfO_2 or ZrO_2. The high k gate dielectrics are usually deposited by the **atomic layer deposition** (**ALD**) method. So the earlier dual

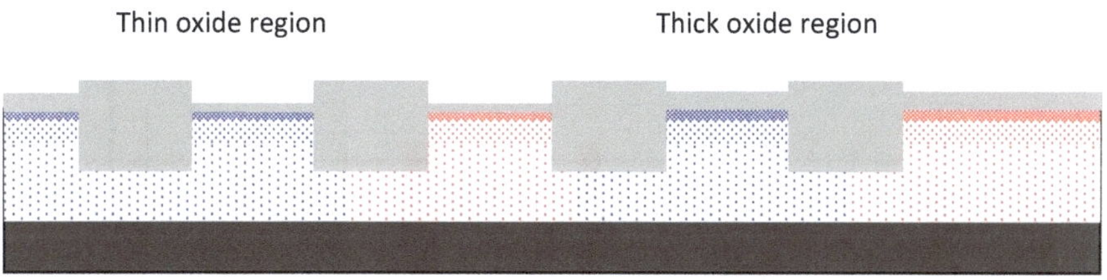

Fig. 3.12 Cross section illustrating the formation of thin and thick oxide regions

Fig. 3.13 Layout illustrating the thick oxide region

oxide growth process only applies to process nodes of 90 nm or above. The complicated process is demanded by the interface quality of the gate oxide, which can change the threshold voltage and operation lifetime if the interface contains too many traps.

After both gate oxides are formed, an undoped polysilicon layer is uniformly deposited over the wafer by CVD. This gate layer is then patterned with the gate mask (GM), often called poly or PO mask by foundries, with the layout shown in Fig. 3.14:

After the polysilicon etch, the wafer cross section will look like Fig. 3.15.

One thing to note is that the actual gate length is often different from the nominal process technology node. Transistors fabricated at the 130 nm process node and above had gate lengths that were about the same as the process node, which is denoted by the half pitch of the most tightly spaced poly. From about the 110 nm to the 65 nm node, the gate lengths shrank faster than the process node, being shorter than the process node. For example, the 110 nm process node may have a nominal physical gate length of about 90 nm. Because of the difficulty in patterning, the gate length again approaches the actual length specified by the technology node for the 45 nm, i.e., 45 nm gate length for the 45 nm process technology node, and below process nodes. For the 20 nm and below processes, the gate length may actually be more than those specified by the process technology node.

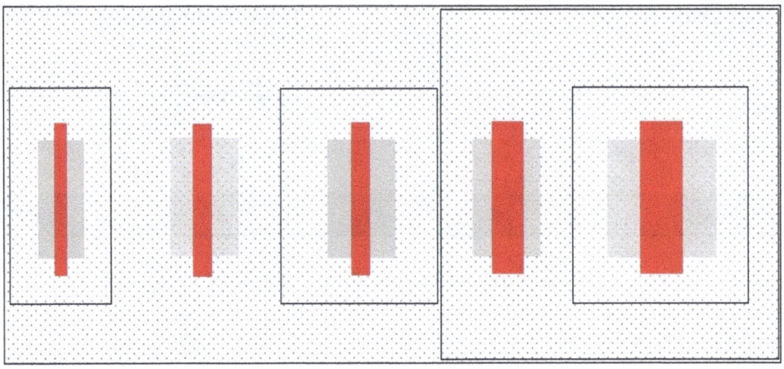

Fig. 3.14 Layout illustrating the gates in *red*

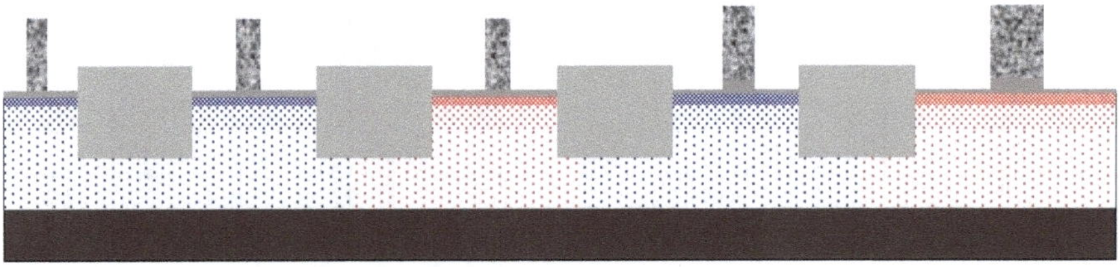

Fig. 3.15 Cross section after gate etch

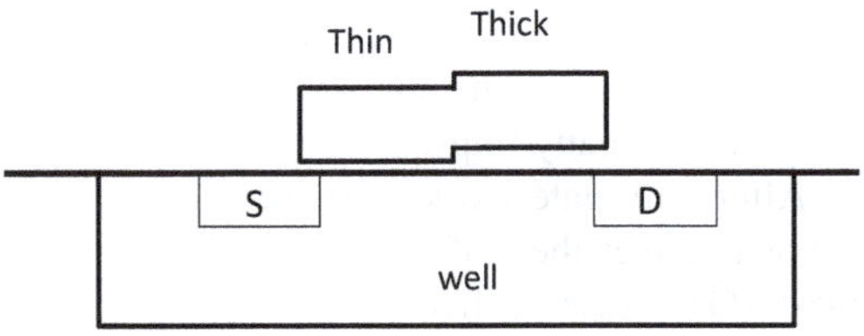

Fig. 3.16 A hybrid MOSFET with a thin gate oxide channel and a thick oxide channel. This device can be used as an antifuse

One important variation of the separate thin and thick oxide devices is that it is possible to make a MOS structure that has a hybrid oxide with both a thin and a thick portion along the same channel (Peng & Fliesler, 2005), as illustrated in Fig. 3.16. The device can be either a MOSCAP or a MOSFET. A practical use of this hybrid device is an antifuse for nonvolatile data storage. As we noted in Chap. 2, the breakdown of the oxide is very sensitive to the thickness of the oxide. A thin oxide, e.g., 3 nm thick, may breakdown at about 5 V while a 5 nm oxide may breakdown at 8 V. To program an antifuse we will need to supply a voltage higher than the breakdown voltage, e.g., 6 V or higher, to ensure all the antifuses will be programmed. But this programming voltage will need to be routed by transistors that can sustain this voltage without breaking down, i.e., with a thick gate oxide. In the hybrid device, the thick oxide portion is used as a select or drive transistor that can tolerate high voltage and the thin oxide device is the antifuse element that actually breaks down after programming. The application of this device is discussed in more details in Chap. 9.

Note that in our hybrid device the boundary between the thin and thick oxide region is drawn and not self-aligned to the polysilicon gate. Since in general the AA2 mask is a noncritical mask, a lower grade mask is usually used and the alignment between AA2 and AA typically is not very precisely controlled. As a result, this hybrid device cannot be a minimal-size device. Design rules set a minimal distance between the thin/oxide boundary to the edge of the polysilicon on both sides. Special care is also needed to ensure the alignment during the manufacturing process.

After the gate formation, the next step is to implant lightly doped drain (LDD) and superhalo extensions if they are employed. LDD and the source–drain usually share the same layout pattern for each of the implant type. The actual masks for the LDD and the source–drain implants may be different. Similarly the core and I/O FETs may have different implants to control their respective short channel effects.

The N-LDD mask for the core FETs can be logically generated from N+ S/D implant layout. The common practice is to generate this mask as a N+ region minus the N-well region and the thick oxide transistor region, e.g.,

$$\text{NLDD\_core} = ((\text{NPlus Not NW})\text{Not AA2}) \tag{3.1}$$

For the N-LDD for the I/O transistors, the mask is generated including the thick oxide region, e.g.:

$$\text{NLDD\_IO} = ((\text{NP Not NW})\text{AND AA2}) \tag{3.2}$$

LDD implant masks for the core and I/O pMOSFETs are generated in the similar ways. Figure 3.17 illustrates LDD layout patterns.

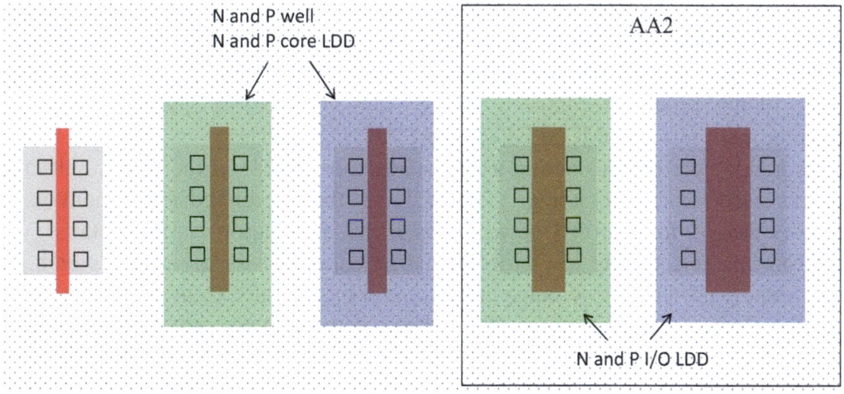

Fig. 3.17 Layout of LDD masks

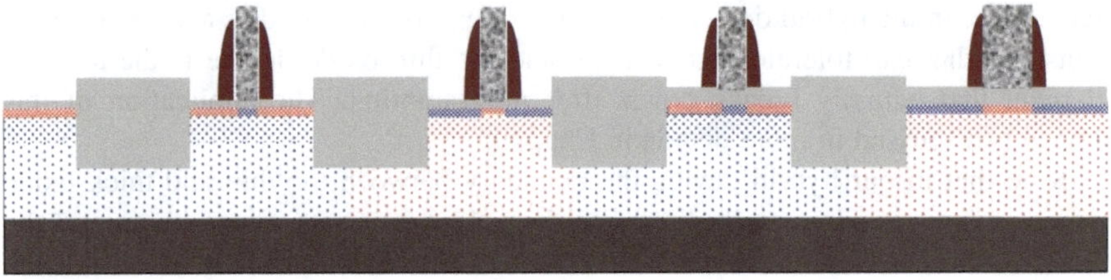

Fig. 3.18 Transistors after spacer formation

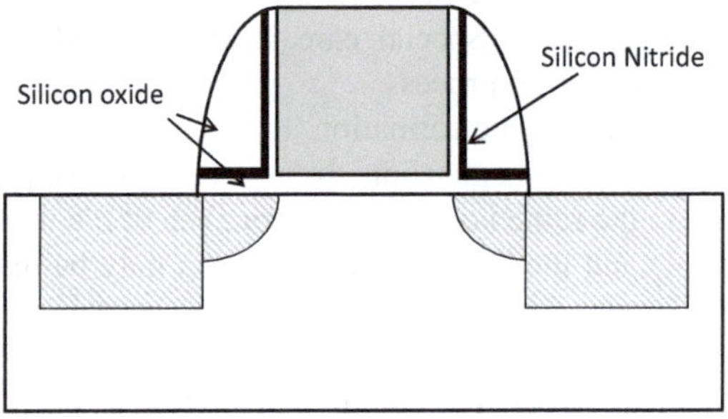

Fig. 3.19 Details of spacer including silicon nitride layers

After the LDD implants, spacers are formed by first uniformly depositing a layer of silicon nitride or a composite layer of oxide/nitride. Due to the nearly isotropic deposition method, the side walls of the poly silicon gate will be covered by approximately same amount of oxide/nitride.

After dry plasma etch, which mostly etch in the vertical direction, the side wall nitride/oxide remains to form the side wall spacer, as illustrated by Fig. 3.18.

The LDD and the associated sidewall spacer structure is one of the most important inventions that enabled the continued scaling of MOSFET. Various spacer structures have been used. A common spacer is a multilayer of thermally grown silicon oxide, silicon nitride, and deposited silicon oxide, as illustrated in Fig. 3.19. The advantage of this ONO multilayer spacer is the reduced gate to source or drain leakage during the silicidation process (discussed later) and the use of nitride layer as the etch stop for contact etch. This type of spacer is used until about 40 nm process technology where strain engineering dictated the use of different materials for pMOSFET (Singer, 2011).

As a side note besides acting as the spacer, the nitride in the ONO multilayer may be used for charge storage. The FET can then be turned into a nonvolatile memory device. In Chap. 10 we will discuss this application in more details.

ONO is a self-sealing insulator against percolation path formation, which has been broadly used in high-voltage insulation for many years. Consider a percolation path has formed and an electron current flowing through the path. ONO contains a potential well and many charge traps inside the nitride layer, and some of these charges will be stored, which in turn cause an electrostatic repulsion to divert further current flow.

After spacer formation, source–drain implants are performed using the N+ or P+ implant masks. This implant step will dope the polysilicon gate of the respective nMOSFET and pMOSFET as well.

After the source and drain doping, the areas of source, drain, and polysilicon are silicided to reduce their sheet resistance. For the area that we do not want to be silicided, e.g., the poly resistor region, a silicide block (SAB) layer is used. Oxide or nitride films are used as the silicide block layer. This region will be covered by PR and patterned by the SAB mask. The silicidation process proceeds with metal deposition, Ti for above 0.25 μm, Co for 0.18 μm to 90 nm technology nodes, and Ni for the 90 and 65 nm processes, annealing which will cause the deposited metal to react with silicon (but not with silicon oxide or nitride) to form silicide over the source, drain, and gate. The unreacted metal, e.g., on the spacer and isolation region, is removed by wet etch in acids afterward. As the silicide will form only on exposed Si and poly regions, it is self-aligned without the need of another critical mask (except for poly resistors). This is often called the self-aligned silicidation or salicide.

The annealing thermal budget for dopant activation during source–drain recrystallization (as the source–drain implant dose often amorphousizes the implanted region) and the salicide formation can be combined, but each process will go through detailed control to optimize the parametric yield.

Note that for more advanced processes that use the finFET transistor structure, silicide is no longer used. Starting when Intel introduced their finFET technology in 2012 in the 22 nm node process, in all finFET processes the silicided contacts were replaced by trenched contact, eliminating the silicide process all together (Das, 2016).

After the salicide formation, a layer of Si oxynitride or Si nitride is deposited to act as an etch stop layer. This is followed by interlayer dielectric deposition, contact patterning, and etch.

From here, multiple layers of metal interconnect will then be deposited and patterned, to make electrical connections between devices. State-of-the-art processes allow up to 11 layers of metal interconnect with different thickness and pitch.

The process before the source–drain–gate metal 1 contact generally is referred to as the front end of the line (FEOL) while the process for fabricating the further metal interconnects is called the back end of line (BEOL) for a given process. We will not discuss BEOL process steps in this book, as only regular metal structures will be used for non-logic devices, even for RF shielding in the waveform-shaping applications.

One final note is that the process flow described in this section is for logic processes using polysilicon gate. Since around 2007, more advanced CMOS processes have moved to metal gates in conjunction with the move to high-*k* gate dielectrics. For example, Intel has revealed some of its 45 nm HKMG process technology (Mistry, Allen, Auth, & Zawadzki, 2007). Intel disclosed that the I/O transistor shares the same gate metallization as the core logic transistor but has a thicker gate dielectric. The I/O transistors use a gate dielectric that is comprised of a 1.9 nm thick HfO_2 layer on top of a 3 nm thick silicon dioxide layer. In contrast, the core logic transistors use a gate dielectric that is a 1.2 nm thick HfO_2 on top of a 0.9 nm thick silicon dioxide layer.

In conjunction with the introduction of metal gate, the gate formation process has moved to a gate last process. One of the reasons for moving to the gate last process is because metal gate is often of a multilayer structure for threshold voltage adjustment. The source and drain formation usually include high temperature anneal process, which could damage the intended gate structure, e.g., the interdiffusion of the different gate metals.

In the gate last process, a dummy gate, e.g., made of polysilicon (Yamasaki, Asai, & Kuramada, 1982) or silicon nitride (Ma et al. 1999), is first deposited and patterned. The source and drain are then formed with the necessary implants and annealing. The IMD is then deposited and CMPed to expose the dummy gate. The dummy gates are then removed using appropriate etch process. In their places, the metal gates are then deposited to maintain the self-alignment with the source–drain structure. The front-end process needs to maintain the tightest critical dimension (CD) with highest parametric yield. The gate last process is a good example on how we trade process complexity with critical dimension.

3.2 Layout and Mask Generation

For a given process technology, the key element to faithfully reproduce the IC design on a silicon wafer is the mask generation. The following table summarizes the key masks needed for the FEOL process steps discussed in the previous section:

In addition to these FEOL masks, many more mask layers are needed to complete the BEOL to produce the final functional IC. Typically, each metal layer will need two masks, one for the patterns and the other for vias. Because CMP is used extensively in BEOL as damascene or dual damascene, dummy patterns are frequently added to enhance the CMP parametric yield. FEOL and BEOL have in total about 40–50 masks to produce an IC in the advanced technology nodes with ten metal interconnect layers.

The mask patterns are based on the circuit schematics by the layout or physical design process. Layout for most digital ICs can be done with automated CAD tools using a cell library of layouts for standard circuit components, like inverters of various nMOSFET and pMOSFET sizes. For analog circuits and some specialized digital ICs that need custom optimization, hand layout is still a common practice. With assistance from the CAD tools, it will be impractical, or unnecessary to layout each mask layer by hand. The common practice is to layout a few essential layers by hand, and then to generate the rest of the masks by logic operations from these hand drawn layers.

For example, the following layers, AA, NW, PO, contact, AA2, SAB, NP from Table 3.1 may be hand drawn. The other layers are then generated automatically from the operations listed in Table 3.2.

Table 3.1 Typical FEOL masks needed for a logic CMOS process

1	AA—active area
2	Reverse AA mask, generated from AA layout
3	NW—n-well mask
4	PW—p-well mask generated from n-well layout
5	AA2—thick oxide area
6	PO—poly gate area
7	NLDD—core nFET area needing LDD
8	PLDD—core FET area needing LDD
9	NLDD2—LDD for I/O nFET
10	PLDD2—LDD for I/O pFET
11	NP—N+ source and drain
12	PP—P+ source and drain
13	SAB—area needing silicide block
14	CO—contact

Table 3.2 Logic operations for making additional masks

Layer	Name	Mask generation
P-WELL	PW	(NOT (NAT OR NW))
Core NLDD	NLDD	((NP NOT NW) NOT AA2)
Core PLDD	PLDD	((PP AND NW) NOT AA2)
IO PLDD	PLDD2	((NP AND NW) AND (AA2))
IO NLDD	NLDD2	((PP NOT NW) AND AA2))

In Table 3.2 the layer NAT is a dummy layer that marks the area for native nMOSFET, which does not get the p-well, including the channel threshold voltage adjusting, implants. In Chap. 4 we will discuss the native FETs in more details. It does not represent a physical mask layer and is only used in the logic operation for the generation of PW mask.

Mask making process is similar to the silicon wafer fabrication process. On quartz blanks Cr/CrO_2 layers are deposited. Then PR is coated and patterned using electron beam lithography. The E-beam lithography and the laser mask maker are direct writers without PR, and the pattern is often 4×–10× of the pattern to be printed on the wafer. A direct write system is much slower than the stepper exposure, as each pattern is written like a line or plane scans. The parts that received sufficient electron or laser exposure are etched away during PR development. The PR is then washed away and glass substrate cleaned.

3.3 Design Rules

In Sect. 2.3.3 we alluded to the need for design rules to ensure devices are functioning correctly, including proper isolation from each other, etc. More generally a set of design rules are imposed on the mask layout designers by the foundries to specify the constraints on layout to guarantee manufacturability of a design. There are two basic types of layout constraints:

1. Resolution—these rules set the size and pitch of the features, e.g., minimal channel length, minimal line width of metal interconnect, minimal separation between devices, etc.
2. Alignment/overlap rules set the minimal or maximum overlap of features such as n-wells and active regions.

The most prominent design rule is the minimum size that can be defined by a process technology, which are usually the channel length of the gate and the size of the contact holes. Next is the minimal spacing (can also be understood as pitch) between two poly or metal lines to guarantee that they are not shorted electrically. The pitch determines the achievable functional density. The technology is often named from the minimal half pitch of the poly gate lines, which determines the maximum number of transistors in a given area. For example, 90 nm technology has the pitch between two poly lines as 180 nm. These resolution rules are determined by the lithography process and ultimately by the wavelength of the light sources used in the lithography machines.

Alignment/overlap constraints specify the alignment between layers such as between a p-well and an n-well, or between a p+ source and drain region to the edge of the n-well. They are usually determined by the quality of the masks, precision of the lithography mask alignment and registry, and process variations.

If the spacing between two regions, e.g., between a p+ drain to the n-well edge, is too small, the drain may short to the p-substrate. This is the punchthrough phenomenon discussed in Sect. 2.3.3.

The alignment rules are important to this book because some of devices, e.g., the high voltage devices discussed in the next chapter, are not self-aligned devices. The alignment accuracy has a big impact on the manufacturability of these devices.

Figure 3.20 illustrates some of the design rules using the example of so-called λ rules (scalable rules). The λ rule set is popular for teaching purposes but not popular in the industry where the direct dimension is used in the "micron" rule. As a

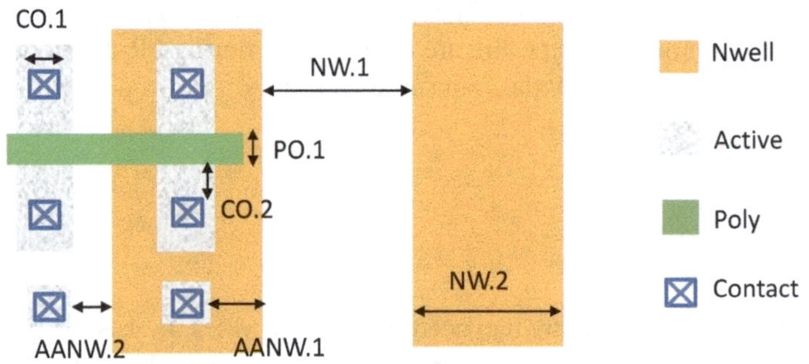

Fig. 3.20 Some exemplary design rules

Table 3.3 Design rules shown in Fig. 3.20

Rule	Description	λ rule example	FreePDK45[6] (nm)
PO.1	Min channel length	2λ	50
CO.1	Min contact size	2λ	65
CO.2	Co to Po spacing	2λ	35
NW.1	NW spacing	20λ	135
NW.2	NW width	20λ	200
AANW.1	NW enclosure of active	4λ	55
AANW.2	NW spacing to active	4λ	35

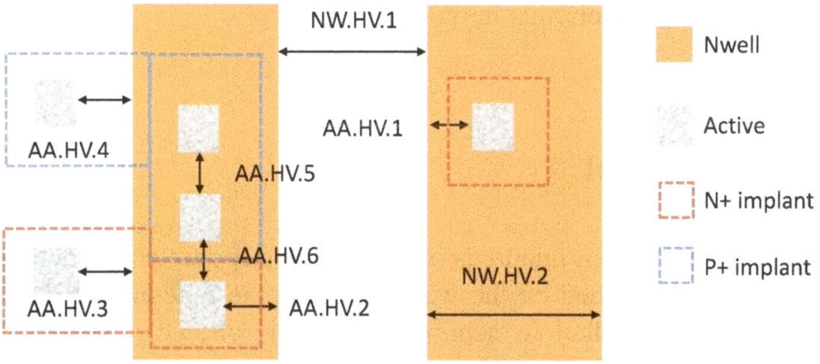

Fig. 3.21 Custom design rules to ensure HV operations in logic CMOS processes

comparison, the last column of Table 3.3 lists corresponding design rules for 45 nm logic process (NCSU EDA Wiki).

There are other design guidelines such as those providing reliability guarantees against electromigration and gate oxide degradation.

As we will discuss in the next chapter, often there are demands for operating devices beyond the specification provided by the foundries. For example, we may take an n-well that was designed for 3.3 V pMOSFET to more than 12 V. In those cases, a custom design rule set may be required to ensure device reliability. Figure 3.21 illustrates some special design rules for high-voltage operations beyond foundry specification.

As we can see, the earlier design rules are all geometric rules. For this reason, some foundries have additional topographical design rules (TDR) which deal with the logical order of mask patterns. Separately there is yet another set of rules that deal specifically with the electrical connections, called electrical design rules (EDR). EDR may include restrictions on the maximum current that can be carried on a wire of certain size to reduce electromigration effects.

3.4 Extension of Logic CMOS Processes

A simple extension of the pure digital logic process presented in Sect. 3.2 is the mixed-signal (MS) process. In the MS process, a few extra layers and a few special devices, such as those shown on the right column in Fig. 3.1, are made available to designers. Also in the MS process, deep n-wells (DNW) are commonly used both for the isolation of substrate noises and to allow negative voltages in radio frequency circuits.

DNW is formed by higher energy implant and a long-time annealing following the implantation. DNWs are deep enough that p-wells and consequently, nMOSFETs, can be placed within them. DNW can be biased to isolate substrate noises from affecting the circuits in the DNW.

3.5 Process Control

Starting with the raw wafer, hundreds of processing steps are performed before the designed circuits are formed on the wafer and become functional ICs. Each step needs to be well controlled to achieve the end result, i.e., the electrical and reliability specifications for the IC, including the added margins that are needed to take into account the intrinsic statistical variations and random defects that can render the fabricated die not unusable.

During wafer processing, we need to assure that each processing step can accomplish the desired results and does not introduce unexpected defects. Since inspection and monitoring costs time and money, only selected critical parameters are monitored. IC fabs may monitor parameters such as film thickness, critical dimensions such as line width, contact hole size, and electrical parameters such as sheet resistance.

When the monitored parameters are outside the production limits the wafers may be reworked or scrapped before more cost is incurred in the subsequent processing steps. **Process control monitor (PCM)** structures are one of the tools used for checking critical dimensions and electrical parameters during processing.

The PCM may contain geometric test structure to allow the measurement of film thickness, critical dimensions such as line width. It may contain structure to test optical alignment. PCM test structures can also have an array of electrical parametric test structures and a small number of functional test devices to allow the measurement of the performance of actual devices, such as threshold voltage of all the FETs, the contact resistance, and via resistance, etc.

These tests are designed to monitor all of the parameters in the technical specifications for each process. Foundries provide standard PCM monitors, which included standard sized transistors, e.g., 10 μm × 10 μm and 10 μm × 28 nm monitors to cover the range of the FET that are used by their customers. Often simple circuits like ring oscillators are included as well.

As the PCM structures usually have no use after the wafers are fabricated and tested, PCMs are usually placed in the scribe lanes between the dice. They are destroyed when the wafers are diced. More complex PCMs that cannot be placed in the dicing lanes, with typical width of 50–80 μm, are placed in the die, see e.g. (Klass, Jain, & Hess, 2009).

The electrical structures are tested after wafers are finished processing and test results are used as part of **wafer acceptance test** (WAT) before shipping to customers.

3.6 Postprocessing

3.6.1 Dicing

After wafer processing is finished and WAT test result accepted, the wafer needs to be diced. The standard technology for separating die is by mechanical sawing. Diamond saw as thin as 10 μm has been developed to use on scribe lane (also called saw street) as narrow as 50 μm.

For die measured in 10s of millimeters across, the silicon area occupied and consequently thrown away during dicing by the saw street is barely noticeable. However, for smaller die, the saw street can be a substantial percentage of the area of the die. For example, for a die of 0.5 mm on a side with 50 μm saw streets, the useful silicon area is

$$500 \times 500/(550 \times 550) = 82.5\ \%\ \text{of the total Si area}$$

The rest, or 17.5 % of the wafer, becomes saw dust and has to be recycled. The cost of each die increases by 21 % as compared to the ideal case of a zero width saw street. If the saw street can be reduced to 10 μm the silicon area loss will be only 4 %.

It is estimated that approximately one trillion semiconductor devices are sold every year, including ICs and optoelectronic chips, sensors, and discrete devices (ICInsights, 2015). Dicing needs to be performed on all of these devices. As the process technology continues to scale to smaller geometries, and with the expected growth of internet-of-things applications which are expected to be served predominantly by small-die ICs, we expect the number of small dies sold each year to grow substantially. Indeed, one example is the RFID tag IC where ten billion RFID chips were sold during a short period of time (Impinj Press Announcement, 2015). So more efficient, including both reduced dicing loss as well as lower dicing cost, dicing methods are needed to address this increasingly important problem.

Many laser dicing technologies, e.g. (Kumagai et al., 2007) which can use saw street as narrow as 20 μm, have been developed and some have been used commercially. The problem with laser dicing is that it is still a serial process, like mechanical sawing, where each saw street are cut individually. It is desirable to use a parallel die separation technology similar to the conventional wafer processing technology.

3.6.2 3D IC Integration and Through Silicon Via

Building in the third dimension has been in practice since the invention of the IC, in the form of adding more layers of interconnects on top of a single layer of active devices. For the most advanced process, up to 12 layers of metal interconnects are common. But adding more active devices such as transistors in the third dimension has been challenging. As we recall, the drive current of MOSFETs is directly proportional to the carrier mobility. High current/carrier mobility as well as low leakage can only be achieved with defect-free single crystal substrate. This fundamental requirement has stymied all previous attempts at building 3D ICs.

Adding more active device layers has been done in the form of die stacking. For example, die stacking has been utilized to add chips of different type, e.g., DRAM stacked on top of flash memory chips. This is more appropriately called 3D package, or 3D **system in package** (SIP), where the dies are connected through flip chip connections or wire bonding, as shown in Fig. 3.22. Both methods are limited by the number of die that can be stacked together or the number of connections that are feasible between different dies.

3D stacked die using **through silicon via (TSV)** is another method to more tight integration. Figure 3.23 illustrates the die stack connected with TSVs. Recent progresses in wafer thinning and shrinking TSV size have made TSV a very promising and feasible technology for large-scale deployment.

Die stacking is attractive for integrating disparate circuits, e.g., MEMS, DRAM on logic circuits, or SRAM on flash. The main issue with die stacking is the heat dissipation. Because of this problem, it is expected that memory on memory will be the first application for the TSV-enabled stacked die technology.

Die stacking, either with wire bonding or TSV, still needs the individual die to be fabricated first. The ultimate 3D chip is monolithic 3D ICs where active circuits are built on top of each other. The first true monolithic 3D IC is the recently released 3D NAND flash memory, first by Samsung in 2014. It turns out that for NAND flash, the speed of the memory transistor is not very critical. As a result, the channel of the transistors can be made of polysilicon that is deposited using chemical vapor deposition.

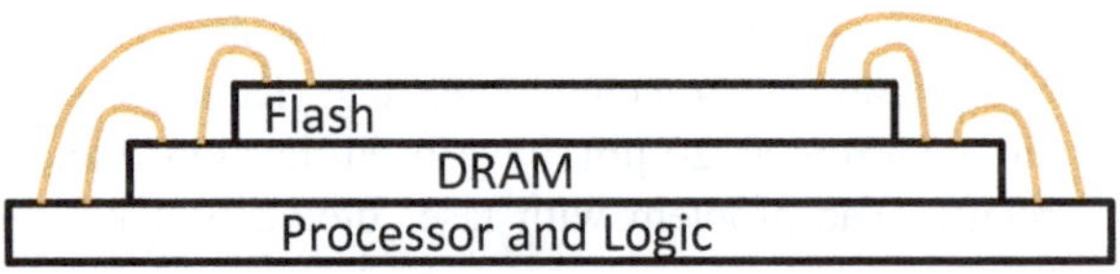

Fig. 3.22 Stacked dice for system in a package (SIP)

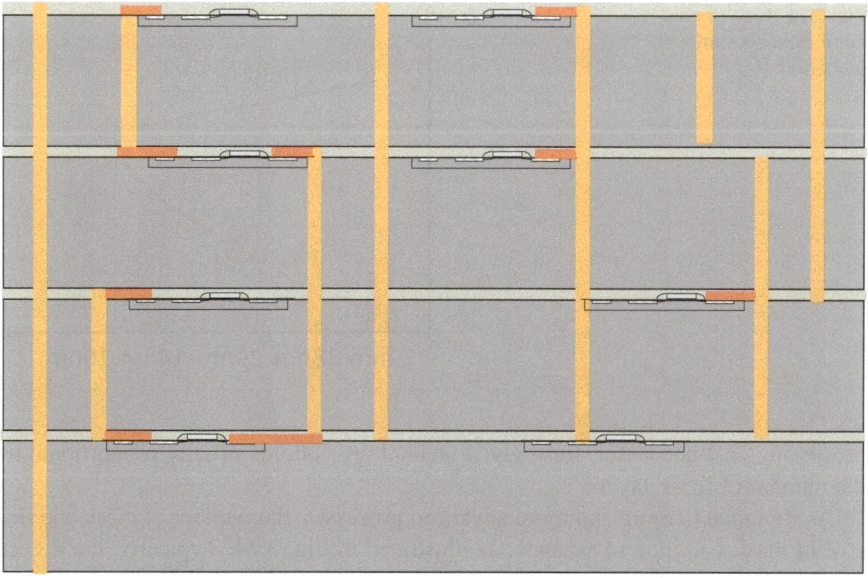

Fig. 3.23 3D chip with through silicon via

3.7 Wafer Processing Economics

3.7.1 Wafer Processing Cost

Moore's law derives from the fact that billions and even trillions of transistors can be packed on a single wafer and fabricated in parallel in a single process sequence. As transistor size scales down exponentially, more transistors are packed on the wafer using newer process technology and at approximately constant wafer processing cost as older generation technologies.

In Sect. 1.2 we briefly discussed the cost of developing and manufacturing ICs. The development cost is a nonrecurring cost which includes both the design and the tooling cost. The cost of a mask set is a nonrecurring cost. A mast set at leading edge can cost on the order of $10 million (Or-Bach, 2014).

The masks in a set are not created equal. The most expensive is the critical mask layers used in the early stages of fabrication, having minimal size and pitch. The masks for patterning polysilicon gate and the source–drain contact are usually the most expensive masks and can cost hundreds of thousands of dollars.

The lowest cost masks are generally at the later stage of BEOL, such as masks for pad opening and top metals.

Once the chip has been developed and in production, the NRE cost decreases and the production cost of the chip becomes more important and recurring. The

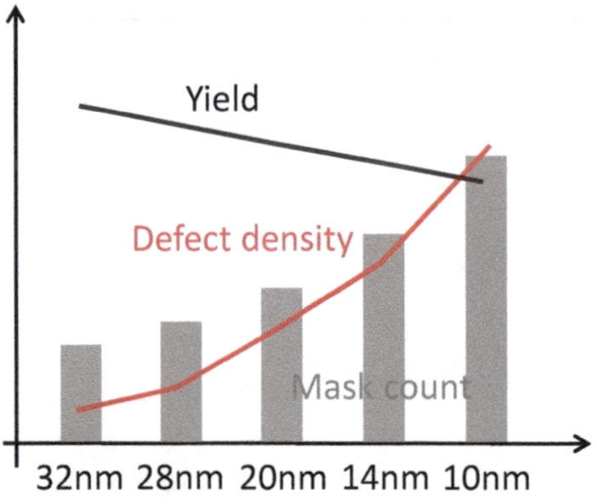

Fig. 3.24 Cost of chip fabrication goes up with process node

processing cost per wafer, for a given technology node, is usually proportional to the number of mask layers.

As we move to more and more advanced processes, the number of mask and the cost of mask continue to escalate, as illustrated in Fig. 3.24. Typically, the defect density goes up and yield goes down.

Cost of fabricating wafers using the leading edge technology can be high when first introduced, often dominated by the lithography steps. As the process matures, equipment cost, especially depreciation cost, goes down and wafers will get cheaper. As more wafers are processed using the new technology, the wafer yield goes up as well. So overall the cost of manufacturing each chip will go down with time.

3.7.2 *Yield*

Wafer yield has a direct impact on the manufacturing cost of each chip. In early days of IC, manufacturing yield is very low even for very small die. It was projected that chip size could never be larger than a few millimeters as the yield would be approaching zero. Early killer defects included thin film pinholes and metal contamination, including sodium drift. Nowadays those sources of yield limiters have all been nearly eliminated and the killer defects are predominately particles. The sources of these particles can be from the air, from the processing equipment, or from the wafer itself.

Based on years of empirical data, several yield models have been developed.

Suppose the mean number of fatal defects per die is λ_0, according to the Poisson probability distribution function, the probability that a die has k defects is given by

$$P(k) = \frac{e^{-\lambda_0}\lambda_0^k}{k!} \tag{3.3}$$

The probability that the die is good is when there is no defect, i.e., when $k = 0$. Therefore, the expected die yield is given by

$$Y = P(0) = e^{-\lambda_0} = e^{-AD} \tag{3.4}$$

where $\lambda_0 = A \times D$, with A being the circuit area of the die and D the defect density.

Another yield model is the Bose–Einstein model, where the yield is given by

$$Y = \frac{1}{(1 + D_1 A)(1 + D_2 A)\cdots(1 + D_n A)}, \tag{3.5}$$

for a process with n critical masking steps, where D_i is the defect density for ith process layer.

For simplicity, assuming defect density is the same for all process steps, the yield model can be simplified to

$$Y = \frac{1}{(1 + D_0 A/n)^n} \tag{3.6}$$

where D_0 is the overall defect density.

There are a number of increasingly more complex and accurate models to describe the wafer yield that is beyond the scope of this book. For our purpose, it is sufficient to note that yield is directly related to the number of mask layers and the die size. This fact is important when we discuss the trade-off between the use of logic-only process vs. specialized processes with additional masking layers.

3.7.3 Cost of Chips

In Sect. 1.2, we discussed the total cost of a chip, taking into account the development cost in terms of NRE (the nonrecurring engineering) cost,

$$\begin{aligned}\text{chip unit cost} = {} & \frac{\text{total NRE cost}}{\text{total expected volume of chips able to sell}} \\ & + \frac{\text{wafer fab cost}}{\text{total yieldable die per wafer}} + \text{test cost per chip} \\ & + \text{packaging cost per chip}\end{aligned} \tag{3.7}$$

The production cost of a fully tested, functional chip is the last three terms of the earlier equation.

$$\text{chip unit production cost} = \frac{\text{wafer fab cost}}{\text{total yieldable die per wafer}} + \text{test cost per chip} + \text{packaging cost per chip} \tag{3.8}$$

As we saw in the last section, the yield, hence the number of yieldable die per wafer, is directly proportional to the size of the die. Adding mask layers, enabling more devices, often will reduce the size of the die, but with a risk of reducing the yield and increasing the cost per wafer. Therefore, it is not necessarily a wise thing to do. The trade-off needs to be evaluated. In Chaps. 8–10 we discussed in detail the case for embedded nonvolatile memory to digital ICs. There a choice can be made by using specially embedded NVM process, which can significantly increase the processing cost of the wafer. An alternative is to build NVM using the basic logic process, where the large NVM cells will make the chip size big. So a question is which approach to use. The answer depends on the size of the NVM. For larger NVMs the more expensive embedded NVM process may be more economical while for small NVM, it may be better to use the larger memory cell with the logic CMOS process.

References

Das, A. (2016). How finFETs ended the service contract of silicide process. *Solid State Technology, 59*(3), 22–26.

Deal, B., & Grove, A. (1965). General relationship for the thermal oxidation of silicon. *J Applied Physics, 36*, 3770–3778.

ICInsights. (2015). *Semiconductor unit shipments to exceed one trillion devices in 2017.* Retrieved from http://www.icinsights.com/news/bulletins/Semiconductor-Unit-Shipments-To-Exceed-One-Trillion-Devices-In-2017/.

Impinj Press Announcement. (2015). *Impinj Sells 10 Billionth Chip as RAIN RFID goes mainstream in retail.* Retrieved from http://www.impinj.com/about-us/news-and-events/press-releases/2015/impinj-sells-10-billionth-chip-as-rain-rfid-goes-mainstream-in-retail/.

Jessing, J. (2007). CMOS fabrication. In W. Chen (Ed.), *The VLSI handbook* (pp. 12–41). Boca Raton: CRC Press.

Klass, F., Jain, A., & Hess, G. (2009). An on-chip process control monitor for process variability measurements in nanometer technologies. In *ICICDT '09. IEEE International Conference on IC Design and Technology* (pp. 203–206). IEEE.

Kumagai, M., Uchiyama, N., Ohmura, E., Sugiura, R., Atsumi, K., & Fukumitsu, K. (2007). Advanced dicing technology for semiconductor wafer—Stealth dicing. *IEEE Transactions on Semiconductor Manufacturing, 20*(3), 259–265.

Ma, Y., Evans, D., Hguyen, T., Ono, Y., & Hsu, S. (1999). Fabrication and characterization of sub-quarter-micron MOSFETs with a copper gate electrode. *IEEE Electron Device Letters, 20*, 254–255.

Mistry, K., Allen, C., Auth, C., & Zawadzki, K. (2007). A 45nm logic technology with high-k +metal gate transistors, strained silicon, 9 Cu interconnect layers, 193nm dry patterning, and 100% Pb-free packaging. *IEDM Technical Digest*, 247–250.

NCSU EDA Wiki. (n.d.). *FreePDK.* Retrieved from http://www.eda.ncsu.edu/wiki/FreePDK45:Contents.

Or-Bach, Z. (2014). *FPGAs as ASIC alternatives: Past & future*. Retrieved from EE Times: http://www.eetimes.com/author.asp?doc_id=1322021.

Peng, J. Z., & Fliesler, M. (2005). *High density semiconductor memory cell and memory array using a single transistor and having variable gate oxide breakdown*. US Patent 6,940,751.

Singer, P. (2011). *Below 22nm, spacers get unconventional*. Retrieved from Solid State Technology: http://electroiq.com/blog/2011/01/below_22nm-spacers-get-unconventional/.

Yamasaki, K., Asai, K., & Kuramada, L. (1982). GaAs LSI-directed MESFETs with self-aligned implantation for n+ layer technology (SAINT). *IEEE Transactions on Electron Devices, 29*, 1772–1777.

[illegible]

Part II
Non-logic Device Design in Logic Processes

In the second part of this book, we will discuss the practices of designing new devices using generic CMOS processes.

In Chap. 4 we explore non-logic MOSFETs that can be made from the generic CMOS process that is originally developed to make only four standard MOSFETs, core n-FET and p-FET, I/O n-FET, and p-FET. We first discuss several ways to obtain MOSFETs with threshold voltages that are not part of the standard offerings without any process modifications.

High-voltage MOSFETs have many applications, including power ICs, LCD display drivers, and HV generators for programming floating-gate nonvolatile memories. They generally require specialized processes to make and are usually not compatible with the generic CMOS process. In the second part of Chap. 4 we will discuss how to build a particular type of MOSFET, LDMOS, that can be made using the basic CMOS process and can operate at voltages much higher than the I/O voltage, e.g., more than 15 V.

In Chap. 5 floating-gate devices are discussed, including floating-gate transistor which forms the foundation for a whole class of circuits—the embedded nonvolatile memories (NVMs) and the neuromorphic circuits. Floating-gate capacitors are also discussed. Lastly a quasi- or semi-floating device as the gain cell for embedded DRAM is briefly reviewed.

In Chap. 6 we discuss bipolar transistors that can be obtained from the CMOS process as well as their uses in the crucial bandgap reference circuits.

In Chap. 7 polysilicon diodes and Schottky diodes that can be made in logic CMOD processes are explored.

Chapter 4
Non-logic MOSFETs in Logic CMOS Processes

Abstract This is a core chapter for this book, covering the design of nonstandard MOSFETs using logic CMOS processes. In Sect. 4.2 we discussed the design of several types of MOSFETs that have different threshold voltages than those provided by the foundries. In particular, the bandgap-engineered FETs have found wide use in voltage reference circuits. Then in Sect. 4.3 we discuss another important class of MOSFETs—LDMOS that can operate at voltage over 15 V and can be manufactured in a generic CMOS process without any process modification. Design, reliability, and manufacturability of LDMOS are reviewed.

Keywords MOSFET • Threshold voltage • Bandgap • LDMOS • High voltage • Junction breakdown • Bandgap reference • Native transistor • Current mirror • Diode connected FET

4.1 Introduction

As we discussed in the last chapter, the most basic, generic logic CMOS processes typically only offer four sets of MOSFETs: core nFET and pFET, and I/O nFET and pFET. Each set of MOSFETs comes with a standard threshold voltage that is optimized for general purpose applications. These V_t's are chosen based on the I_{on} to I_{off} trade-off in logic circuits.

Variations of V_t at different W and L due to narrow width and short channel effects, respectively, are minimized but usually could not be eliminated. The SPICE models provided by the foundries usually include such relationship and are able to model the V_t changes with W or L with reasonable accuracy. These FETs are sufficient for most digital applications.

However, very often transistors with other V_t are required. This kind of demand is most prevalent in ICs that integrate analog, mixed signal, RF, and digital functions, i.e., system-on-a-chip (SOC) ICs. With the growing demand for ultra low power ICs to implement the internet of things, the need for MOSFETs with multiple, and custom designed V_t will tend to increase. But unless a customer has a huge wafer volume, a foundry usually will not offer customized FETs with dedicated V_t.

Y. Ma, E. Kan, *Non-logic Devices in Logic Processes*,
DOI 10.1007/978-3-319-48339-9_4

When faced with strong demand, the foundries sometimes offer derivative processes targeting large segment of the market, as illustrated by Fig. 3.1. For example, a derivative lower power process is usually available where the transistor V_t's are higher than those in the generic logic CMOS process. As a result, the FETs have lower leakage and the resulting ICs have low standby power consumption. Most of these processes will require more mask levels than the basic CMOS process. As a result, the wafer processing cost will be higher. In this chapter, we will discuss several methods to design of MOSFETs with nonstandard V_t without the additional mast or process modifications.

The transistors in the logic processes are also specified to operate at either the core or I/O supply voltages. The core voltage is typically 1.0 V for the more advanced processes while I/O can be up to 1.8, 2.5, or 3.3 V to fit the **printed circuit board (PCB)** standards. In Sect. 4.3 we study several transistor designs that can operate beyond the highest voltage offered in a given process.

4.2 MOSFETs with Nonstandard Threshold Voltages

In this section, we discuss several methods to obtain nonstandard V_t using the generic CMOS process without modification. These transistors can be used for special circuits to save power and/or area. Examples of extremely low power and small bandgap voltage reference circuits are discussed in Sect. 4.2.1.

As discussed in Sect. 2.4.1, the threshold voltages of a MOSFET for given oxide thickness, channel, and well doping are given by

$$V_t = V_{fb} + 2\phi_B + \frac{\sqrt{4qN_a\varepsilon_s\phi_B}}{C_{ox}} \tag{4.1}$$

For a given process with predefined gate oxide thickness, we have the following knobs that can be tuned for varying the threshold voltage:

1. Varying substrate (well) doping, N_a
2. Varying the flatband voltage, V_{fb}
3. Through short channel and narrow width effects.

Varying the substrate or well doping is possible. As described in Chap. 3, the p-well doping is usually different than the p-substrate. As a consequence, we have the possibility of obtaining two more nFETs by not using the p-well. These nFETs are built directly on p-substrate and are called native nFETs—one with the core gate oxide and one with the I/O gate oxide. Since the doping of the substrate is about two to three orders of magnitude less than the p-well, the last term of Eq. 4.1 can be usually neglected, resulting a threshold voltage that is near zero for both core and I/O native nFETs.

The native nFETs are usually available for all logic processes and good models are provided by the foundries. Since the native nFETs do not have the usual high p-well doping and usually lacks the anti-punchthrough doping as well, the minimal channel length specified by the foundries is usually much larger.

In the following sections, we discuss the other two possibilities.

4.2.1 Bandgap-Engineered MOSFETs

A method for varying the flat band voltage is by bandgap (BG) engineering of the polysilicon gate. The bandgap-engineered MOSFETs have found applications in circuits such as bandgap voltage references.

The standard MOSFET uses p+ doped polysilicon gate for PMOS and n+ doped gate for NMOS. By changing the type of polysilicon gate, we can change the flatband voltage by the bandgap, i.e., ~1.1 eV for silicon and obtain corresponding MOSFETs with V_t shifted by the bandgap, respectively.

Figure 4.1 shows the cross sections of a normal n+ gate nFET and a bandgap-engineered nFET, where the n+ gate is replaced by a p+ gate. The bandgap-engineered FET is actually a composite of three FETs: two edge FETs of the normal type on the ends of the channel and one bandgap FET in the center. The edge nFETs are protective devices, necessary to prevent the p+ implant from encroach onto the n+ source and drain. If p+ dopant happens to get implanted in the source and drain area, especially near the edges, the source–drain will short to the substrate. We will end up with either a highly leaky or nonfunctioning MOSFET.

The V_t of the bandgap FET is much higher, shifted by the silicon bandgap (~1.1 V) from the normal FET. Table 4.1 listed all the possible bandgap-engineered FETs from a generic CMOS process that offer two standard sets of FETs and the corresponding threshold voltages. We listed the typical threshold voltage in a hypothetical 0.13 μm technology.

The channel length of the edge FETs is kept at a minimal to reduce its contribution to the overall transistor characteristics. Because of its high V_t and longer channel length, the middle FET usually dominates the overall characteristics, including the drive current, of the whole BG FET. The effect of the edge nFETs can be usually ignored. Thus, the composite FET can be, to first-order approximation, modeled by a nominal nFET with channel length L_p, and with its V_t shifted by the bandgap voltage, i.e., 1.1 V. In reality, as we will see later, the shift is smaller due to other factors.

Figure 4.2 shows the corresponding layouts for these two FETs.

The minimal channel length of the edge FETs, L_n, is determined by the minimal overlay resolution between the gate mask and the N+ implant mask and the lateral scattering of the implanted dopants or the lateral migration of the dopants during the annealing process. As a result, the BG FET is not a self-aligned device. If L_n is too small or negative, the p+ doping can encroach on the source–drain region, which will make them partially p+ doped and short to the p-well or p-substrate.

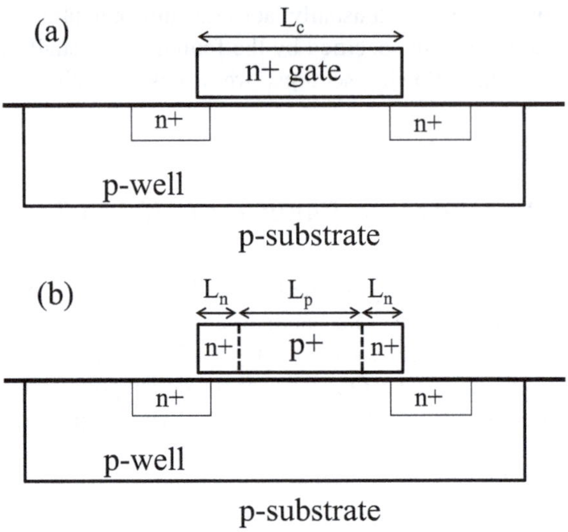

Fig. 4.1 Standard nFET (**a**) and bandgap-engineered nFET (**b**). BG-engineered FET has two edge normal FETs to protect the implant of p+ dopant in the n+ source and drain

Table 4.1 Bandgap-engineered nFETs

	Nominal V_t	Bandgap V_t	Example −0.13 μm	0.13 μm BG FETs
Core nFET	VTN	VTN+VBG	0.3	1.4
I/O nFET	VTN	VTN2+ VBG	0.5	1.6
Native nFET	VTNative	VTNa+ VBG	0	1.1
I/O native nFET	VTNative2	VTNa2+ VBG	0.1	1.2
Core pFET	VTP	VTP− VBG	−0.3	−1.4
I/O pFET	VTP2	VTP2− VBG	−0.5	−1.6

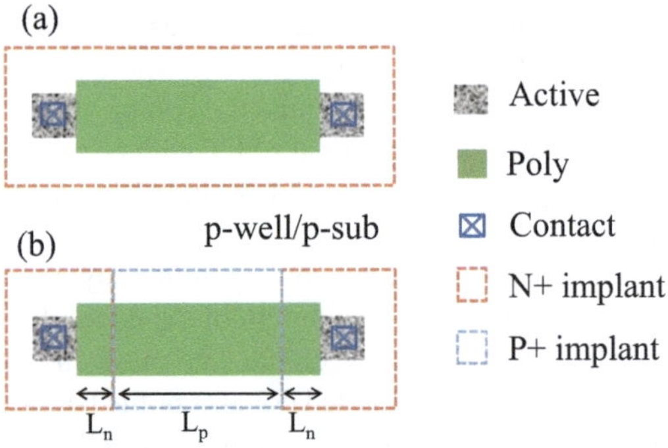

Fig. 4.2 Layout for standard nFET (**a**) and BG nFET (**b**)

Thus, a new design rule specifying a minimal L_n needs to be implemented to assure that the bandgap device is manufacturable. Usually L_p is set at several times of L_n, e.g., $L_p = 5L_n$, to ensure that we can obtain a consistent V_t on the middle region. If L_p is too short, the n+ implant regions on the edge of the polysilicon gate may merge. So for $L_n = 0.1$ μm, the overall length of the FET is then $0.5 + 2 \times 0.1 = 0.7$ μm. In other words, the BG FETs are necessarily long channel devices. They are not meant to be used as current drivers.

Despite of such limitations, the bandgap FETs have found use in voltage reference circuits, as discussed in the next section.

4.2.2 Bandgap FET Application: Voltage Reference Circuit

One of the applications of the bandgap-engineered FET is in current mirror circuits that provide voltage reference. Voltage reference circuits are found in many IC circuits. In particular, most memory ICs use voltage references as part of the precision voltage converters, either downconverter or charge pumps. A very popular class of voltage reference is the **bandgap reference** (**BGR**). BGR output is a function of the bandgap of silicon, a physical parameter which, unlike parameters like threshold voltage, does not vary from process to process, fab to fab, and have a well-known, small temperature dependence.

Figure 4.3 shows a typical CMOS bandgap voltage reference circuit. In this figure PMOS transistor P1 is a BG pFET with an n+ doped polysilicon gate, based on the similar principle as the p+ gate nFET discussed earlier. The threshold voltage of the P1 transistor is $V_{tp1} = V_{tp} -$ bandgap (the negative sign is because the threshold voltage of pFET is negative). P2 is a normal p+ poly pFET with threshold voltage V_{tp}. The current source provides a small I_{bias} current, e.g., 100 nA, with a voltage drop of V_{cs}.

In this design, pFETs P1 and P2 and nFETs N1 and N2 are transistors with the same W and L. Transistors N1 and N2 then form a current mirror with the same V_{gs}. Thus, the current flow through N1 and N2 is the same and equals to half of I_{bias}. As a

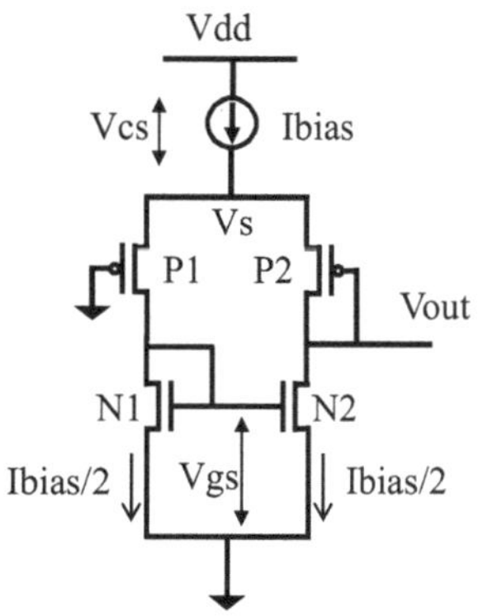

Fig. 4.3 Bandgap reference circuit using a special n+ gate PMOS transistor

result, the current through P1 and P2 is also the same. N1 and P2 are diode-connected MOSFETs. As discussed in Sect. 2.4.1, due to the small current being fed by I_{bias}, the voltage drop across the two pFETs is the corresponding threshold voltage and given by the following.

V_{gs} for P1 is given by $V_{gs1} \cong 0 - V_s$ and V_{gs} for P2 is given by $V_{gs2} \cong V_{out} - V_s$ where $V_s = V_{dd} - V_{cs}$ is the source voltage of P1 and P2. As P1 and P2 are forced to have the same current, ignoring the output resistance in the subthreshold or saturation region, they will have the same gate drive: $V_{gs1} - V_{tp1} = V_{gs2} - V_{tp2}$. Hence,

$$V_{out} = V_{tp} - V_{tp1} = \text{VBG} \tag{4.2}$$

The output voltage is then silicon bandgap voltage, VBG ~1.1 V. The output voltage will have the same temperature dependence as that of the silicon bandgap as well. We can see that I_{bias}, V_s, and V_{cs} will not affect $V_{out} = \text{VBG}$ as long as all transistors are in saturation (in subthreshold, $V_{ds} > 3kT/q$ is usually sufficient to saturate the drain current).

The required supply voltage can be estimated as

$$V_{dd} = V_{out} + |V_{gs}| + V_{cs} \sim 1.1 + 0.3 + 0.2 = 1.6 \text{ V} \tag{4.3}$$

The advantage of this bandgap reference circuit is that this circuit is very small and power efficient, as compared to the conventional bandgap reference circuits found, e.g., in (Baker, Li, & Boyce, 1998), and in examples in Sect. 6.3 using parasitic bipolar transistor. The design is also very robust from process variations—as long as the BG transistor is fabricated correctly, which is easy to do in the long-channel case, the output is very insensitive to mismatching.

As can be seen easily from the circuit, this bandgap reference has very limited output current capability. The analysis indicates that the maximum useful current is approximately one-half of the bias current, I_{bias}.

The circuit in Fig. 4.3 can be adapted to extract the difference of threshold voltage of any two transistors with the same type, as long as the transconductance prefactor $\mu_n C_{ox} W/L$ can be matched. However, the temperature insensitivity may not be inherited.

Even though the earlier circuit works well at a low supply voltage, a design with even lower supply voltage is possible. Note that in Eq. 4.3, the voltage drop V_{gs} is the threshold voltage of the pFET. This voltage drop can be reduced by using a FET with lower or even zero V_t, which leads to the idea of using the native FET. An implementation is shown in Fig. 4.4 (Kuhn, 2010). This circuit uses a couple of native nFETs to enable the use of even lower supply voltage. The key of this circuit is the BG nFET (Nat1 in the figure) where the n+ polysilicon gate of the usual native nFET is replaced by a p+ gate. The threshold voltage of the normal native nFET is $V_{tn} \sim 0$. The bandgap-engineered native nFET has a threshold voltage of approximately, $V_{tn} + \text{VBG} \sim 1.1$ V.

For the BG-engineered native nFET Nat1 with a small bias current, $V_{gs1} = V_{ref} \sim V_{BG}$. For Nat2, its $V_{gs2} = V_{ref} - V_{out}$. Or $V_{out} = V_{ref} - V_{gs2}$. With

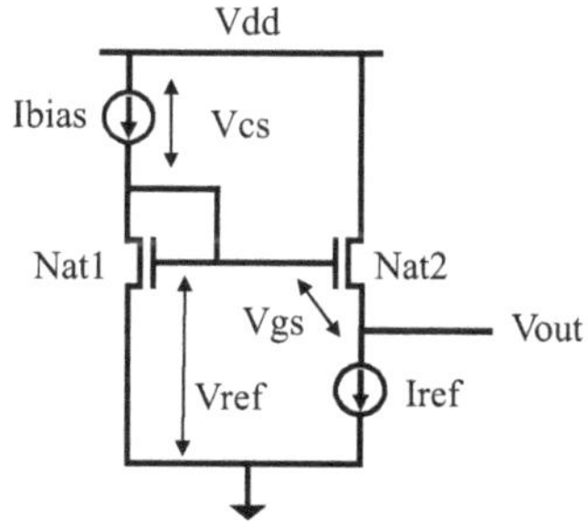

Fig. 4.4 Low-voltage bandgap voltage reference circuit, adapted from a patent by J. Kuhn (2010)

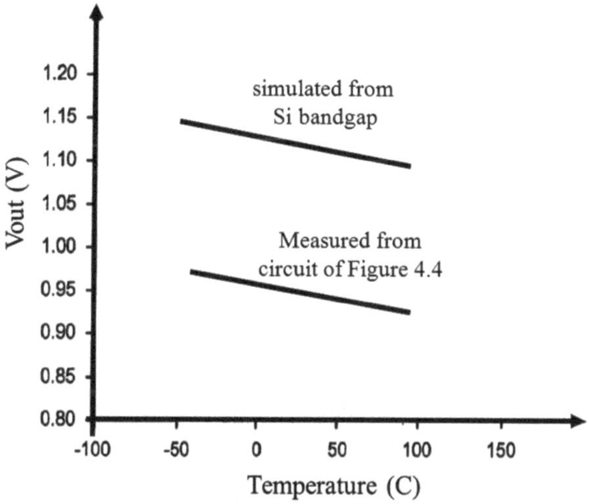

Fig. 4.5 Measured temperature dependence of output voltage of a bandgap reference circuit, using a native NFET with p+ gate, compared to a simulated silicon bandgap, adapted from Kuhn (2010)

appropriate sizing of the transistor for Nat2 which is a normal native nFET, we get $V_{gs2} \sim V_{tn} \sim 0$. Therefore, $V_{out} = V_{ref} \sim V_{BG}$, same as the circuit in Fig. 4.3.

The normal native FET Nat2 acts as an output buffer. Assuming that the effective *W*/*L* of Nat2 is *n* times that of the *W*/*L* of Nat1, the output should be able to output *n* times the current of I_{bias}.

The supply voltage is lowered to

$$V_{dd} = V_{gs} + V_{tn} + V_{cs} = 1.2\ \text{V}$$

With the reduced supply voltage, this circuit consumes less power. Because of the simplicity and the lower power consumption, it is useful for applications such as RFID and wireless sensors.

Figure 4.5 shows the measured output voltage for the circuit shown in Fig. 4.4 as a function of temperature from −40 to 85 °C (Kuhn, 2010). Two curves are shown: one is the theoretical simulated results from the temperature dependence of the silicon bandgap. The second curve illustrates the measured output voltage from a test chip. It is seen that the measured voltages exhibit the same temperature dependence as the theoretical prediction but are 200 mV lower. The 200 mV may be explained by the polysilicon depletion effect near the interface between polysilicon and the gate oxide, i.e., the dopant concentration at the interface is lower than that corresponding to fully degenerated doping. As a result the difference is smaller than the actual bandgap.

4.2.3 Source–Drain-Engineered MOSFET

The BG FETs change the threshold voltage by around 1 V, which sometimes is too large for some applications. For example, in Table 4.1 we can see that the threshold voltages of the nFET are 0, 0.3, 0.5, ~1.1, ~1.4, ~1.6 V. There is a big gap between 0.5 and 1.1 V.

We can use the short channel and the narrow width effects discussed in Sect. 2.4.3 to obtain more V_t variations. There are two ways to do this. The first way is to use the variations of V_t from the channel length and/or width dependence already built in a given process and specified in the models that foundries provide, e.g., Figs. 2.20 and 2.21.

As an example, suppose the design requires an nFET with $L = 0.5$ μm and $W = 1$ μm and V_t of 0.4 V. And 0.3 V is the nominal V_t for such a device. We can replace the 1/0.5 nFET with either two 1/1 nFET in parallel, or two 2/0.5 nFET in series. The net result may emulate a device with V_t of 0.4 V and drive strength of a 1/0.5 device.

Another way to take advantage of the short channel effect (SCE) is to physically change the factors affecting SCE without modifying the process. Recall that the SCE is mainly due to the effect of source and drain on the channel region and that LDD and the associated halo implants are used to a large extent to control the SCE, we can then modify the LDD for a MOSFET which will change the short channel behavior.

There two possibilities for doing this, including:

1. A MOSFET without the LDD implant
2. Swap the LDD implants of core and I/O transistors.

In Fig. 4.6a we show a standard MOSFET with the normal LDD implant. An example maybe the core nFET of a generic logic process, with its normal short channel behavior. The V_t may have substantial V_t roll-off, as shown in Fig. 2.20.

In Fig. 4.6b the LDD of the core nFET is replaced by the LDD implant of the I/O nFET. This changes the short channel behavior. As a result, the V_t roll-off curve as well as the narrow width effect may be different, and different V_t variation can be obtained.

In Fig. 4.6c the LDD implant is skipped altogether. This device will function in most processes. But due to the stronger short channel effect, the V_t roll-off will be sharper than that of the MOSFET with LDD (Liu et al., 1993).

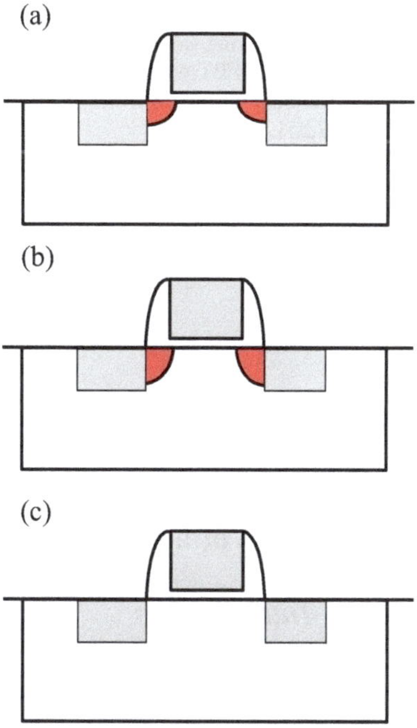

Fig. 4.6 Source–drain engineering for V_t variation: (**a**) normal FET, (**b**) FET with a different LDD implant, and (**c**) FET with no LDD implant

These special FETs can be obtained from the standard CMOS process without process modification. The only needed changes are the way the masks for the LDD implants are generated. For example, to generate Fig. 4.6c, we may replace the standard LDD mask logic generation

$$\text{NLDD} = (\text{NP NOT NW}) \tag{4.4}$$

By, defining a dummy mask layer covering the region of the special FET, where LDD is to be skipped and

$$\text{NLDD} = (\text{NP NOT NW})\text{NOT}(\text{Dummy}) \tag{4.5}$$

There are interesting applications with those special devices (Hsieh et al., 2004). Besides the V_t changes associated with different short channel effect, the removal of the LDD implant and the halo implant will substantially change the electric field distribution in the source and drain region. As a result, the hot carrier injection will be stronger than normal. Therefore, this type of FETs may be used as an injection device for programming floating-gate memories.

Figure 4.7 illustrates a patent by (Lin et al., 2009) that uses a single polysilicon gate pFET with no LDD implant as the storage device. In the figure the spacer (20) is comprised of oxide (22), nitride (24), and oxide (26) layers or ONO and is used as charge storage. The LDD layer is removed to improve the high carrier injection close to the drain junction.

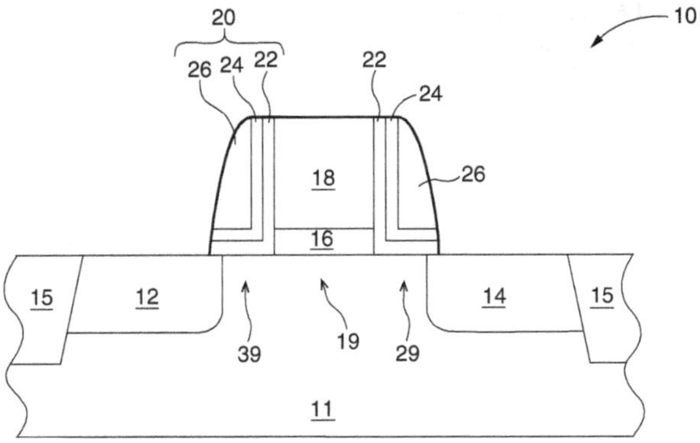

Fig. 4.7 A nonvolatile memory device using single polysilicon gate pFET with no LDD implant. From US Patent # 7,551,494 (Lin, Chen, Shen, King, & Hsu, 2009)

4.3 High-Voltage MOSFET

4.3.1 Introduction

The previous section covered several methods for making FETs with nonstandard V_t. In this section, we discuss building FETs that can tolerate high voltages.

As we discussed in Sect. 2.7, most digital ICs operate with two supply voltages: a core voltage that usually is around 1 V or less and an I/O voltage of typically either 1.8, 2.5, or 3.3 V.

Many circuits can benefit from or work only at higher voltages. As will be discussed in Chaps. 8–10, to program and erase nonvolatile memory devices such as floating-gate transistors, often voltages exceeding 3.3 V are needed. For example, to erase a floating-gate memory cell with a 7 nm thick tunnel oxide using Fowler–Nordheim tunneling, a voltage of about 10 V is required. Another example is the use of hot carrier injection to program a NOR type of flash memory, where more than 5 V is needed at the drain of the floating gate.

High-voltage (HV) transistors are needed to handle, including both the generation and the switching of these voltages. A transistor that is reliable over the desired operating condition, voltage and temperature ranges, is the essential element of all high-voltage circuits. The high-voltage devices are used in two main components of the high-voltage systems:

1. Voltage generators capable of taking the input of core or I/O voltages and pump up to the desired voltage. Different names are used for these types of circuits that include charge pumps, voltage doublers, and voltage generators.
2. Switching circuits where the high voltage is routed from the voltage generator to the place of use. These types of circuits include high-voltage switches and row and column decoders.

There are two classes of solutions to obtain devices that can operate at higher than I/O voltage:

1. Using additional process steps to build custom transistors with thicker oxide, lightly doped source and drain, and wells that can sustain the desired high voltage. This approach usually incurs additional cost—both development cost and higher cost of the wafer processing. In addition, the additional processing steps generally have a negative impact on both the performance of the other devices and the overall wafer yield.
2. Use design and layout solutions to achieve high-voltage devices and circuits without modifying existing logic processes, but with area penalty.

In this section, we only explore the second approach. As discussed in Chap. 2, the main barriers for a MOSFET to operate at high voltage are the breakdown of the gate oxide and the breakdown of the source–drain junctions.

We first show how to design the transistors to avoid oxide breakdown. The gate oxide breakdown usually occurs at an oxide electric field of about 15–20 MV/cm. For the usual gate oxides, the breakdown occurs at much lower voltages than 10 V, e.g., ~3 V for 1.5 nm oxide, ~5 V for 3 nm oxide (for 1.8 V I/O), ~8 V for a 5 nm oxide (for 2.5 V I/O), ~11 V for 7 nm oxide (for 3.3 V I/O), etc.

One way to operate oxide at higher voltage is to reduce the operating time at the desired voltage. Recall from Sect. 2.4.3 that gate oxide breakdown voltage is related to how long it is biased at this voltage, as measured by the time-dependent dielectric breakdown (TDDB) characteristics of a given gate oxide.

Therefore, designers of a high-voltage FET can use two approaches to address the oxide breakdown problem:

1. Reducing the electric field experienced by the gate oxide by voltage divider techniques.
2. Reducing the required time or duty cycle under the maxim stress. The time a MOSFET is used for high-voltage operations is usually application dependent. For example, the programming circuit of an embedded NVM needs HV transistors. If the NVM is specified for 1,000,000 programming operations and each operation takes only 5 ms to finish, the oxide of the HV transistor may only need to have a lifetime of 5 s at the required voltage.

The source–drain junction breakdown usually occurs at around 5–7 V. Therefore, they cannot be used for HV operations over ~7 V. However, for another junction in the p-substrate-based CMOS process, e.g., the n-well to p-substrate, the breakdown voltage is usually higher than 10 V. This observation led to the invention of the LDMOS discussed in the following sections.

4.3.2 LDMOS

One commonly used high-voltage (HV) transistor is the so-called LDMOS, which stands for **Laterally Diffused MOSFET**. If special high-voltage process is used, LDMOS can sustain hundreds of volts and has been used extensively in power electronics such as inverters for motor drives. However, the high-voltage processing steps often involve prolonged anneals which may adversely affect the performance of core, low-voltage transistors. Hence traditional LDMOS cannot be easily integrated with standard CMOS devices.

In our version of the LDMOS, we use an n-well as the laterally diffused drain. It may be more appropriately called lightly doped drain MOS (Ballan & Declercq, 1999). This LDMOS implementation can be fabricated in the standard CMOS process and provides relatively high-voltage (~15 V) capability without any extra process steps. Applications for such devices include in HV generators and switches for programming nonvolatile memories (Pesavento et al., 2004), liquid crystal display (LCD) drivers, RF power amplifiers, power management circuits (Moscatelli, Merlini, Croce, Galbiati, & Contiero, 2000), and avalanche photodiodes (Lee & Abshire, 2013).

A cross section of a LDMOS is shown in Fig. 4.8. The device can be built using the standard CMOS process of Chap. 3. The gate oxide can be either the core oxide or the I/O gate oxide. The source and gate are laid out and fabricated as the conventional nFET device. The drain has a more complex structure. It is an n-well, where the drain contact is just an n-well tap. The n-well is used as the drain extension that overlaps with the channel.

This device can sustain high voltage on the drain side only. The maxim drain voltage is limited by the n-well breakdown voltage. Even at high drain voltages that far exceed the gate oxide breakdown voltage, the gate oxide will not breakdown. This is because most of the high voltage is dropped across the depletion region in the n-well. This can be seen in the simulation plot in Fig. 4.9 (Wang et al., 2005). The lightly doped N-well region also reduces the electric field and increases the avalanche breakdown voltage at the drain.

The simulation was performed for a 0.25 μm process with STI. In this simulation, the source V_s and gate V_g are biased at 0 V and drain V_d at 14 V. As shown by the equipotential lines at 2 V intervals, the voltage under the gate oxide is close to zero, with all the voltage drops occurring in the n-well. The nFET is in the OFF state, and therefore, it provides a high block voltage for the drain bias. If there is no such drain extension region, the oxide at the drain–gate junction will break down at much lower bias.

Figure 4.10 shows a sample layout and lists the important design rule parameters for the LDMOS. In addition to the usual FET dimensions such as channel length (L_c) and width (W), the design parameters of the LDMOS involve several key dimensions that affect its functionality and reliability: (1) the L_w parameter is the overlap of the n-well with the active area, (2) extension of poly to the STI in the drain (L_o), (3) width of STI in the drain (L_{dp}).

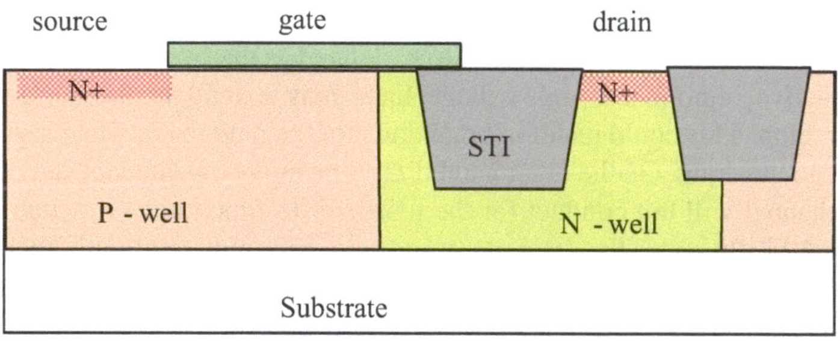

Fig. 4.8 Cross section of an LDMOS in a standard CMOS process

Fig. 4.9 Simulation of electrical potential for a 0.25 μm LDMOS with V_d at 14 V (Wang et al., 2005)

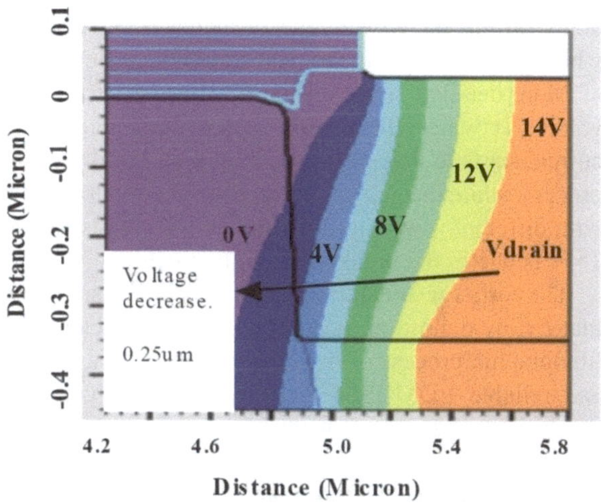

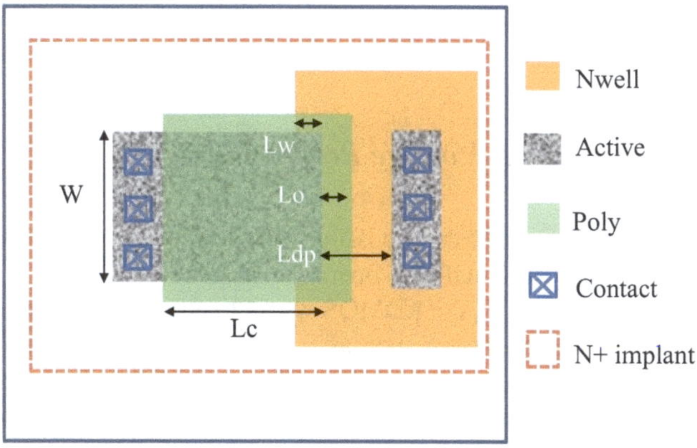

Fig. 4.10 Layout and related design rule parameters of an n-type LDMOS with n-well as the diffused drain. The p-well, by default, covers anywhere that is not n-well

L_w is the most critical parameter for the design of LDMOS. As can be expected from Figs. 4.8 to 4.10, when L_w is too large, defined as the overlap of n-well and the channel active region, the high-voltage lines may extend to the surface of the channel region. This could result in breakdown of the gate oxide when high voltage is applied to the drain. On the other hand if L_w is negative, no channel can be formed and the channel will not conduct for the transistor to function as a switch.

Figure 4.11 illustrates the drain, well, and source current commonly seen in these LDMOS FETs. The figure shows that the drain and source currents equal each other, as expected from a conventional MOSFET at small drain voltage. The drain current then increases rapidly, starting at about 5 V. The increase, however, is not accompanied by an increase in source current. The increase in drain current is associated with the increase of n-well current. This observation suggests that the drain current increase is due to increased reverse leakage at high drain voltage, which generates significant substrate hot carrier currents.

More detailed study on the L_w dependence is to look at BVdss, which is the voltage at which the reverse-biased drain junction breaks down and significant current starts to flow between the source and drain by the avalanche multiplication process, while the gate and source are shorted together. BVdss of a series of devices with different n-well to diffusion active area is shown in Fig. 4.12. Breakdown at 15 V is due to the breakdown of the n-well.

The earlier result shows that L_w has a very small manufacturing window. Notice that L_w, by definition, is an alignment parameter that is determined by the photolithographic process of n-well and the active area masks. Thus to yield functional and reliable LDMOS devices, the mask alignment has to be controlled fairly precisely, to less than 0.1 μm.

The other parameters, especially L_o and L_{dp}, which also affect the field electric field distribution, are important for LDMOS built on LOCOS process (Griffith et al., 2000). Only minimal effects are observed on the performance and reliability of the LDMOS for the STI process (Wang et al., 2005).

4.3.3 Improving High-Voltage Performance

The previous section exposed two possible problems with the LDMOS: small manufacturing window and small HV operating range.

As can be seen in Fig. 4.12, the LDMOS works to about a drain voltage of 15 V for typical mature logic processes such as 0.25 μm and 0.18 μm processes. This upper voltage limit is set by the junction breakdown of the n-well. Unfortunately, it appears that this breakdown voltage is smaller for more advanced processes, i.e., going from about 15 V for a 0.18 μm process to about 13 V for a 65 nm process. This is due to the increased n-well doping for the more advanced processes. To enable operation at higher voltage we have to find ways to increase this breakdown voltage.

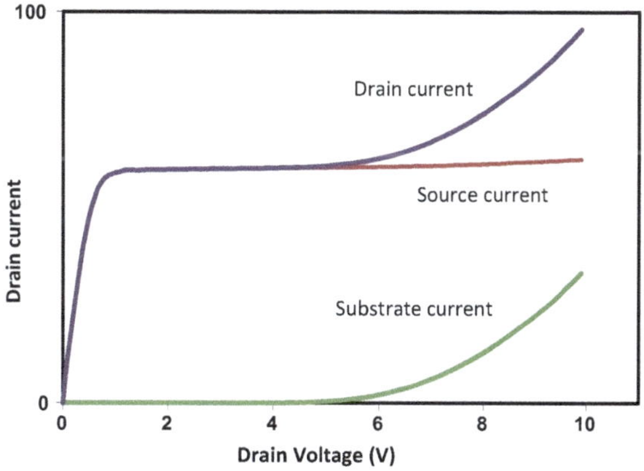

Fig. 4.11 Drain, source, and well current of an LDMOS

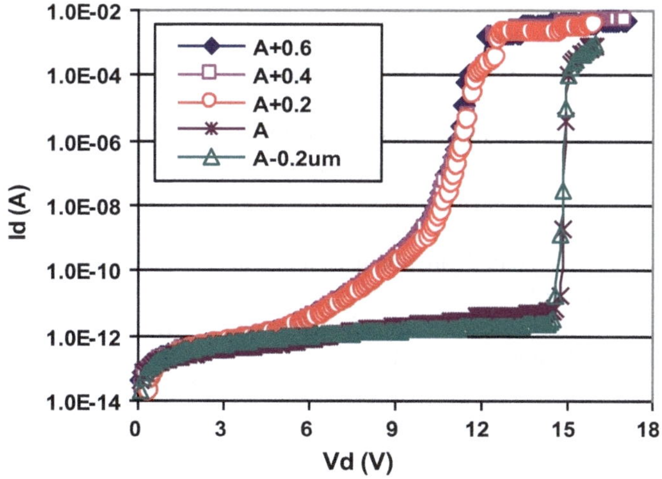

Fig. 4.12 Breakdown behavior of LDMOS as a function of the L_w parameter (Wang et al., 2005), where *A* is a process and foundry dependent parameter

As discussed in Chap. 2, the junction breakdown is inversely related to the doping concentrations at both sides of the junction. Analysis of the breakdown voltage for the n-well in the LDMOS suggests that the weakest point is at the junction between the n-well and p-well. This explains the decrease of the breakdown voltage at more advanced processes as both n-well and p-well doping increases. This is especially true as more advanced processes uses retrograde wells with nonuniform dopant concentration. The weakest point of the junction is then where the n-well and p-well concentration is the highest.

This analysis also points to a way to increase the breakdown voltage. The solution is from the realization that the n-well and p-well do not have to touch each other. Recall from the process flow and the mask generation discussion in Chap. 3, the p-well is usually generated, by default, as the complement of n-well, i.e., everywhere that is not defined by n-well mask is covered by p-well implant. Fortunately, this does not have to be the case. The p-well implant is much higher than the doping level of the p-substrate, e.g., $\sim 10^{17}$ dopant/cm^3 vs. $\sim 10^{15}$ dopant/cm^3. So the solution is to draw the p-well separately to allow a separation of native substrate region between the n-well and p-well. This will result in a "graded" junction around the n-well, as illustrated by Fig. 4.13. An additional design parameter, L_{sub}, is introduced, which is the drawn separation between the n-well and p-well masks.

Indeed, as shown in Fig. 4.14, the breakdown voltage of the graded junction can be increased from ~15 V to as high as 40 V in a 0.18 μm logic CMOS process (Liu et al., 2011). The figure shows that the breakdown voltage is dependent on the separation between the n-well and p-well in a near square law.

Based on this result, we can design a graded junction that can go to much higher voltage. The cross section and layout of a graded junction LDMOS is shown in Fig. 4.15.

Figure 4.16 shows the drain current trends for the new LDMOS, where the drain n-well is surrounded by $L_{sub} = 0.5$ μm native region. We can see that these two types of LDMOS behave very similarly up to about 15 V. At which point the conventional LDMOS shows much fast rise in drain current, indicating the n-well breakdown, while the graded junction n-well did not show any breakdown behavior even at more than 20 V.

The measured transistor characteristics of graded junction LDMOSes are shown in Fig. 4.17. In the figure the LDMOS with L_{sub}=0 is the old LDMOS, while the LDMOS with $L_{sub} \neq 0$ is the new graded junction LDMOS. Note that the old LDMOS shows breakdown at around 14, while the new LDMOS do not show such breakdown.

4.3.4 Manufacturing Challenges

As discussed in Chap. 3, modern MOSFETs are made with many self-aligned steps where the gate to source–drain overlap, and the silicidation of gate and source–drain, are all self-aligned. This is the key to our ability to shrink the transistor size to less than 10 nm, which needs nanometer accuracy, and to manufacture billions of transistors on a single die with high yield.

Unfortunately LDMOS is not a self-aligned device as the alignment of the drain to the channel region is defined by lithography rather than by the process itself. As discussed in Sect. 4.2.2 the critical parameter L_w, which is the overlap between the n-well drain and the channel, is the most critical parameter defining the functionality of the LDMOS device. It also affects the drive current significantly. Thus, the

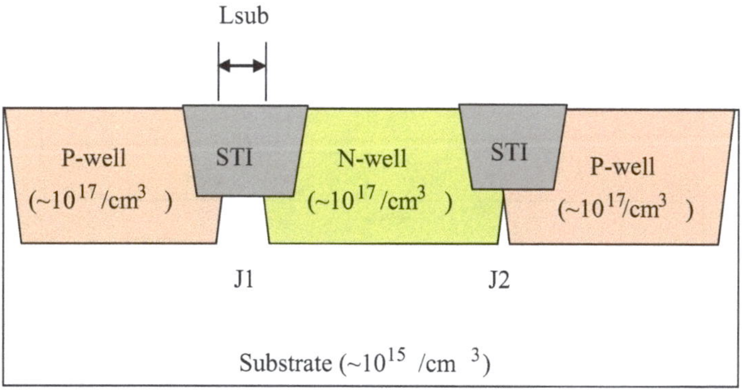

Fig. 4.13 Graded junction J1 where n-well is separated from p-well by a moat of lower doping p-substrate. J2 is the regular n-well/p-well junction

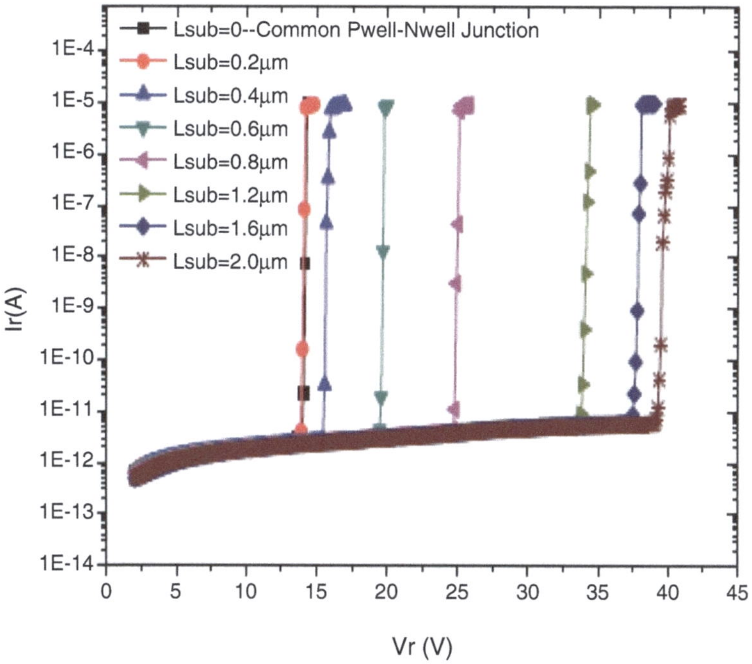

Fig. 4.14 Breakdown, as measured by the reverse bias current, of n-wells surrounded by native moat of different width (Liu, Wang, & Wu, 2011)

alignment between the n-well and the channel masking steps is a critical processing concern for the manufacturing of the LDMOS device.

Another consequence of the LDMOS layout and the non-self-alignment is the discovery of orientation dependence. Wang and coworkers, including one of the

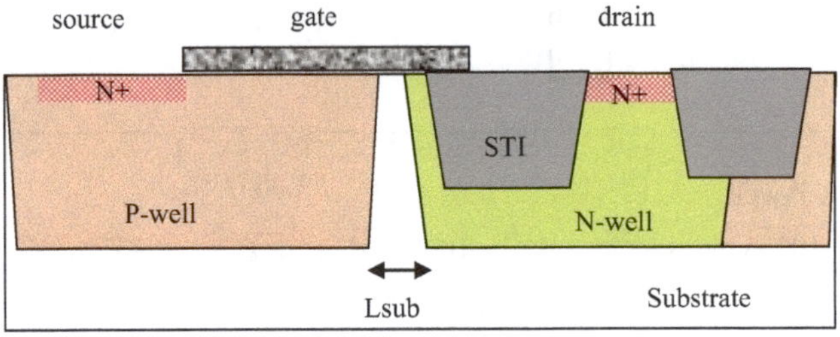

Fig. 4.15 Cross section of a LDMOS with n-well drain surrounded by a native moat

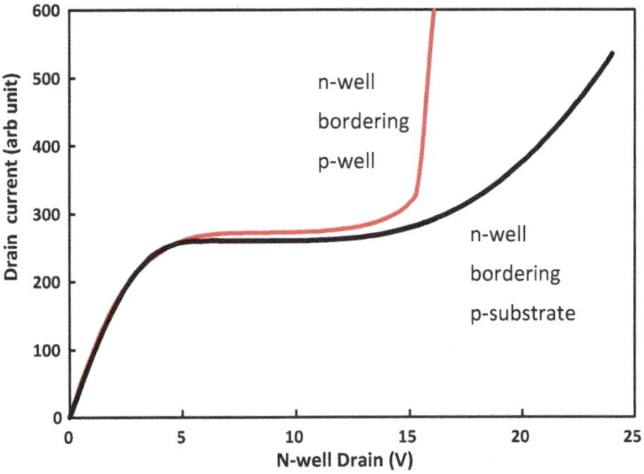

Fig. 4.16 Illustration of drain current for LDMOS with drain made of n-well surrounded by p-wells and n-well surrounded by p-substrate moat

authors (YM), have shown that the drain current of LDMOS is strongly dependent on the orientation of the N-well to channel direction (Wang et al., 2005). In addition, hot carrier reliability of the LDMOS is also dependent on layout orientation for 0.25 μm N-LDMOS. It suggests that layout orientation might have some impact on the lifetime of I_{dsat}. The device lifetime is longer at an orientation when the I_{dsat} is also the largest.

The observed layout orientation dependence may be explained by (1) misalignment between gate oxide mask and n-well mask; (2) the fact that some of the n-well implants are typically done with a 7° angle (Wolf & Tauber, 1990). As a result, the effective L_w for different orientations will be different from drawn dimension.

The LDMOS manufacturability is affected by the layout orientation dependence. Lack of restrictions on the orientation of the N-LDMOS device will result in even smaller design window. This requires carefully designed process monitoring devices as part of the wafer acceptance and wafer sort testing. Examples of the

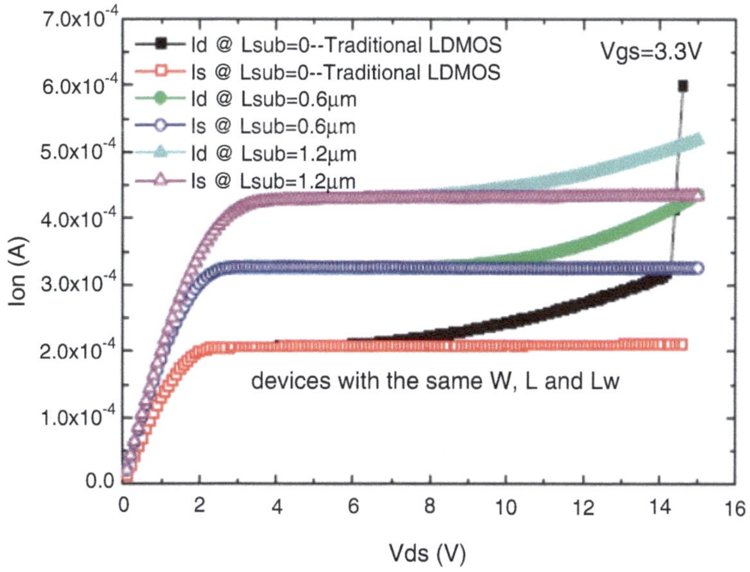

Fig. 4.17 Drain characteristics of LDMOS with different native moat width, from $L_{sub} = 0$ which is the case of the conventional LDMOS to $L_{sub} = 1.2$ μm (Liu et al., 2011)

device may include LDMOS devices with different orientation, different L_w, and other device parameters. Lot to lot, wafer to wafer, die to die, and within die variations should all be monitored.

Figure 4.18 illustrates the I_{dsat} variations of a production LDMOS in the process control monitor structures across about 1000 wafers at a 0.25 μm process node (Wang & Ma, 2008). Large variations of I_{dsat} of up to 50 % are seen. This suggests that LDMOS is not optimal for use in precision circuits. However, LDMOS has been demonstrated commercially in many applications where the precise drive current is not necessary. For example, we will show in Chaps. 8 and 10 that LDMOS has been used in high-voltage switch circuits for embedded nonvolatile memories.

The rectangular layout illustrated in Fig. 4.8 has sharp corners which might introduce edge breakdown and reduce the device performance and/or manufacturability. Circular structures are implemented to eliminate edge effect as illustrated (Lee & Abshire, 2013) in Fig. 4.19.

Finally, since the minimum dimension of n-well is usually large in all logic processes, the LDMOS is usually even larger, several times the size of the n-well. As can be seen from the layout in Fig. 4.12, due to the presence of the native moot around the n-well and associated design rules, the graded junction LDMOS is even larger.

Since LDMOS takes up much larger area than a high-voltage transistor built with dedicated process, LDMOS in the logic process may only be economically used when only a few of such devices are needed in a design. If large number of

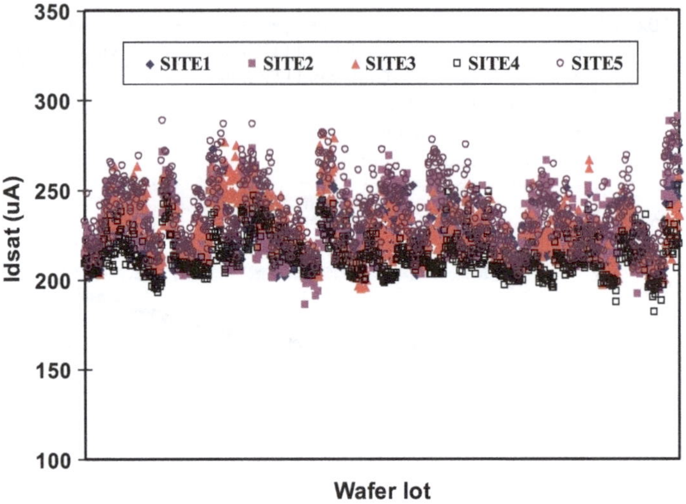

Fig. 4.18 I_{dsat} for LDMOS inside the production PCM from 1000 wafers at a 0.25 μm process node (Wang & Ma, 2008)

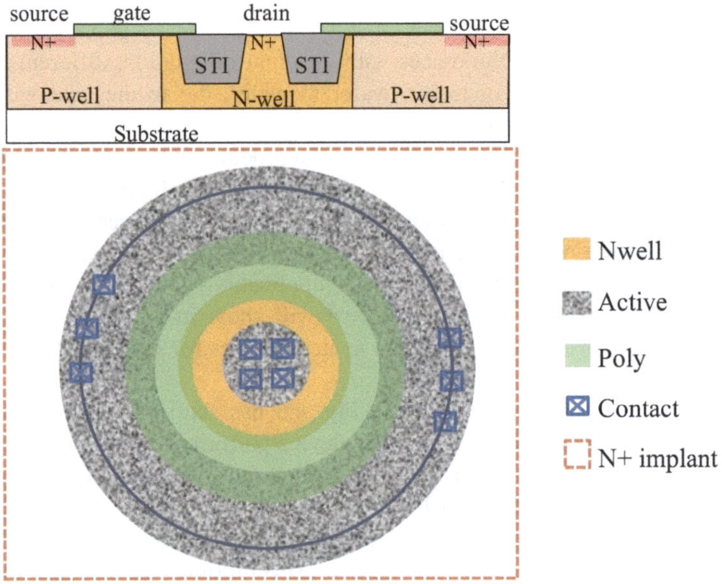

Fig. 4.19 Circular layout for increasing breakdown voltage of LDMOS

high voltage transistors are needed in the circuits, it is often more economical to use special high-voltage processes. One needs to evaluate on a case-by-case basis which approach is most appropriate. A rule of thumb is that if the high-voltage devices are only used by a small percentage of the circuits for a small percentage of the time, then the LDMOS approach introduced here may be appropriate.

References

Baker, R. J., Li, H., & Boyce, D. (1998). *CMOS—Circuit design, layout, and simulation*. New York: IEEE Press.

Ballan, H., & Declercq, M. (1999). *High voltage devices and circuits in standard CMOS technologies*. Berlin, Germany: Springer.

Griffith, E. C., Power, J. A., Kelly, S. C., Elebert, P., Whiston, S., Bain, D., et al. (2000). Characterization and modeling of LDMOS transistors on a 0.6 μm CMOS technology. In *Proceedings of the 2000 International Conference on Microelectronic Test Structures* (pp. 175–180). IEEE.

Hsieh, C. S., Kao, P. C., Chiu, C. S., Hon, C. H., Fan, C. C., Kung, W. C., et al. (2004). NVM characteriscs of single-MOSFET cells using nitride spacers with gate-to-drain NOI. *IEEE Transactions on Electron Devices, 51*, 1811–1817.

Kuhn, J. (2010). *Voltage reference circuit with low-power bandgap*. US Patent #7,808,387.

Lee, T.-H., & Abshire, P. (2013). Design and characterization of high-voltage NMOS structures in a 0.5 μm standard CMOS process. *IEEE Sensors Journal, 13*, 2906.

Lin, C., Chen, H., Shen, S., King, Y., & Hsu, C. (2009). *Single-poly non-volatile memory device and its operation method*. US Patent #7,551,494. Washington, DC: US Patent and Trademark Office.

Liu, Z., Hu, C., Huang, J. H., Chan, T. Y., Jeng, M. C., Ko, P. K., et al. (1993). Threshold voltage model for deep-submicrometer MOSFET's. *IEEE Transaction on Electron Devices, 40*, 86–95.

Liu, Y., Wang, B., & Wu, H. (2011). An N-channel graded-junction lateral diffused MOS transistor in 0.18μm low-power logic CMOS process. In *2011 International Semiconductor Device Research Symposium*. College Park: IEEE.

Moscatelli, A., Merlini, A., Croce, G., Galbiati, P., & Contiero, C. (2000). LDMOS implementation in a 0.35 μm BCD technology (BCD6). In *12th International Symposium on Power Semiconductor Devices and ICs* (pp. 323–326). IEEE.

Pesavento, A., Gilliand, T., Lindhorst, C., Srinivas, S., Bernard, F., Salazar, S., et al. (2004). Embedded non-volatile memory in logic CMOS. In *Non-volatile Semiconductor Memory Workshop* (p. 49). Monterey: IEEE.

Wang, B., & Ma, Y. (2008). Opportunities and challenges in multi-times-programmable floating-gate logic non-volatile memories. In *2008 International Memory Workshop* (pp. 120–125). IEEE.

Wang, B., Nguyen, H., Mavoori, J., Horch, A., Ma, Y., Humes, T., et al. (2005). Effect of layout orientation on the performance and reliabiltiy of high voltage N-LDMOS in standard submicron logic STI CMOS process. In *43rd IEEE International Reliability Physics Symposium* (pp. 654–655). IEEE.

Wolf, S., & Tauber, R. N. (1990). *Si processing for the VLSI era* (Process integration, Vol. 2). Sunset Beach, CA: Lattice Press.

Chapter 5
Floating-Gate Devices in Logic CMOS Processes

Abstract In this chapter, we discuss another important class of devices, devices containing floating gate including floating-gate transistors and capacitors, that can be produced in logic CMOS processes. We describe some examples of making floating-gate memory cells in a logic process and several methods of programming, including tunneling and hot carrier injection, floating-gate devices. The importance of coupling ratio and its implications to the NVM cell design is discussed extensively. In the last section, we review a semi-floating-gate device, that of gain cell which have found application as an embedded DRAM cell.

Keywords Nonvolatile memory • MOSFET • Floating-gate transistor • Floating-gate capacitor • Tunneling • Fowler–Nordheim tunneling • DRAM • Gain cell • Semifloating gate • Coupling ratio • Threshold voltage • Program • Erase

5.1 Introduction

Measured by pure number produced by the semiconductor industry, floating-gate (FG) transistor is perhaps the most widely used semiconductor device. FG transistor is the basic storage element in virtually all flash memories.[1] As an example, a single 16Gbit NAND flash memory chip, at one bit per cell, has over 16 billion float gate transistors on a single die. For the more commonly seen **multilevel cell (MLC)** chip, with two bits per cell, a 128Gbit chip has more than 64 billion floating-gate transistors.

These transistors are densely packed and fabricated using special flash memory processes. The process is very different from the standard logic process described in Chap. 3. Because of packing density requirement, all present planar floating-gate flash memory processes use two layers of polysilicon gate, e.g., see Fig. 5.1. The first poly layer is the floating gate which is separate from the silicon substrate by a tunneling oxide, usually 7–10 nm thick. The second polysilicon layer is the control gate (CG) which sits on top of the floating gate, separated by a deposited ONO (SiO–SiN–SiO) layer about 10 nm thick.

[1] More recently, some 3D flash memories started to use charge trapping transistor, not floating gate, to store charge.

Y. Ma, E. Kan, *Non-logic Devices in Logic Processes*,
DOI 10.1007/978-3-319-48339-9_5

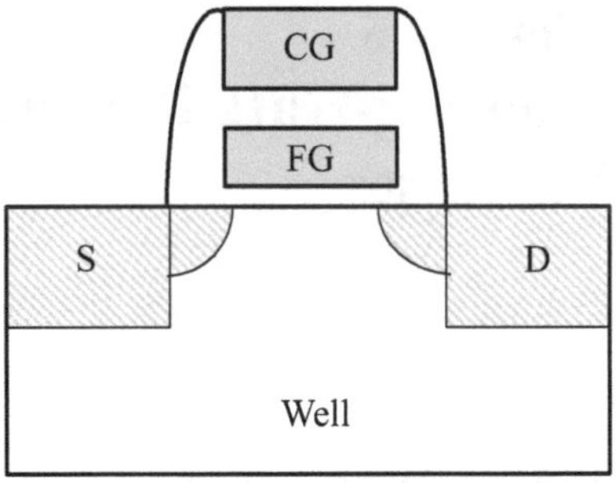

Fig. 5.1 Double polysilicon floating-gate transistor used in flash memory chips

To program and erase these double poly floating-gate cells, programming voltages as high as 20 V are necessary. To make the double polysilicon floating-gate transistors and the high-voltage transistors, many additional mask layers and processing steps are needed, in additional to the steps for making the basic logic CMOS transistors. These special processing steps, often involving long anneal times, are not compatible with the requirements for making high speed and high density transistors for high performance logic ICs. For these reasons, the flash memory processes are usually not compatible with logic CMOS processes.

The primary use of floating-gate devices is for data and code storage. Besides data storage, floating-gate transistors have found many embedded applications, including trimming for analog circuits (De Vries & Ma, 2007) and as synaptic transistors to model neuromorphic systems (Mead, 1989). For these embedded applications, it is desirable to use floating-gate devices that are compatible, and ideally can be made with the standard logic CMOS processes without process modifications.

In this chapter, we will present techniques for making floating-gate devices using standard logic CMOS processes with a single polysilicon layer and for the operation of these devices. The floating-gate transistor is the main focus but we also discuss another use of floating gate, i.e., in capacitors. In Chaps. 8–10, we will discuss the uses of these floating-gate devices as nonvolatile memory and in Chaps. 11 and 12 we will discuss analog and other nondata storage applications such as biosensors.

Floating-gate transistor was first invented by Kahng and Sze at Bell Labs in 1967 (Kahng & Sze, 1967). Based on the floating-gate concept, Intel's Frohman invented the EPROM (erasable programmable read-only memory) in 1971, where erasure is done using UV irradiation. Electrically erasable version of EPROM was invented in 1978 by Perlegos, also at Intel, resulting in a floating-gate EEPROM (electrically erasable programmable read-only memory). In 1980, Masuoka at Toshiba developed the modern version of flash memory, now used in everything from USB sticks to digital cameras, smart phones, and solid-state disk drives.

The central figure of floating-gate transistors and capacitors is the "floating gate" where the normal polysilicon or metal gate in the MOSFET is not contacted and is entirely isolated by surrounding insulators. The isolated gate can be used to store charges which in turn can be used to vary the threshold voltage of the FG transistor.

The threshold voltage of the floating-gate transistor is then used to store information in nonvolatile memories.

The key quantity, the charge on the floating gate, can be varied by a number of methods. The earliest attempt, which is still being used, is by using ultraviolet radiation. This is the mechanism behind the EPROM memory. UV can be used only to remove charge from the floating gate and, as a result, is useful only for one-time programmable memories. The erasure process is also slow and very inconvenient in stand-alone end product. As a consequence, electrical methods using electron or hole tunneling or injection have taken over to enable the tremendous growth in the EEPROM. NAND and NOR flash memories are special types of EEPROM.

One of the main performance metrics of the floating-gate device is the programming speed—how fast the charge level can be changed on the floating gate. Typically one millisecond or less is desired. A related metric is, once the charge is placed on the floating gate, how long can they stay there. For standard memory applications, a retention period of 10 years within the storage temperature range is usually required. Even longer, e.g., 50 years, retention period may be needed for some applications such as archival storage.

Using a representative programming time of 30 μs and a retention time of 10 years, these two modes of operation differ in time scale on the order of

$$10\,\text{year}/30\,\mu\text{s} \cong 10^{12}.$$

Therefore, it appears to be a great challenge to be able to perform two operations with such vastly different time scales.

We translate these numbers to electrical quantities. The floating gate typically has a capacitance of 1 femto-farad or less, depending on the technology node and the size of the floating-gate cell. To be able to reliably distinguish the information store on the floating gate, the FG voltage change is usually on the order of 1 V between the programmed and erased state. Thus, we need to move total charge, $Q = \text{C–V} = {\sim}1$ femto coulomb (about 6000 electrons), each time we program a floating-gate memory cell. To do this in ~ 30 μs means the programming current is about ~1fC/30 μs = 10^{-11} Amp. To keep the data in 10 years, the leakage current will need to be less than ~$10^{-11}/10^{12} \sim 10^{-23}$ A.

Fortunately the tunneling and injection phenomena discussed in Chap. 2 have strong, exponential voltage dependence. This voltage dependence has been used as a means to accelerate the charging or discharging of the floating gate in the required time interval. On the other hand, when the voltage is removed, the charge can be retained under normal operating conditions. The hot carrier injection can be used to inject charge (either electrons or holes) to the floating gate while FN tunneling has been used for both erase and program floating-gate-based memory cells.

These operating conditions are illustrated graphically in Fig. 5.2. In the figure, the current is plotted in logarithmic scale against voltage in linear scale. The two principal operating regions are represented by the two bubbles. The first is the storage and read region on the lower left. In this region, the floating gate is at a lower voltage and is read repeatedly or in storage for a long time. Thus, the current,

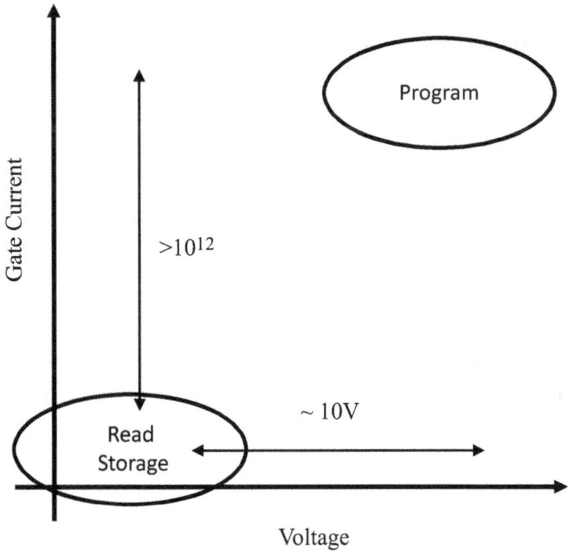

Fig. 5.2 The regions of IV domain for programming and storage of floating-gate transistors

classified as the leakage current, from the floating gate needs to be less than 10^{-23} A as mentioned in the previous paragraph. During the program/erase phase, in the bubble on the top right, the cell needs to be programmed in ~1 ms or less, necessitating a program current about 12 decades higher than that during the storage phase. As we recall from the discussion on hot carrier injection and FN tunneling, the current-to-voltage characteristics of both processes are exponential and can be used as the mechanism for the necessary voltage acceleration.

One further requirement is that, as we discussed in Chap. 4, it is difficult to handle high voltage in the CMOS process, it is therefore desirable to be able to accomplish the ~12 decade current acceleration with a voltage range of about ~10 V or less.

5.2 Floating-Gate Transistors

In this section, we discuss floating-gate transistors that can be obtained from the single poly CMOS process of Chap. 3. Figure 5.3 shows a basic floating-gate cell including a conventional pFET as a select transistor and a floating-gate pFET. The select transistor can be either an nFET or a pFET. The select transistor is needed when the cells are arranged in a row and column array. Since one of the states of the floating-gate transistor is an "On" state, the select transistor is turned off when the memory cell is not selected in order not to disturb the memory cells in the same row or column.

The floating gate usually comes out of fabrication in an uncharged state and the FG transistor is turned off. To ensure this is the case, often an UV erasure step is

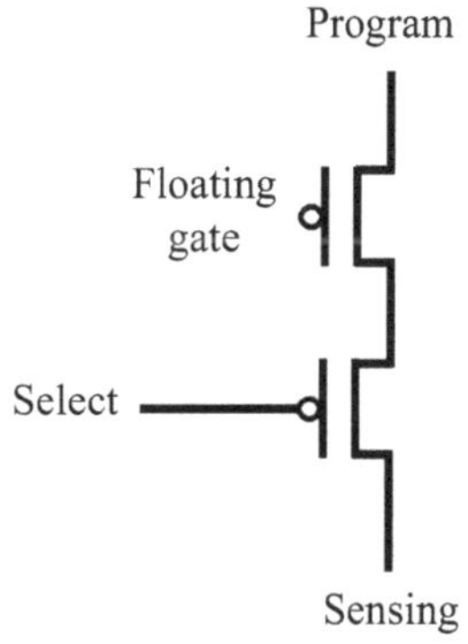

Fig. 5.3 Floating-gate transistor with a select transistor

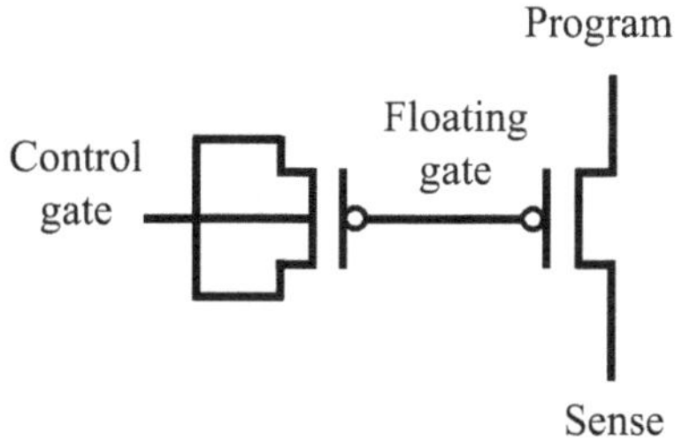

Fig. 5.4 An electrically erasable floating-gate cell. The select transistor is omitted

used at the end of the chip fabrication. Electrons can be injected onto the floating gate from the "Program" node using hot carrier injection. When the floating gate accumulates enough electrons, the pFET may turn on. To read, external circuitry enables the select transistor (at the "Select" node), and a sense amplifier connected to the "Sense" node detects the state via presence or absence of current flow.

To be able to erase the electrons on the floating gate electrically, an additional device is needed to couple to the floating gate, for example, a capacitor. Figure 5.4 shows a basic two-transistor floating-gate cell that can be used both to add and remove charge on the floating gate. The select transistor is omitted for simplicity. The cell can be implemented using standard CMOS process in a single poly process without extra masks or process steps.

An example cross section of how this is achieved is shown in Fig. 5.5. Two of the largest structures of this cell are the n-wells. One is used by the floating-gate pFET. The second n-well hosts the control gate device, which can be either a MOSFET or a MOSCAP. Since only one node, the control gate node, is necessary, the second device only needs an n-well contact.

Three basic operations can be performed on this floating-gate memory cell: erase, program, and read. Pulling the n-well-labeled control gate to high voltage while keeping other nodes at low voltage or ground can cause Fowler–Nordheim (FN) tunneling which removes electrons from the floating gate. This operation is called "erase." By biasing the "Program" node to high voltage, hot carrier injection

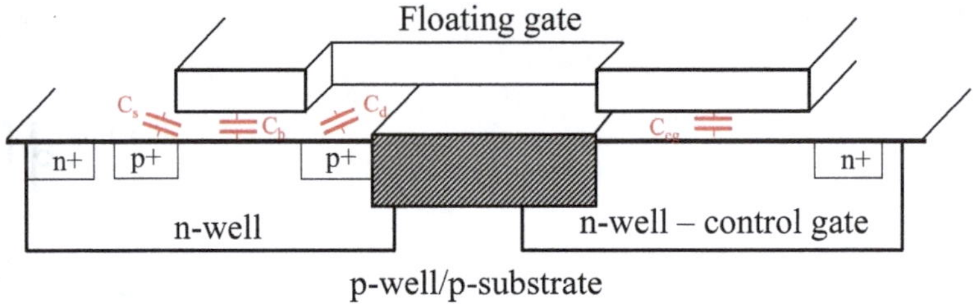

Fig. 5.5 Floating-gate transistors with single poly gate. The middle section of the floating gate is out of the plane. The control gate node needs only an n-well contact. Some built-in capacitors are also illustrated in the figure

(through the impact-ionized hot electron injection process discussed in Sect. 2.4.4) can occur to inject electrons onto the floating gate, which is named "program." To read or sense the charge on the floating gate, the control gate is biased at ground to allow current to flow through the program and sense nodes, as detected by a sense amplifier. An external selected transistor may be added, e.g., similar to Fig. 5.3 to isolate the memory cell from others in an array.

To estimate the voltages needed for these operations, it is important to introduce one of the most important concepts of floating-gate device physics: the so-called ***coupling ratio (CR)***. The voltage of a floating gate is not determined by the charge on the gate alone. It is also affected by the voltages on the nearby nodes through capacitive coupling. The capacitive analysis is simplest to understand using the example of Fig. 5.5. In Fig. 5.5 the most important capacitances are illustrated in red, including capacitance between the control gate and the floating gate, C_{cg}; the capacitances between the floating gate to the source and the drain (C_{s}, and C_{d}, respectively), and the capacitance to the bulk, C_{b}. There are many other parasitic capacitances, for example, the floating gate to the substrate through the STI region, that are usually much smaller and can be neglected in this simple analysis. On the whole these parasitic capacitances can have a significant contribution and warrant detailed analysis in actual floating-gate memory cell design.

The voltage on the floating gate can be found to be

$$V_{\mathrm{fg}} = \frac{Q_{\mathrm{fg}}}{C_{\mathrm{tot}}} + \alpha_{\mathrm{s}} V_{\mathrm{s}} + \alpha_{\mathrm{d}} V_{\mathrm{d}} + \alpha_{\mathrm{cg}} V_{\mathrm{cg}} + \alpha_{\mathrm{b}} V_{\mathrm{b}} \cdots = \frac{Q_{\mathrm{fg}}}{C_{\mathrm{tot}}} + \sum \alpha_{\mathrm{i}} V_{\mathrm{i}} \tag{5.1}$$

where Q_{fg} is the charge on the floating gate, the total capacitance C_{tot} seeing from the floating gate is given by

$$C_{\mathrm{tot}} = \sum C_{\mathrm{i}} \tag{5.2}$$

and the coupling ratio for the *i*th capacitor is defined by

$$\alpha_{\mathrm{i}} = \frac{C_{\mathrm{i}}}{C_{\mathrm{tot}}} \tag{5.3}$$

For simplicity and for the purpose of illustrating the most important dependences, we assume that the source and drain capacitances are negligible. As a consequence, there are only two remaining coupling coefficients, α_{CG} and α_b, which are related,

$$\alpha_{CG} + \alpha_b = 1 \tag{5.4}$$

This is because $C_T \cong C_{cg} + C_b$. The voltage across the gate oxide on the left n-well in Fig. 5.5 can then be shown to be

$$V_{ox} = V_{fg} - V_b = \alpha_{CG}(V_{cg} - V_b) + \frac{Q_{fg}}{C_{tot}} \tag{5.5}$$

As we have seen in Chap. 2, the FN tunneling current is exponentially dependent on the electric field in the oxide, $E_{ox} \sim V_{ox}/t_{ox}$, where t_{ox} is the thickness of the tunnel oxide. We then can see from Eq. (5.5) that the tunneling current is closely correlated to the applied voltage and the coupling ratios. As will be discussed in detail in Chaps. 8–10, the coupling ratio affects the required voltage for programming operations and is one of the most important considerations in designing the memory cell for nonvolatile memory.

We will discuss in more details the available methods for changing the charges on the floating gate in Chap. 8. With a change in the charge on the floating gate, the apparent threshold voltage of the floating-gate transistor also changes and is given by, see (Shen et al., 2007),

$$\Delta V_{th} = -\frac{\Delta Q_{fg}}{C_{tot}} \tag{5.6}$$

This is the basis for virtually all floating-gate nonvolatile memories—by electrically varying the charge on a floating gate we can change the threshold voltage of a floating-gate transistor.

Figure 5.6 illustrates the cell current of a floating-gate cell with the floating gate at neutral and at a charged condition. The shift in threshold voltage depends on the type of MOSFET (n vs. p) and the sign of the net charge. For n-type floating-gate transistor, the neutral floating-gate condition corresponds to an "on" or "1." When electrons are injected on to the floating gate, the apparent threshold voltage of the FG transistor is shifted higher, corresponding to an "off" or "0" state. For p-type FG transistor, negatively charge floating gate is the on state. As a result, if the control gate is biased at an intermediate voltage as indicated by the arrow in the figure, the cell current will be either on or off, corresponding to a "1" or a "0" state. This is the method for "reading" the floating-gate cell.

Intermediate cell current levels can be explored to store more states, leading to multilevel cell (MLC) NAND flash memory. Up to eight levels or three bits are routinely stored on one flash memory cell.

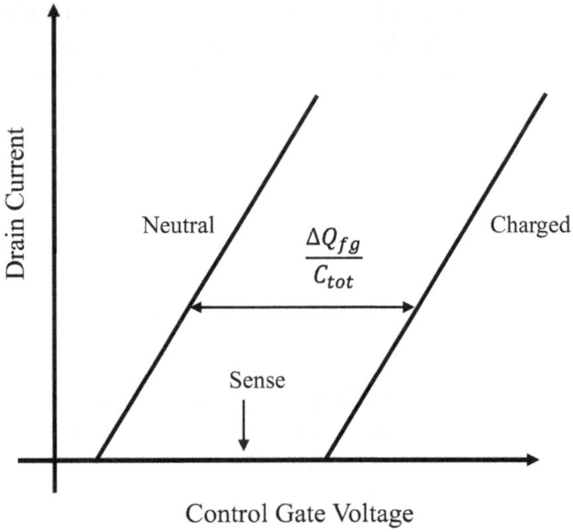

Fig. 5.6 Illustrated drain current vs. control gate voltage for a generic MOSFET with neutral and charged floating gate

5.3 Floating-Gate Capacitor

Integrated variable capacitor (varactor) is a valuable device in integrated circuit design and has found wide applications in, among others, switched capacitor and voltage controlled oscillator (VCO) circuits (Sze, 1981). A typical varactor is a MOS capacitor using voltage controlled depletion depth to vary the junction capacitance. As the depletion width adjustment is by the majority carrier motion with dielectric relaxation time, MOS varactors can be operated in very high frequency (e.g., around 50 GHz, limited mainly by gate line resistance).

Floating gate offers another possibility to produce a variable capacitor (Hyde & Ma, 2004). The floating variable capacitor is based on back-to-back series-connected MOS capacitors with a shared floating gate. By changing the charge on the floating gate, the total capacitance of this composite capacitor can be varied by close to 100 %.

The schematic for one type variable capacitor is shown in Fig. 5.7, which is implemented with back-to-back connected pMOS capacitors, C_1 and C_2. A small programming MOSFET is also shown, which is used to remove or add electrons to the floating gate. The capacitance of this small FET is neglected when calculating the coupling ratio at node V_1 and V_2.

The capacitance across the V_1-V_2 nodes can be expressed in the following equation, assuming that the tunneling device is much smaller and its capacitance can be ignored,

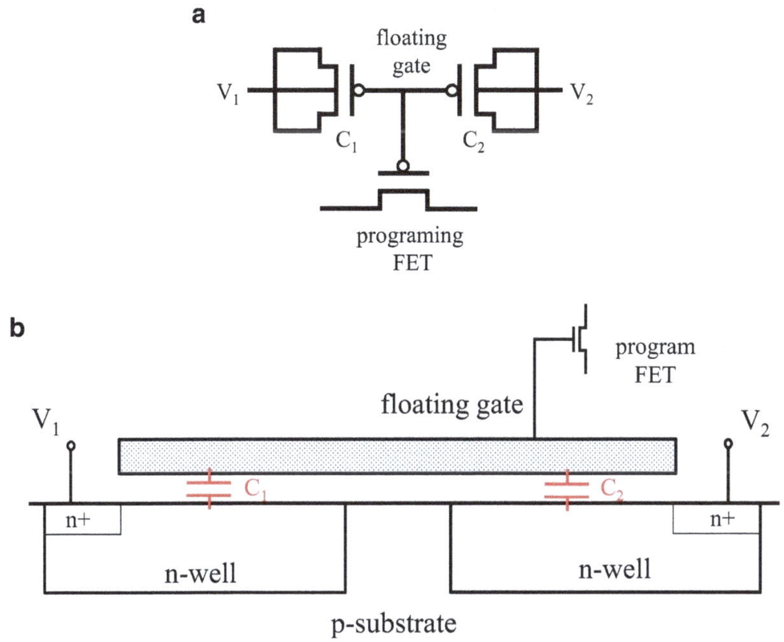

Fig. 5.7 Schematic and cross section of a variable capacitor implemented using three pMOS floating-gate devices—two large capacitors and a small programming FET. For simplicity, we assume C_1 and C_2 have the same size

$$C = \left(\frac{1}{C_1} + \frac{1}{C_2}\right)^{-1} \tag{5.7}$$

The MOSCAP capacitance C_1 and C_2 depend on the gate-to-well biases, which in turn are related to the coupling ratio and the charge on the floating gate, and are given by

$$\begin{aligned} V_{gb1} &= \frac{Q_{fg}}{C_{tot}} + \alpha_2(V_2 - V_1) \\ V_{gb2} &= \frac{Q_{fg}}{C_{tot}} - \alpha_1(V_2 - V_1) \end{aligned} \tag{5.8}$$

where C_{tot} is the total capacitance of the floating gate, ~$(C_1 + C_2)$, and α_1 and α_2 are coupling coefficients of terminal V_1 and V_2 to the floating gate, respectively. By changing the charge on the floating-gate Q_{fg} we can change the floating-gate-to-well biases and achieve the desired accumulation/depletion capacitance variations.

As an example, let us examine a couple of simple cases. First is when electrons are removed from the floating gate to leave it positively charged and V_{fg} at a value corresponding to $Q_{fg}/C_{tot} \sim 3$ V. In this case, when $V_2 - V_1 <$ ~2 V both MOSCAPs

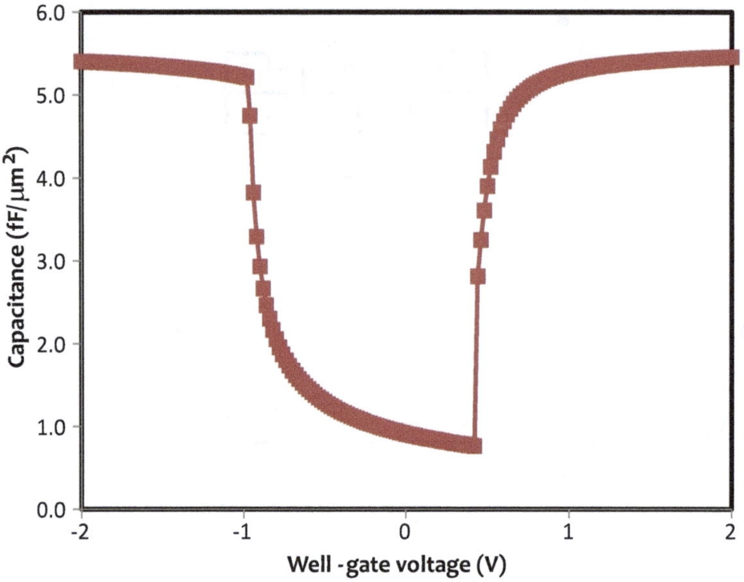

Fig. 5.8 Simulated high frequency C–V curve of a typical 3.3 V pFET

are in the accumulation state and $\alpha_1 \sim \alpha_2 \sim 0.5$. The total capacitance is then $C_1/2 = C_2/2 = C_{ox}/2$, with $C_{ox} = WL\ \varepsilon/t_{ox}$. W and L are the width and the length of the MOSCAP.

When there is no charge on the floating gate, the two capacitors will have opposite biases. Near the bias condition $V_2 - V_1 \sim 0$ we will have both MOSCAPs in the depletion state and the overall capacitance is minimized, $C = C_2/2 = C_1/2 = C_{dmin}/2$, where C_{dmin} is the depletion capacitance of the MOSCAP.

The earlier two cases are illustrated by simulations. First, the C–V curve of a stand-alone MOSFET is simulated. Figure 5.8 plots the C–V curve for a pFET with a 6.2 nm thick gate oxide, a flatband voltage of 1 V, and a threshold voltage of 0.5 V. The parameters are picked to emulate a 3.3 V I/O pFET available from foundries. Note that the voltage at the X axis is the well-to-gate voltage. As a result the C–V curve is flipped from the normal C–V curve of a pFET and similar to the C–V curve of the nMOSCAP shown in Fig. 2.9.

When two of these pFETs are connected back to back, as in Fig. 5.7, maximum capacitance is half of the single capacitance shown in Fig. 5.8. As shown in Fig. 5.9, we can see that the C–V curve indeed looks like two back-to-back connected capacitors of Fig. 5.8. Here in the simulation we placed a high positive charge on the floating gate, which shifted the threshold voltage of both pFETs. One C–V curve has a $V_t \sim -3$ V and the other, being in the opposite orientation, has $V_t \sim 3$ V.

As Q_{fg} is reduced, these two curves move toward each other and start to merge, reaching a minimum $C = C_{min}/2$ where C_{min} is the minimum capacitance of one of the MOSCAP. Figure 5.10 shows such a series of C–V curves with the last one being with floating gate at neutral state, $Q_{fg} \sim 0$.

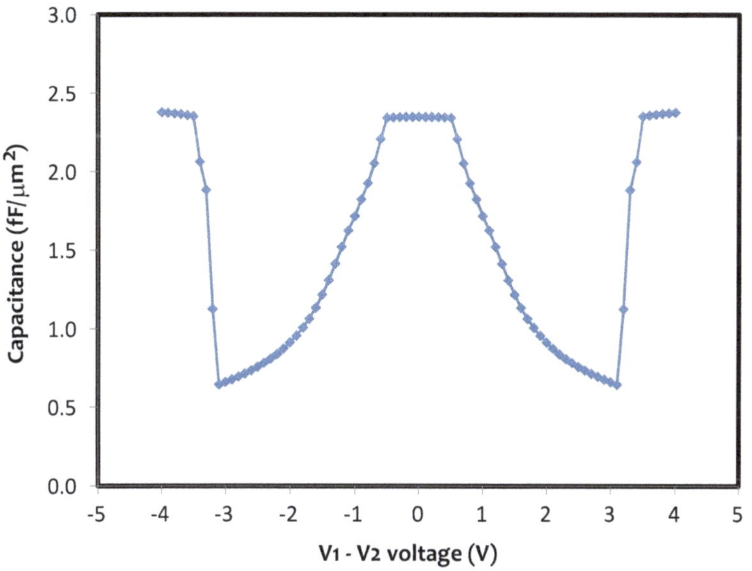

Fig. 5.9 Simulated C–V curve of two back-to-back connected pFETs of Fig. 5.8

Continuing in the same rationale, now adding negative charge to the floating gate, in Fig. 5.11 we can see that capacitance of the two floating gates will start to cross over each other and shift to the opposite directions.

The experimental C–V curves of two back-to-back connected 10 μm × 10 μm pFETs are shown in Fig. 5.12 (Hyde & Ma, 2004). The device was fabricated in a logic CMOS process using the 3.3 V I/O FETs with 7 nm gate oxide. The observed results are similar to simulation results described earlier.

Using the bandgap engineering concept discussed in the last chapter, we can get a drastically different version of the variable capacitor. By changing the dopant of floating gate in Fig. 5.8 from p+ to n+, the flatband voltage of each MOS structure is shifted by ~1 V. This drastically changed the C–V characteristics. Simulated C–V curves of a back-to-back connected pFETs with n+ floating gate as a variable capacitor are shown in Fig. 5.13.

As we will show in Chap. 10, the floating-gate transistors with 7 nm oxide have good charge retention characteristics, (Ma et al., 2004). Thus, the variable capacitors built of the 3.3 V I/O oxide of a logic CMOS process are expected to hold its value. For increased capacitance, thinner oxide devices, e.g., 1.8 V FET with 3 nm gate oxide may be used. But a more frequent calibration process may be necessary to maintain the desired capacitance to compensate for charge leakage.

Notice that even though the charging and discharging of the floating gate can be slow, the capacitance response to V_1 and V_2 are very fast with cutoff frequency up to 50 GHz when the back-to-back FETs are in the accumulation and depletion regions, as only the majority carrier participates in the capacitance adjustment. However, when either one of the FET has to be in or out of the inversion state, the C–V

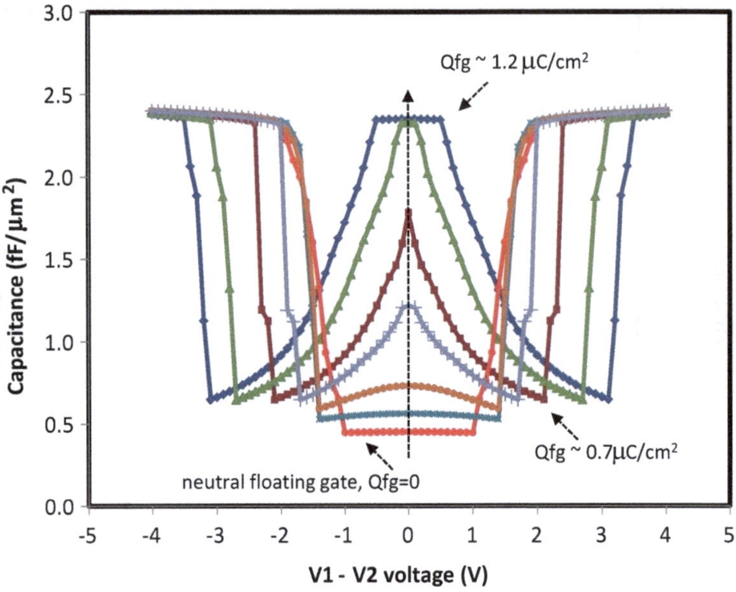

Fig. 5.10 Simulated C–V curves with varying floating-gate voltages

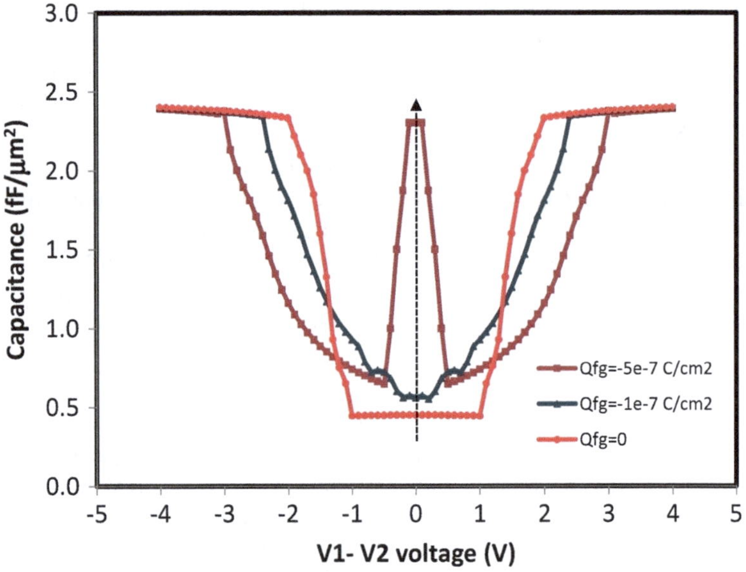

Fig. 5.11 Simulated C–V curves of back-to-back connected pFETs with ~ neutral to negative charge on floating gate

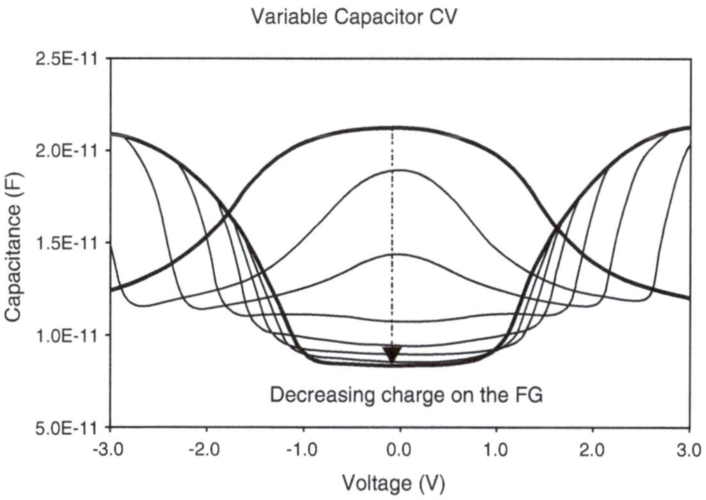

Fig. 5.12 Inversion-mode MOSCAP C–V curves with different amount of charge on the floating gate. Capacitance measured at 100 KHz. The device is comprised of two back-to-back connected 10 μm × 10 μm 3.3 V pFETs and a small tunneling device (Hyde & Ma, 2004)

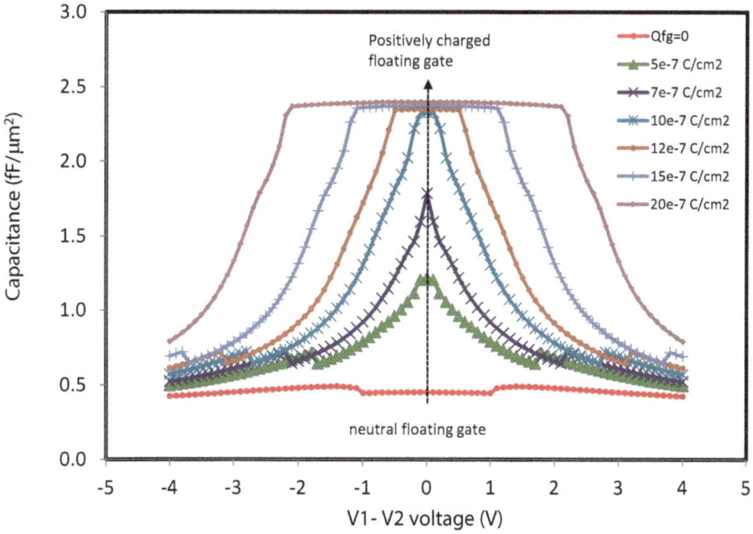

Fig. 5.13 Simulated C–V curves for back-to-back connected n+ gate pFET with different floating-gate charge

response can be slow (tens of Hertz) if no deliberate channel injection structure is built. At the inversion state, the minority population change is slow when no source of minority carriers is implemented.

The earlier two examples show the feasibility of creating many types of novel variable capacitors using floating-gate technology, which has the advantage of nonvolatility, easy to program, and continuous variability as compared to specialized, digitally controlled varactors and capacitor banks. These nonvolatile variable capacitors may find applications in switched capacitor, digital-to-analog (DAC), or analog-to-digital (ADC) converter circuits.

5.4 Semifloating Embedded DRAM Cell

An alternative method for charging and discharging a floating gate is by directly connecting it to a MOSFET, as shown in Fig. 5.14. Strictly speaking the floating-gate node here is only semifloating as all MOSFETs usually have substantial leakage, either through the junction to substrate or through the subthreshold leakage. Thus, the floating node can only hold the charge for a very short period of time, usually on the order of micro to millisecond. Still this type of semifloating node can be used to store charge and has been used in as DRAM cells for embedded DRAMs.

Recall that the conventional DRAM cell is a one transistor one capacitor (1T1C) cell, as shown in Fig. 5.15. The basic cell can be addressed by the word line (WL) and sensed at bit line (BL) when arranged in a dense row and column array. The cell stores a "0" when the capacitor is not charged and "1" when it is charged. Since the charge can leak, mostly through the access transistor, refresh (recharge of the capacitor to maintain the information stored) periodically is needed. And the capacitor needs to have a large capacitance to avoid frequent

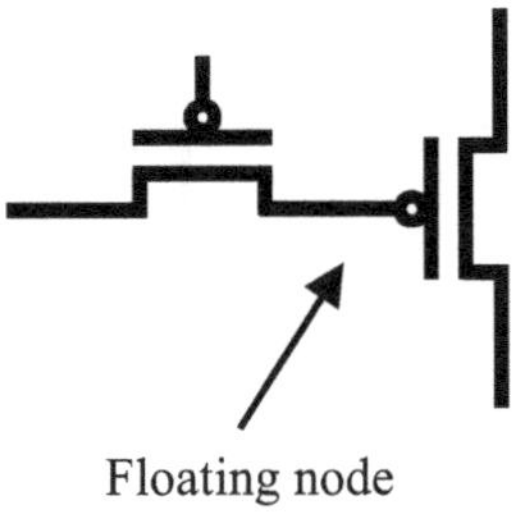

Fig. 5.14 Arrangement for making a semifloating gate

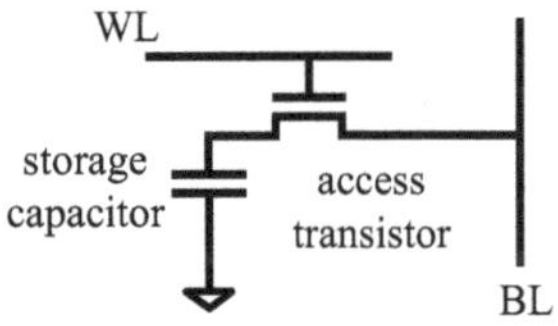

Fig. 5.15 Standard 1T1C DRAM cell

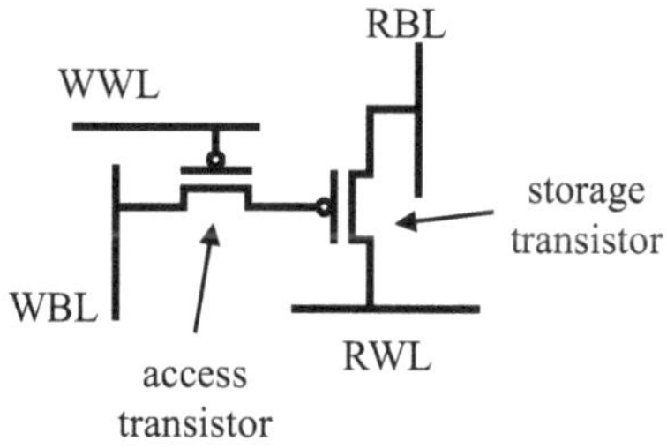

Fig. 5.16 Use of semifloating device as an embedded DRAM cell

refresh. As a result, high density DRAM chips are usually made using special processes with very complex process steps for making the capacitor in the trench isolation or in the back end.

Figure 5.16 shows the use of the semifloating device of Fig. 5.14 as a DRAM cell. Since these types of cells can be made using the standard CMOS transistors, they have been used in embedded DRAM applications for many years. Note that although the nFET version of the device has also been used, pFET is favored here because pFET usually has lower subthreshold leakage than that of an nFET of same size. Therefore, the retention of pFET-based cell is usually better than nFET-based cell. A review of this and a number of similar cells can be found in (Teman et al., 2012). These cells can be classified by the number of transistors used in a cell, e.g., 2T, 2T + 1C, 3T cells.

For the cell in Fig. 5.16, during write, the write bit line (WBL) is driven to high or low, depending on the data to be written to the cell. The write word line (WWL) is then enabled (driven low) to store the data to the floating node.

During read, the read bit line (RBL) is precharged to Vdd. The read word line (RWL) is then driven low. If the storage node is low, e.g., the storage transistor is on and RBL discharge. If data "1" is stored at the storage node, the storage pFET remains off and the RBL voltage will be maintained. This voltage change or the lack of the RBL line is detected by a sense amplifier connected to the RBL.

The read process explains why these types of cell are also called gain cells because the storage transistor also acts as an amplifier. This is fundamentally different from the DRAM cell shown in Fig. 5.15. In the usual DRAM cell, the storage state is distinguished by whether the storage capacitor is charged or not. During read, the bit line is first charged to Vdd/2. The sensing of the stored charge of a conventional DRAM cell is by turning on access transistor, and let the charge stored on the storage capacitor to be shared with the charge on the precharged bit line. The resulting bit-line voltage change therefore is dependent on the capacitance ratio between the storage capacitor and the total bit-line capacitance. As a result, the storage capacitance has to be large enough, compared to the bit-line capacitance to be detectable by the sense amplifier.

On the other hand, the storage capacitance in the gain cell only determines whether the storage transistor is in an on or an off state. The bit line is charged or discharged to a separate word line. In this regard, the gain cell is much more like a floating-gate transistor of a nonvolatile memory device discussed in Sect. 5.1.

The cell of Fig. 5.6 uses the standard logic transistors (either core or I/O transistor). Therefore, embedded DRAMs based on this type of cell can be manufactured in standard CMOS processes. The drawback is that due to the off/leakage current, and the small capacitance of the storage node, these cells need to be refreshed often, typically on the order of 1–10 ms. Since the subthreshold leakage current of a MOSFET is directly determined by its threshold voltage, it is often desirable to deploy some of the tricks discussed in the previous chapter to obtain pFET of higher threshold voltage and lower leakage current. Another solution to increase the storage capacitance, and reducing the refresh rate, is by adding an additional MOSCAP to the storage node. This solution, called a 3T or 2T + 1C cell, however, increases the cell area.

5.5 Concluding Remarks

In this chapter, we discussed several devices based on the floating-gate structure, including floating-gate transistors and associated memory cells, capacitors, and resistors.

Floating-gate transistors have found wide application, principally as the storage element in nonvolatile memory devices. In Chaps. 8–10 we will discuss in detail how these memory transistors and cells are arranged in arrays and coupled with supporting circuits to form functional blocks to be integrated with other circuits.

Floating-gate variable capacitors are used less often but have found niche applications in special circuits like analog-to-digital, digital-to-analog converters, trimming circuits, etc. These novel devices provide nonvolatile and continuous adjustable RC delay capabilities in standard CMOS processes.

Embedded DRAM is also possible using semifloating transistors in standard CMOS processes. The concept has been deployed in the so-called gain cells and used in embedded applications.

References

De Vries, A., & Ma, Y. (2007). A logical approach to NVM integration in SOC design. *Electrical Design News, 52*, 73–80.

Hyde, J., & Ma, Y. (2004). *Metal dielectric semiconductor floating gate variable capacitor*. US Patent Application #20040206999: US Patent Office.

Kahng, D., & Sze, S. (1967). A floating gate and its application to memory devices. *Bell System Technical Journal, 46*, 1288–1295.

Ma, Y., Gilliland, T., Wang, B., Paulsen, R. P., Wang, C., Nguyen, H., et al. (2004). Reliability of pFET EEPROM with 70A tunnel oxide manufactured in generic logic CMOS processes. *IEEE Transactions on Device and Materials Reliability, 4*, 353–356.

Mead, C. (1989). *Analog VLSI and neural systems*. Reading, MA: Addison-Wesley.
Shen, R.-J., Lin, F.-L., Chou, A.-F., Yang, E.-S., & Hsu, C.-H. (2007). Flash memories. In W.-K. Chen (Ed.), *The VLSI handbook* (p. 54-1). Boca Raton, FL: CRC Press.
Sze, S. (1981). *Physics of semiconductor devices* (2nd ed.). Hoboken, NJ: Wiley.
Teman, A., Meinerzhagen, P., Burg, A., & Fish, A. (2012). Review and classification of gain cell eDRAM implementations. In *27th Convention of IEEE* (pp. 1–5). New York: IEEE.

Chapter 6
Bipolar Transistors in Logic CMOS Processes

Abstract In this short chapter, we review bipolar junction transistors (BJTs) that can be obtained in the basic CMOS processes. The parasitic BJTs have been extensively discussed, mostly in the context of avoiding the latch-up risk for CMOS ICs. We here focus on a couple of applications, especially in the voltage reference circuits, of the BJTs. We also discuss a special punchthrough transistor that can be obtained from the CMOS process.

Keywords BJT • Bipolar junction transistor • Lateral NPN • Vertical PNP • Lateral NPN • Bandgap reference

6.1 Introduction

The workhorse of CMOS circuits is the MOSFET. However, bipolar transistors are often finding use in CMOS chips as well. As is well known, the parasitic bipolar transistors generated in the CMOS process, both pnp and npn, can cause latchup and interrupt the original circuit functions. There are plenty of literature on the latchup process and how to avoid it, see e.g. (Wolf & Tauber, 1990).

In this chapter, we review the basics and some beneficial uses of parasitic npn and pnp bipolar transistors available in the basic CMOS process. Since they are parasitic devices, most characteristics of these transistors are not ideal. In particular, the parasitic transistors are not designed for their current drive capability and are not generally used that way, e.g., as power amplifiers or in precision circuits. One common use of bipolar junction transistor (BJT) is in ESD protection circuits. Another important use of the parasitic bipolar transistors is in current or voltage reference circuits.

6.2 Parasitic Bipolar Transistors

In Fig. 6.1 we reproduce the cross section of pFET and nFET MOSFET transistors in the twin-well CMOS process on p-type substrate, discussed in Chap. 3. Three parasitic transistors—a lateral npn and pnp, and a vertical pnp transistor—are

Y. Ma, E. Kan, *Non-logic Devices in Logic Processes*,
DOI 10.1007/978-3-319-48339-9_6

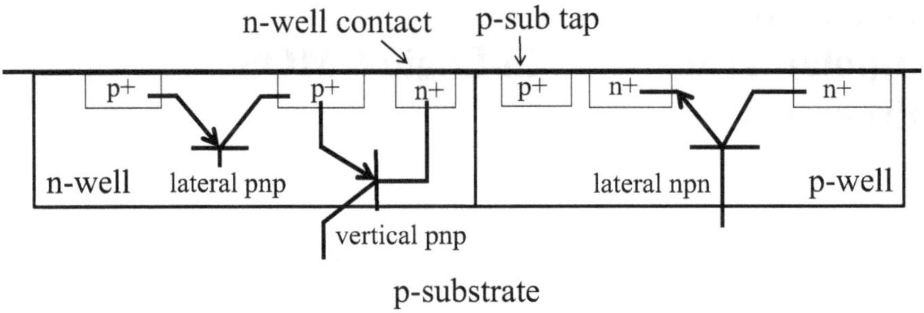

Fig. 6.1 Parasitic vertical pnp, lateral pnp, and lateral npn bipolar transistors from a CMOS process

highlighted. Notice all of these BJTs conduct current away from the oxide–silicon interface.

For the lateral npn transistors in Fig. 6.1, current gain strongly depends on n+-to-n+ layout spacing because it represents the width of the base. P-well doping and the well depth are other parameters that affect the current gain. Notice that the collector doping is as high as the emitter doping, and therefore the base width modulation at high collector bias is expected to be much stronger, which will result in a small output resistance and small voltage gain. In other words, the collector current will depend on the collector voltage in the active mode.

In the parasitic vertical pnp bipolar transistor, current gain depends on n-well depth, well concentration, and built-in field in the well. For a typical CMOS technology, the well depth is usually ~1.0 μm and is retrograde doped at around 1×10^{17} cm^{-3}. A current gain (β) of approximately 10 is usually obtained. The collector doping comes from the p-substrate, so large output resistance and high voltage gain can be expected.

The BJT current gain is controlled by reverse injection and recombination in the base region. Since the n-well is fixed in doping and depth, there is no design variable that can be changed to modify the current gain. Because the vertical pnp is a parasitic device and not optimized by process tuning, the current gain can vary by more than 30 %. Note that the collector of the vertical pnp is the p-type substrate and is most often grounded. Due to the lower base resistance and higher output resistance, vertical pnp transistors are more commonly used.

6.3 Reference Circuits Using Parasitic BJTs

An interesting application of parasitic bipolar transistors is to simulate silicon cochlea (van Schaik, Fragnière, & Vittoz, 1996). Other more common applications of the parasitic bipolar transistors are in current or voltage sources.

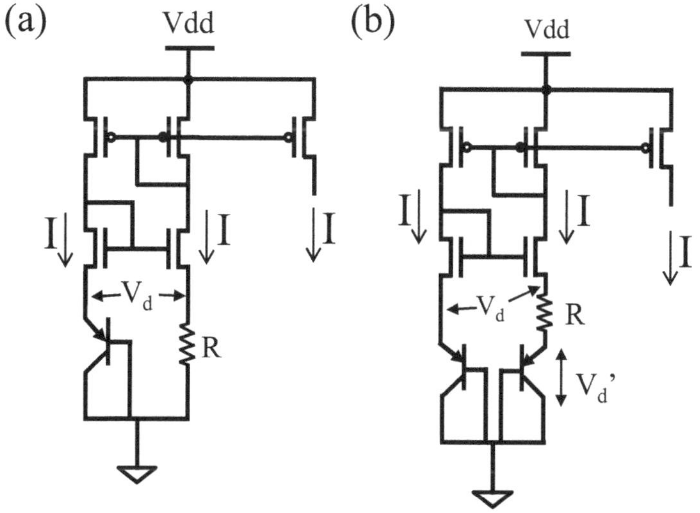

Fig. 6.2 (**a**) Current source; (**b**) diode temperature sensor based on parasitic bipolar transistors

Current sources are widely used in analog circuits to provide biasing and to provide load for amplifier stages. The key to current sources are matching characteristics for manufacturability and sensitivities to power supplies and temperatures.

Some CMOS current source designs use current mirrors and rely on the threshold voltage of the MOSFET to provide current standard. The drawback is the matching characteristics due to the threshold voltage variations. Also threshold voltage usually has strong temperature dependence which is hard to compensate.

Figure 6.2 shows reference circuits based on diode-connected parasitic vertical pnp BJT (Baker, Li, & Boyce, 1998). In both implementations, the current mirrors use identical and matched transistors, and the currents on the vertical branches are all equal.

For Fig. 6.2a, because of the equal current, the voltage drop across the pnp and across the resistor will be equal, the current I then is given by,

$$I = \frac{V_d}{R} \tag{6.1}$$

where V_d is the voltage across the diode-connected pnp. For small current, $V_d \sim 0.6$ V and is nearly independent of layout and process variations, except for the temperature dependence of the resistor.

The circuit in Fig. 6.2b is a thermal voltage ($= k_B T$, where k_B is the Boltzmann constant and T is the ambient temperature in Kelvin) referenced circuit. The diode on the right may be larger by a factor of K, as measured by the emitter area ratio between pnp1 and pnp2.

Again because of the equal currents, we have,

$$V_{\mathrm{d}} = V_{\mathrm{d}}^{'} + \mathrm{IR} \tag{6.2}$$

For the diode-connected pnp on the left, the diode current equation is given by the following relation,

$$I = I_0 \mathrm{e}^{qV_{\mathrm{d}}/kT} \tag{6.3}$$

For the diode-connected pnp on the right branch, the emitter current is,

$$I = \mathrm{KI}_0 \mathrm{e}^{qV_{\mathrm{d}}^{'}/kT} \tag{6.4}$$

From the previous three questions, we obtain,

$$I = \frac{\ln(K)}{qR} kT \tag{6.5}$$

The current is proportional to the absolute temperature T. The additional parameter K for the right circuit makes the *resistor* implementation easier, as there are often limited choices of *resistors available in the CMOS process*.

Voltage reference can be constructed as well. In practical circuits, more cascode stages are usually added to increase the output resistance. Figure 6.3 shows an example of a bandgap reference circuits using the bipolar transistors.

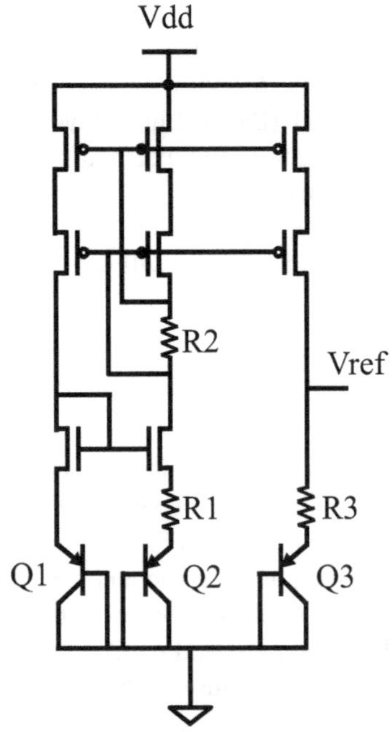

Fig. 6.3 Bipolar transistor bandgap reference circuits with cascoded FETs

One drawback with the cascode bandgap reference circuits discussed earlier is the high supply voltage needed. As subthreshold MOSFET has more serious matching issues and subject the slow recovery from disturbance, the biasing and mirror FET are often set at above V_t. Due to the cascode MOSFETs, each FET needs to operate with at least a V_t drop, as a result, V_{dd} needs to be around four times V_t plus an overhead voltage to operate properly.

6.4 Punchthrough Transistor

As discussed in Sect. 2.2.3, punchthrough is one of the most fundamental limitations in semiconductor device physics (Sze, 1981). It is one of the dominant factors determining the device size and packing density in all semiconductor manufacturing processes. Significant process development efforts have been expended to control punchthrough effects in advanced CMOS processes, including the use of many antipunchthrough implants such as super-halo doping. Even so, punchthrough still limits the packing density devices through many minimum spacing and device dimension rules. In fact, the scaling of NOR-type flash memory has essentially stopped in part because the channel length of the programming transistor cannot be scaled down further due to punchthrough (Lai, 2008).

However, the punchthrough phenomenon can be used to construct bipolar transistors in CMOS. Punchthrough transistors include both lateral pnp or npn transistors, similar to the ones in the previous section. The main difference is that while the parasitic transistors in the last section are mostly based on the source or drain as the terminal, here at least one terminal is the well, either p-well or n-well. These transistors are similar to static induction transistor (Nishizawa, Terasak, & Shibata, 1975).

Figure 6.4 shows three types of punchthrough transistor structures. First two devices are n-well to n-well spacing, with a base in between and outside the n-wells, respectively. The third is n-well to n+ diffusion spacing structure. The operation principles of punchthrough transistors can be found, e.g. (Li, Pan, & Wilamowski, 2003). Punchthrough transistors can be explained by a two-step model: in order for the current to flow between the two terminals, carriers need to first escape the barrier at the emitter and transport through the merged depletion region to reach the second terminal. The substrate (base) region can modulate the punchthrough current through biasing and change the barrier height at the emitter junction. The authors showed that for small current the current is exponentially dependent on the emitter voltage. I–V relationship turns quadratic at medium current and finally into a linear relationship at high current limit. These I–V characteristics are illustrated in Fig. 6.5.

The well/emitter current can be controlled by both gate and emitter/well voltage. Also by changing the geometry spacing of the terminal, transistors with different voltage multiplication factor can be fabricated. Figure 6.6 illustrates punchthrough current vs. well bias for the punchthrough transistor structure of Fig. 6.4c in a

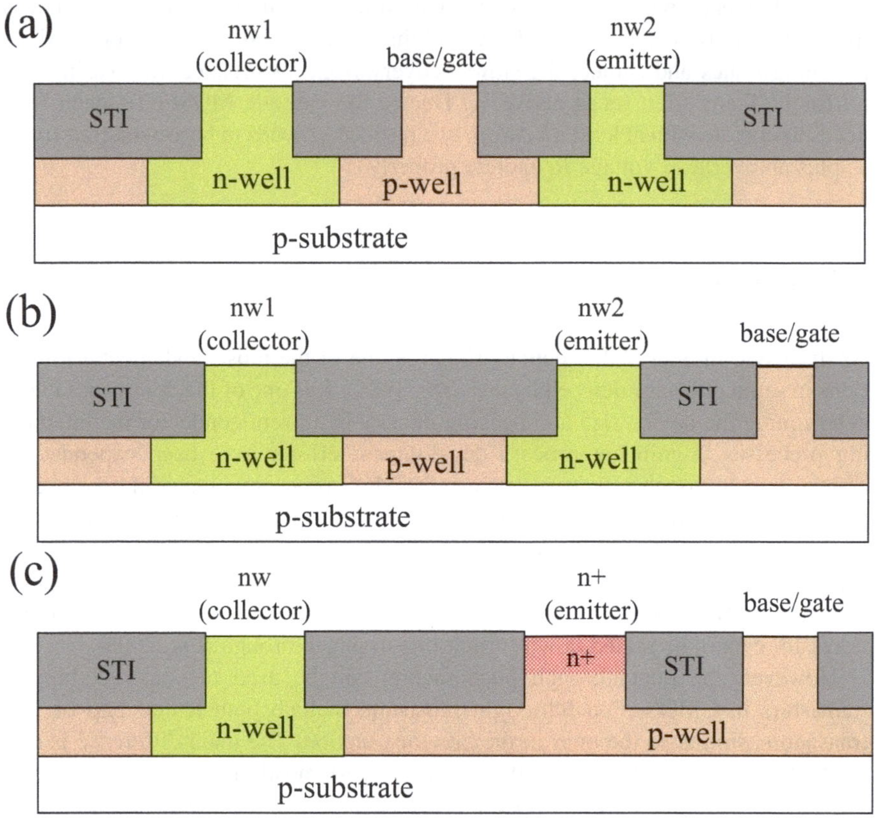

Fig. 6.4 Punchthrough transistor structures that can be implemented in CMOS process—(**a**) n-well to n-well with base in between, (**b**) n-well to n-well with base outside, and (**c**) n+ diffusion (np) to n-well

CMOS process. In this figure the gate is at ground, the n-well current is plotted vs. n-well voltage for a few emitter n+ voltages. With p-substrate at ground, the punchthrough current between the n+ and n-well starts at about 3 V. With n+ at 1 V the punchthrough gets pushed out to about 7 V. There is a voltage multiplication of about 4 in this case.

Thus, punchthrough is not a simple function of the voltage difference between the two n-wells. Rather, the punchthrough voltage is increased by a large voltage amplification factor when the first n-well is biased. The amplification can exceed 10.

With the previous example, we can use the principles for operating punchthrough transistors to reduce punchthrough in CMOS layout design. Because the voltage amplification factors are found to be typically 5–10, it suggests that the spacing design rule can be relaxed if the neighboring nodes are biased with nonzero voltages. In fact, one can deliberately reduce the spacing rule if circuit design techniques are used to bias neighboring nodes. This practice can result in sizable

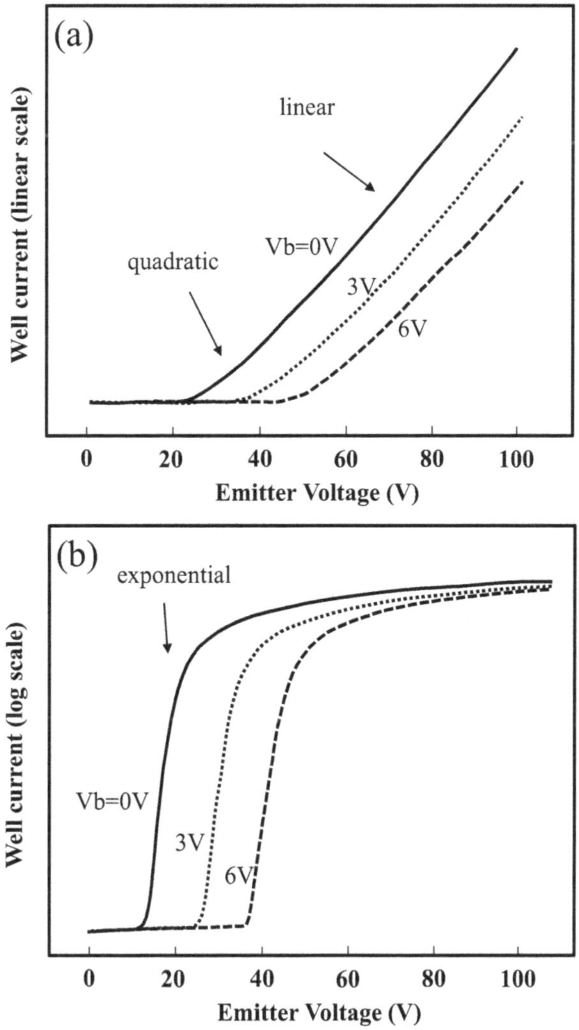

Fig. 6.5 Simulated I–V relationship for a punchthrough transistor in linear (**a**) and log scale (**b**)

reduction in chip size, especially in high voltage processes and analog circuits, such as display drivers and voltage converters. Similarly, it may be possible to reduce the channel length of NOR memory transistors further if the source is biased at a nonzero voltage.

The PT transistors can be used in high voltage power electronics and also have been proposed for high frequency operations.

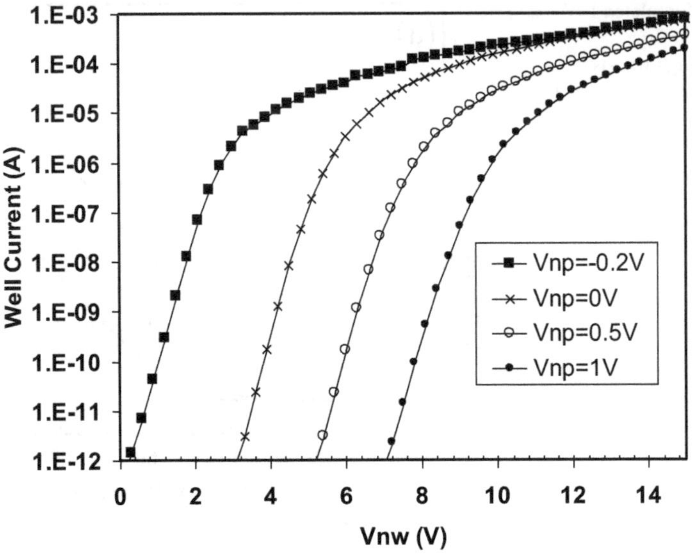

Fig. 6.6 Punchthrough current vs. well bias for the punchthrough transistor structure of Fig. 6.4c in a CMOS process

References

Baker, R. J., Li, H., & Boyce, D. (1998). *CMOS—Circuit design, layout, and simulation.* New York: IEEE Press.

Lai, S. (2008). Flash memories: Successes and challenges. *IBM Journal of Research and Development, 52*, 529–535.

Li, X., Pan, H., & Wilamowski, B. M. (2003). Gate-controlled punch through transistor. In *Proceedings of the 15th Biennial University/Government/Industry Microelectronics Symposium, 2003* (pp. 226–229). New York: IEEE.

Nishizawa, J., Terasak, T., & Shibata, J. (1975). Field-effect transistor versus analog transistor (static induction transistor). *IEEE Transactions on Electron Devices, 22*, 185–197.

Sze, S. (1981). *Physics of semiconductor devices* (2nd ed.). Hoboken, NJ: Wiley.

van Schaik, A., Fragnière, E., & Vittoz, E. (1996). Improved silicon cochlea using compatible lateral bipolar transistors. In *Advances in neural information processing systems* (pp. 671–677).

Wolf, S., & Tauber, R. N. (1990). *Si processing for the VLSI Era* (Process integration, Vol. 2). Sunset Beach, CA: Lattice Press.

Chapter 7
Diodes in Logic CMOS Processes

Abstract In this chapter, we discuss special diodes that can be made in standard logic CMOS processes. We first discuss the polysilicon diode and a close relative, the polysilicon resistor. In particular, the different temperature coefficients of resistivity (TCR) of p and n polysilicon make it possible to obtain a resistor that is temperature independent. Then we showed several examples of making Schottky diodes in logic processes and discussed their properties and applications.

Keywords MOSFET • Diode • Polysilicon diode • Schottky diode • Zener diode • Polysilicon resistor • TCR

7.1 Introduction

There are many kinds of diodes in CMOS processes, including well diodes and polysilicon diodes. Most of the diodes are integral part of the CMOS device and circuits. Dedicated diodes can be useful as well. In addition to the diode-connected MOSFET and BJT described in the previous chapters, examples are the use of Zener diode as a voltage limiter, well diodes in ESD protection circuits, and Schottky diodes in high-speed circuits and in charge pump circuits.

In this chapter, we discuss polysilicon diodes and Schottky diodes that can be fabricated in standard logic CMOS processes.

7.2 Polysilicon Diodes

The structure of the polysilicon p–n diode is a polysilicon layer with p-type and n-type doping regions. Undoped polysilicon layer is deposited on top of oxides, either on gate oxide over the channel region or on the field oxide on top of STI. Then, the p-type (n-type) highly doping region on the polysilicon is doped during the p+ source and drain formation of the pMOS. To form p–i–n diode, an undoped region is covered by photomask to block both the p-type and n-type implants. A layout and a cross section of p–i–n poly diode are shown in Fig. 7.1.

Y. Ma, E. Kan, *Non-logic Devices in Logic Processes*,
DOI 10.1007/978-3-319-48339-9_7

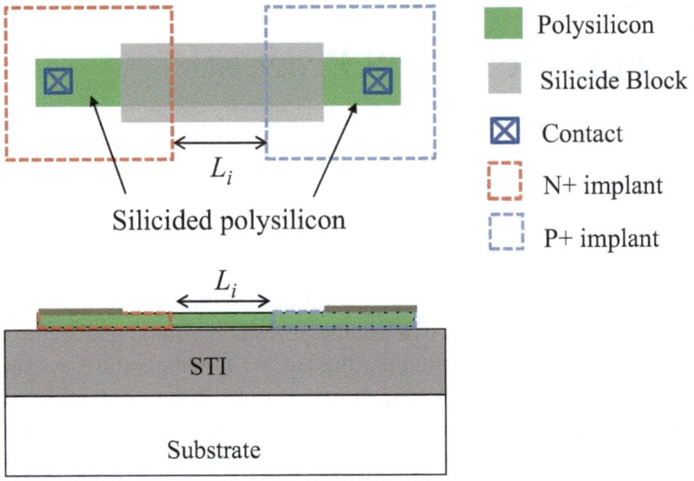

Fig. 7.1 Layout and cross section of a polysilicon diode, where L_i is the drawn width of the intrinsic region

In CMOS processes, polysilicon is usually silicided by default and cannot make a p–n junction. Therefore, a salicide block (SAB) layer is needed to prevent the silicidation of the polysilicon and produces p–n or p–i–n poly diodes. SAB is usually a standard offering by the foundries.

During the layout of a p–i–n (PIN) diode, a dummy layer needs to be added to mark the intrinsic region. The PP and NP masks will then be generated to exclude the region covered by the dummy layer from getting any implants, either p+ or n+ type.

Figure 7.2 shows the IV curves for several PIN diodes with varying length of intrinsic region, labeled L_i in Fig. 7.1 and Lc in Fig. 7.2 (Ker & Chang, 2003). The thickness of the polysilicon layer is 200 nm. Normal diode I–V curves are seen at positive biases. A striking feature from these curves is the breakdown of the reverse biased junction that is strongly dependent on the width of the intrinsic region.

In Fig. 7.3 we illustrate the breakdown voltage as a function of the width of the intrinsic region. This feature can be used to make Zener-type diode with adjustable breakdown voltages by changing L_i.

The I–V characteristics of p–i–n diodes are similar to the standard diode at low frequencies. At higher frequencies, the diode looks like an almost perfect resistor with good linearity. This behavior is due to the stored charge in the intrinsic region as the polysilicon has many grain boundary traps that can hold the static charge. At low frequencies, the stored charge can be removed and the diode turns off. At higher frequencies, there is not enough time to remove the stored charges. The diode never turns off. In other words, the poly PIN diode has a poor reverse recovery time.

Compared to well diodes which are necessarily adjacent to the substrate, polysilicon diode has low parasitic capacitance and no concerns for junction

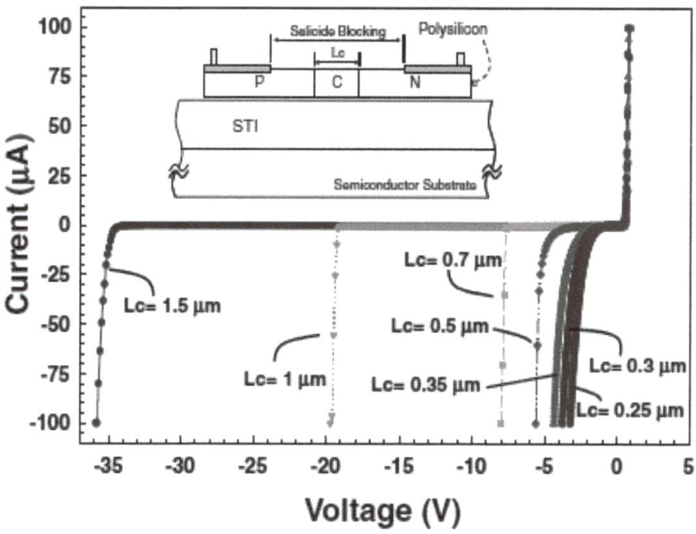

Fig. 7.2 I–V curves of polysilicon diodes, the width of the intrinsic region is labeled L_c (Ker & Chang, 2003)

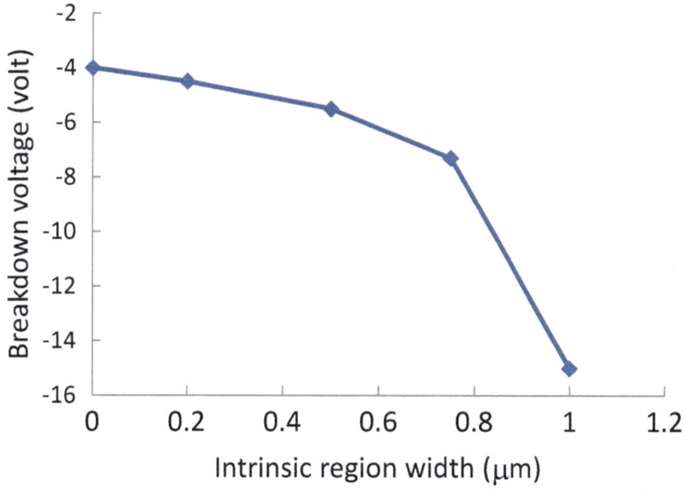

Fig. 7.3 Reverse breakdown voltage as a function of the intrinsic region width

breakdown. In addition, the isolations between polysilicon traces are much more compact, allowing diodes with much smaller area than well diodes.

Another advantage of the polysilicon diode is the negligible substrate leakage current.

A drawback of the poly diode is the current limitation in the forward direction due to the high series resistance in the intrinsic region, as well in the doped but unsilicided polysilicon regions. In comparison to bulk, crystalline silicon at the same doping, polysilicon mobility is often 5–10 times smaller.

PIN diode can be used as attenuators in RF circuits, and in fast switches, photodetectors, and high voltage power electronics applications. Commercial stand-alone PIN diodes are available.

7.2.1 Polysilicon Resistors

By a simple modification of the polysilicon PIN diode layout, we can obtain polysilicon resistors with designed temperature coefficients of resistance (TCR). As we recall, TCRs of polysilicon resistors can be either positive, whereas resistivity increases with temperature, or negative, where resistivity decreases with temperature. In fact, negative and positive TCR values of R_{bulk} are found in n^+ and p^+ polysilicon resistors in the 0.18 μm process, respectively (Chuang et al., 2003).

The opposite temperature dependence makes it possible to design a composite resistor that have no temperature dependence. For two resistors with different TCRs, α_1 and α_2, in series, the total resistance is given by:

$$R = R_1(T) + R_2(T) = N_1 R_{s1}(1 + \alpha_1 \Delta T) + N_2 R_{s2}(1 + \alpha_2 \Delta T), \tag{7.1}$$

where R_s and N denote the sheet resistance and number of squares for the individual resistors, respectively.

The earlier relation can be rewritten as follows:

$$R = N_1 R_{s1} + N_2 R_{s2} + (N_1 R_{s1} \alpha_1 + N_2 R_{s2} \alpha_2)\Delta T. \tag{7.2}$$

The first two terms have no temperature dependence. For two materials with known and opposite temperature coefficients, we can obtain a resistor with no temperature dependence in its total resistance when proper N_1 and N_2 are chosen to meet the required R and zero TCR.

For Eq. (7.2), the temperature dependence of the total resistance is eliminated when

$$N_1 R_{s1} \alpha_1 + N_2 R_{s2} \alpha_2 = 0,$$

or,

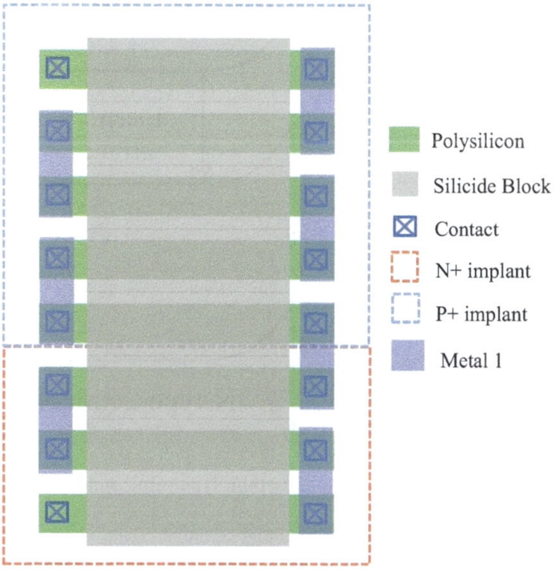

Fig. 7.4 Composite polysilicon resistor strip comprised of two n+ and p+ poly with different number of squares and temperature coefficients, α_1 and α_2

$$\frac{N_1}{N_2} = -\frac{\alpha_2}{\alpha_1}\frac{R_{s2}}{R_{s1}}. \tag{7.3}$$

Since the n+ and p+ unsalicided polysilicon cannot be in contact with each other, lest diodes are formed, the n+ and p+ segments need to be connected by silicided sections and metal. Figure 7.4 illustrates a possible layout of the composite polyresistor, with N_1/N_2 ratio of 3/5.

As we noted in Chaps. 4 and 6, resistors are essential elements in many voltage or current reference circuits. The ability to obtain a resistor with no or little temperature dependence is of great use to these circuits.

7.3 Schottky Diodes

In Fig. 7.5 we replotted the band diagram of Fig. 2.8 of a metal–semiconductor junction. Because of the energy barrier, this structure exhibits diode behavior. This type of diode, generally called Schottky diode is an important device for many applications.

The barrier height is a function of the work function of the metal, the doping level, and type (n or p) of the semiconductor. The barrier height is given by the difference between the metal work function, ϕ_M, and the silicon electron affinity, χ,

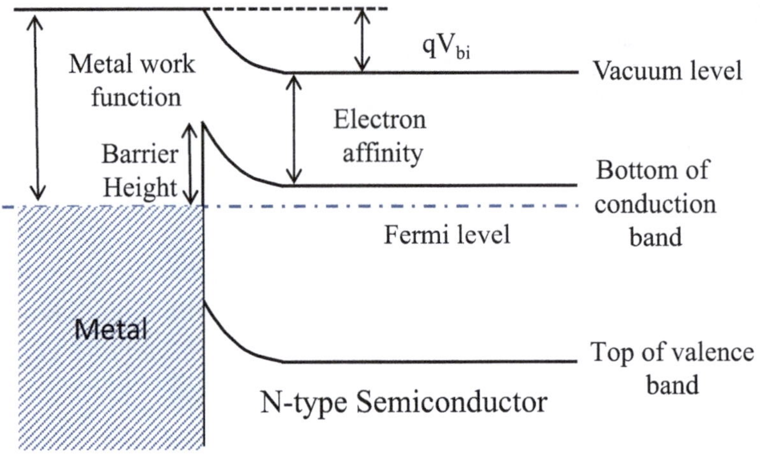

Fig. 7.5 Energy diagram of metal/n-type semiconductor interface

$$\phi_{BN} = \phi_M - \chi \tag{7.4}$$

For n-type silicon substrate, the barrier height has been found for various metals and metal silicides. For example, for Ti silicides the barrier height is on the order of 0.6 V (Sze, 1981).

To make a Schottky diode using a standard CMOS process, one possibility is to remove the highly doping source and drain implant in the contact region. In older CMOS process where aluminum metallization is directly made to contact with the silicon, Al–Si Schottky diodes were made in a 2 μm CMOS process (Milanovic et al., 1996).

In more advanced processes, when the contact is made with silicided wells, Schottky diodes can be formed between a silicide and an n-well. The type of silicide is Ti–Si for 0.25 μm and the following process, Co–Si for 0.18 μm to 90 nm process, and Ni–Si for 90 nm and more advanced process. Silicide to p-well diode can be formed in the same way in CMOS processes with the constraint that all the semiconductor nodes are shorted due to the sharing of the p-substrate. However, in the mixed-signal process where p-well may be isolated by a deep n-well, independent silicide-p-Si diodes can be formed.

Figure 7.6 shows the I–V curves for $CoSi_2$ to n-Si (n-well) and p-Si (p-well) junction diodes that is manufactured in a 0.13 μm logic CMOS process without any process changes (Sankaran & O, 2005). The p-type junction has a higher reverse leakage current, indicating a lower barrier.

Extraction of the barrier height from the I–V curves yielded a value of 0.35 eV for CoSi/p-Si and 0.52 eV for the n-type. The reverse breakdown voltage is ~15 V.

The performance of a Schottky diode can be measured in terms of drive current, area, and speed. The current characteristic of a Schottky diode has the same form as that of a p–n diode, e.g., Eq. (2.8), where I_0 is directly related to the barrier height.

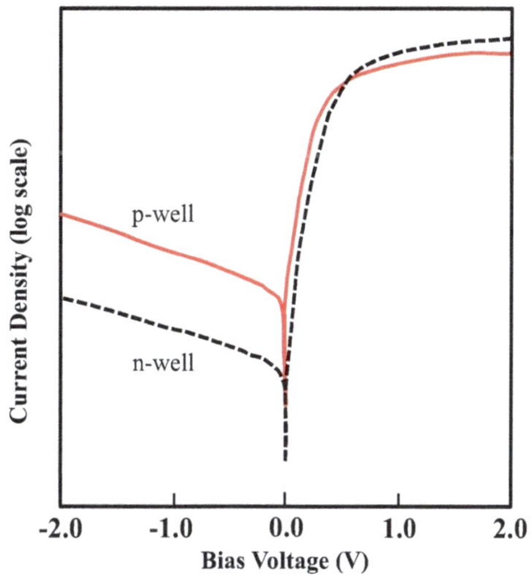

Fig. 7.6 I–V curve for a $CoSi_2$ to n-well and p-well silicon junction manufactured in a logic CMOS process, adapted from (Sankaran & O, 2005)

Table 7.1 Estimated and measured barrier height for different materials

Junction	Barrier height to n-Si	To p-Si	Process
Al-Si	0.78 eV		2 μm (Milanovic et al., 1996)
TiSi2/Si	0.45 eV		0.35–0.25 μm (Jamali et al., 2004)
CoSi2/Si	0.52 eV	0.35 eV	0.13 μm (Sankaran & O, 2005)

The barrier height is determined by the materials used in constructing the diode. For a given CMOS process, the diode barrier height is then fixed. Table 7.1 shows some estimated barrier heights. As the metal or silicide work function can vary around 0.5 eV depending on the grain orientation and the interface trap pinning, these Schottky diodes are subject to larger variations in parametric yields.

In almost all cases, parasitic factors such as series resistance play an important role in reducing the drive current and need to be carefully considered during layout. For example, Fig. 7.7 shows the basic layout for a Schottky diode in a standard CMOS process. To reduce the series resistance, we may use interdigitated fingers (Rivera, 2001), as illustrated in Fig. 7.8.

In the Schottky diode of Fig. 7.8, the anode and cathode are separated by the shallow trench. The current path is longer as electrons have to go around the STI, increasing the series resistance. Two ways to reduce the series resistance are to eliminate the STI directly or separate the anode and cathode using polysilicon gates (Han et al., 2011). These two methods are illustrated in Fig. 7.9.

The most important difference between Schottky and p–n diodes is the large-signal switching speed. When the diode switches from reverse to forward conducting state or vice versa, the recovery time of this reversal can be on the order of hundreds of nanoseconds for p–n diodes, due to the process of

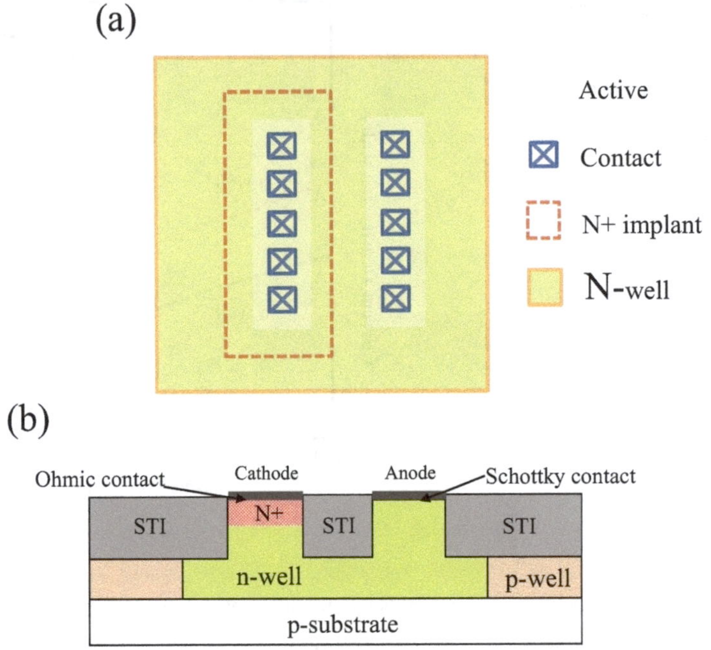

Fig. 7.7 Layout of an n-well Schottky diode (**a**) and corresponding cross section (**b**)

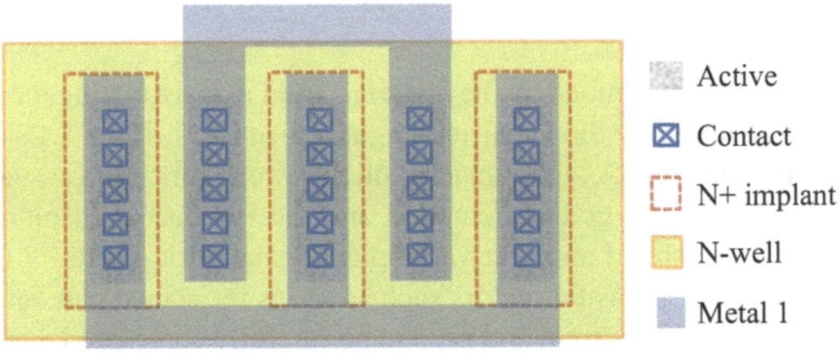

Fig. 7.8 A Schottky diode with interdigitated fingers to reduce series resistance

recombination of majority and minority carriers. On the other hand, the Schottky diode is a majority carrier device, where there is only one type of carrier, e.g., electron for n-type semiconductors in Figs. 7.7 and 7.9. The majority carriers are quickly injected into the conduction band of the metal contact by thermionic emission to become free moving electrons. So the switching between forward and reverse can be virtually instantaneous, limited only by the dielectric relaxation time. As a result, the Schottky diode can be very fast. The cutoff frequency is

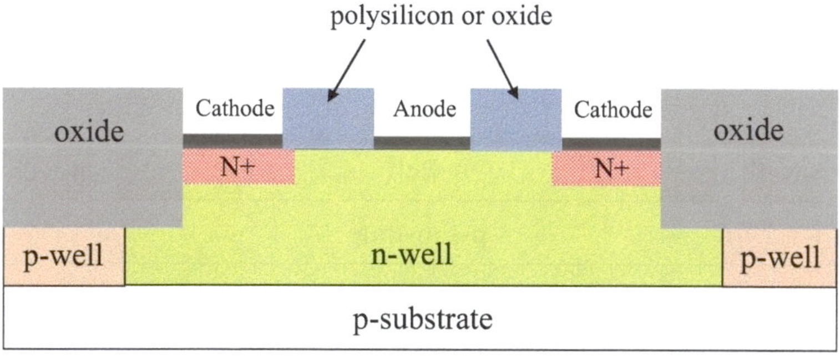

Fig. 7.9 Schottky barrier diode structure with polysilicon or oxide separation

related to the resistance and capacitance of the junction and given by (Sankaran & O, 2005),

$$f_{\text{cutoff}} = \frac{1}{2\pi R_s C_{\text{tot}}} \tag{7.5}$$

Thus, for higher speed, both the series resistance, R_s, and total capacitance, C_{tot}, which is comprised of both the junction capacitance and other parasitic capacitances, should be reduced. The junction capacitance is given by the following equation:

$$C_J = A\sqrt{\frac{qN_D\varepsilon_{Si}}{2(V_{bi} - V_a)}} \tag{7.6}$$

where A is the area of the junction, V_a is the applied voltage, N_D is the doping of the silicon well, and ε_{Si} is the dielectric constant of the silicon. The area is determined by the required drive current, while ε_{Si} is the material property and cannot be changed. It is, however, possible to reduce the parasitic capacitance by going to a lower N_D.

Parasitic capacitance between the n-well and the p-substrate can be reduced by the same trick we used to increase the voltage of LDMOS. By inserting a native region between the n-well and the surrounding p-well, we can reduce the parasitic capacitance at the cathode (Ma et al., 2010), as illustrated in Fig. 7.10.

The minority injection transport in p–n diodes and the thermionic emission in Schottky diodes actually happen in both diodes, just which mechanism is more dominant depending on the barrier height. For p–n diodes where the barrier height is larger than 0.8 eV, minority injection is much larger than thermionic emission, and hence the transport is bipolar. To establish or eliminate the minority carriers at the turn-on stage, the transport is slow. For Schottky diodes with less than 0.6 eV barrier, minority injection can be ignored.

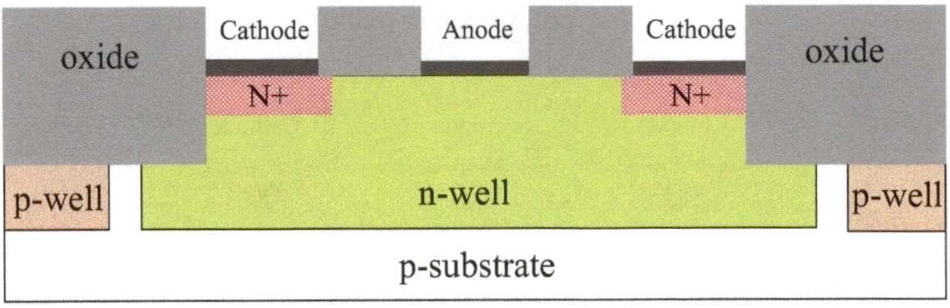

Fig. 7.10 Modified Schottky diode with reduced cathode parasitic capacitance

Comparing with the p–n diodes, Schottky diode has low turn-on voltage, fast recovery time, and low junction capacitance. As a result Schottky diodes are popular for RF applications, e.g., at frequency of up to the terahertz domain and RF power detection circuits (Rivera, 2001). For example, Schottky diodes have been used in RF detectors and mixers at frequencies up to 50 GHz. Terahertz detectors are designed and fabricated using 0.13 μm CMOS process (Han et al., 2011). To achieve the cutoff high frequency, the designers used polysilicon gate-separated Schottky diodes with a 16 unit-cell structure, each with area of $0.4 \times 0.4\ \mu m^2$, with a total series resistance of 8 Ω and junction capacitance of 10 fF, yielding a cutoff frequency of 2 THz. The Schottky diode is integrated with on-chip patch antenna and preamplifier to achieve 280 GHz detector operation that performs better than other competing technologies.

Schottky diodes are also used as rectifiers in switched-mode power supplies. The low forward voltage and fast recovery time lead to increased efficiency which is especially important for RFID application (Jamali et al., 2004).

References

Chuang, H., Thei, K., Tsai, S., & Liu, W. (2003). Temperature-dependent characteristics of polysilicon and diffused resistors. *IEEE Transactions on Electron Devices, 50*(5), 1413–1415.

Han, R., Zhang, Y., Coquillat, D., Videlier, H., Knap, W., Brown, E., et al. (2011). A 280-GHz Schottky diode detector in 130-nm digital CMOS. *IEEE Journal of Solid-State Circuits, 46* (11), 2602–2612.

Jamali, B., Cole, P., Ranasinghe, D., & and Zhu, Z. (2004). Design and optimisation of Schottky diodes in CMOS technology with application to passive RFID system. In *Smart Structures Devices and Systems II, Proceedings of SPIE, vol. 5649*. Bellingham, WA: SPIE.

Ker, M.-D., & Chang, C.-Y. (2003). High-current characterization of polysilicon diode for electrostatic discharge protection in sub-quarter-micron CMOS technology. *Japanese Journal of Applied Physics, 42*, 3377–3378.

Ma, Y., Oliver, R., Humes, T., & Mavoori, J. (2010). *Schottky junction diode devices in CMOS*. US Patent 7,732,887.

Milanovic, V., Gaitan, M., Marshall, J., & Zaghloul, M. (1996). CMOS foundry implementation of schottky diodes for RF detection. *IEEE Transactions on Electron Devices, 43*, 2210–2214.

Rivera, B. R. (2001). Design and layout of Schottky diodes in a standard CMOS process. In *2001 International IEEE Semiconductor Device Research Symposium* (pp. 79–82). New York: IEEE.

Sankaran, S., & O, K. (2005). Schottky barrier diodes for millimeter wave detection in a foundry CMOS process. *IEEE Electron Device Letters, 26*, 492–494.

Sze, S. (1981). *Physics of semiconductor devices* (2nd ed.). Hoboken, NJ: Wiley.

Part III
Selected Applications

This part of the book illustrates the use of the non-logic devices described in Part II in selected applications.

The most used class of applications is probably the embedded non-volatile memory based on floating gate transistors discussed in Chap. 5. Chapters 8–10 explore the design of several types of nonvolatile memories in logic CMOS processes.

In Chap. 8 we review basic floating-gate memory cell designs and several methods for programming and erase of these memory cells. We then briefly discuss the basic circuit modules of embedded nonvolatile memory, including charge pumps and sense amplifiers.

Chapter 9 discusses one-time programmable (OTP) memories using floating gate devices or fuse and antifuse devices. The advantage of the OTP memory is that the memory cell is usually smaller than that of the multiple-time programmable (MTP) memory. Also OTP memory is more easily implemented in most advanced logic CMOS processes.

Several logic MTP memories are described in Chap. 10. Some of these memories have seen commercial deployment. Typically due to the added program and erase functionality, the memory cells of MTP memories are much larger, limiting its use to small density applications.

In Chap. 11 we selected a few non-data-storage applications of the non-logic devices, especially the floating gate based devices. These include applications in trimming resistors or capacitors in precision analog circuits, digital to analog converters, RFID chips, and in neuromorphic computing.

Chapter 12 presents a treatise on CMOS biosensors in both Faradic (current sensing) and non-Faradic (field sensing) operations. Examples of isolation and compensation techniques to interface with the biological saline buffers are presented. One of the main reasons of IC clean-room manufacturing is to eliminate the influence of alkali ions in the semiconductor wafers, but the CMOS biosensors have to be operated with close proximity to this electrolyte environment with high variability.

In Chap. 13 we describe the distributive waveform shaping circuits in high-frequency applications by employing embedded waveguides and varactors.

Finally Chap. 14 presents a brief overview of recent developments in CMOS processing, especially with regard to the emergence of FinFETs, 3D NAND flash memory, and emerging nonvolatile memory technologies.

Part III
Selected Applications

Chapter 8
Logic Nonvolatile Memory

Abstract Chapter 8 is the first of three chapters that form a core application for the devices we discussed in Part II, in embedded NVMs. In this chapter, we give an overview of the embedded NVM design using logic CMOS process. We review the basic memory cell design, the programming and erasing methods, and array structure of embedded NVMs. High-voltage circuits, including charge pumps and high-voltage switches, are also discussed. Reliability issues associated with embedded NVMs are introduced.

Keywords Nonvolatile memory • Logic nonvolatile memory • NVM • Embedded NVM • Floating-gate transistor • Memory cell • Memory array • Reliability • Data retention • Endurance • Disturb • Tunneling • Fowler–Nordheim tunneling • HCI • Hot carrier injection • Band-to-band tunneling • Coupling ratio • Threshold voltage • Program • Erase • High-voltage switch • Sense amplifiers • Row and column decoders • Readout buffer • Charge pump • High-voltage generator • Dickson charge pump • OTP • One-time programmable • MTP • Multiple-times programmable

8.1 Introduction to Embedded NVM

Long before nonvolatile memory came to the public's awareness from the wide use of flash memory in digital cameras and USB drives, embedded nonvolatile memories (eNVMs) have found extensive use in a wide array of integrated circuits. Currently, eNVM technology is one of the most important differentiating features in many segments of the semiconductor market. While high density NVM, principally NAND flash for data storage, is more easily recognized by the consumer, eNVM is driving many of the fastest growing applications in automotive and industrial control, home entertainment, Smartcards, and **radiofrequency identification** (RFID) and is expected to play a critical role to enable the **internet of things** (IoT).

Some application scenarios of eNVM are as follows:

1. A few bits for storing analog trim information in analog or RF ICs. These bits are not frequently updated;

Y. Ma, E. Kan, *Non-logic Devices in Logic Processes*,
DOI 10.1007/978-3-319-48339-9_8

2. Dozens of bits as chip ID for authentication (e.g., serial numbers), or functional settings such as repairs and redundancy rows or pages in DRAMs. These bits are often programmed once and not changed;
3. Hundreds to thousands of bits for storing security keys in encryption, data or code in RFID chips, and smart cards. These bits may need frequent but not constant change;
4. Megabits for data or code storage in microcontrollers. The code storage is not changed often, less than 1000 times over the life of the chip. But data could need frequent change, e.g., over one million times.

The main attraction of eNVM is its ability to retain data when the power is turned off. Nonvolatile storage is important to most IoT devices which are battery powered and have to be powered off when not in use. In addition, the code stored in an on-chip NVM can be accessed much more quickly and are generally more secure than off-chip options.

In Chap. 5 the history of floating-gate-based memory was briefly discussed, including EPROM and EEPROM which are still used widely as the storage device in most eNVM applications. To recap, key milestones in the development of NVM include

- 1967—Floating-gate memory invented at Bell Labs (Kahng and Sze)
- 1971—Floating-gate avalanche-injection MOS (FAMOS) (Frohman–Bentchkowsky)
  - EPROM
- 1974 EEPROM (SAMOS)
  - First use of stacked polysilicon gate
- 1981 EEPROM
  - First use of bidirectional FN tunneling for program and erase
- Late 1980s—Flash memory (Toshiba and others)
- 2013—3D NAND flash memory (Samsung)

As with all ICs, the main driver for the continued scaling of nonvolatile memory ICs, especially stand-alone NVM data storage chips, is to reduce the cost per bit. On the other hand, by integrating NVM with other functions, the cost of the overall solutions needing NVM can be driven down further. Embedded NVM technology and design choices are therefore driven by achieving the lowest overall cost including the NVM and the logic and analog functions on a system on chip (SoC), while providing the features and performance needed by the system.

One of the largest markets for eNVM is for code and data storage in microcontrollers (MCUs). A review of the history of eNVM in microcontroller applications is illustrated in Fig. 8.1. The evolution of integrating NVM on MCUs followed several milestones. Initially stand-alone ROM chips are used. An example of such approach is the BIOS boot program in earlier PCs. The total solution in that case

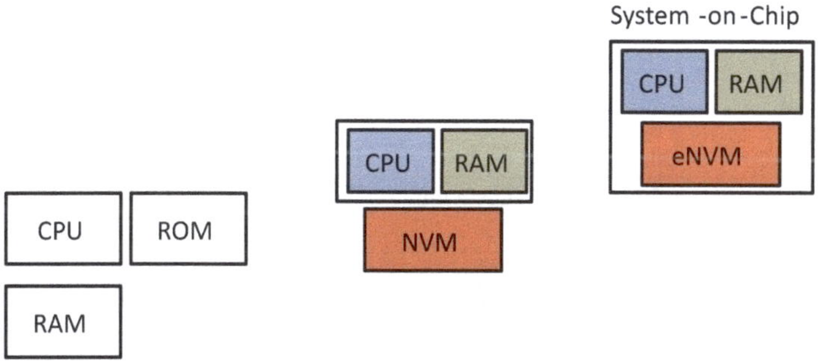

Fig. 8.1 Evolution of flash and eNVM memory technologies for microcontroller applications

consisted of multiple chips, including CPU, SRAM, or DRAM chips for data and scratchpad memory, and ROM chip for code storage. Then SRAMs began to be integrated with the CPU, reducing the number of chips in a system. Later on, mask-programmable ROM, programmable EPROM, and eventually to full function NVMs are integrated with RAM and CPU to make up single-chip flash MCUs. Nowadays state-of-the-art flash MCUs have megabytes of flash memory on chip.

There are many options for designers to integrate NVM on chip, including mask-programmable ROM, One-Time-Programmable (OTP) memories using fuse or antifuse, EEPROM, or Flash (De Vries & Ma, 2007). Table 8.1 compares a number of embedded NVM technologies.

OTP memory such as the original EPROM and the more recent fuse or antifuse-based memories can only be programmed once. Types of multiple-times programmable (MTP) eNVM include EEPROM, which can be programmed and erase at byte level, and eFlash, where erase can only be done by a block of many kilobytes, while programming can be done on a page of kilobytes. EEPROM excels at storing a few bytes that may need to be quickly updated at the byte level.

Because special processes are used, the bit cell of embedded flash memories usually is the smallest among the memories in Table 8.1. EEPROM is somewhat larger to allow byte-level program/erase ability. On the other hand, OTP memories including fuse and floating-gate-based OTP memory can be made with logic-only process and usually have larger memory cells. Logic NVM MTP memories that will be discussed in Chap. 10 have the largest memory cells.

In the rest of this chapter, we will discuss some basic concepts, including programming methods and reliability concerns, for all NVMs. Fuse and antifuse-based OTP memories will be discussed in Chap. 9 and MTP memories are discussed in Chap. 10.

Table 8.1 Embedded nonvolatile memory technology comparison, adapted from De Vries and Ma (2007)

	MTP			OTP	
	EEPROM/ flash	SONOS flash	Logic NVM	Fuse/ antifuse	Logic NVM
Storage medium	Floating gate	ONO film	Floating gate	Metal/ oxide film	Floating gate
Process	2–3 Generation lag	2–3 Generation lag	Leading but requires 3.3 V I/O	Leading	Leading but requires 3.3 V I/O
Extra masks from baseline CMOS processes	5 or more	2–3	0	0	0
Cell area	Small	Small	Large	Medium	Small
Program	HCI or FN	HCI or FN	HCI or FN	Current/ voltage stress	HCI
Program speed	μ to ms	ms	ms	μ to ms	ms
Program power consumption	Medium	Low/ medium	Low	High/ medium	Medium
Erase mechanism	FN	FN	FN	N/A	N/A
Erase speed	ms/page	ms/page	ms/page	N/A	N/A

FN Fowler–Nordheim tunneling, *HCI*, hot carrier injection

8.2 Logic NVM

Although there are many technology choices for embedding NVM in logic ICs, most presently available technologies need process modifications and additional mask steps, such as thicker oxides and additional implants, from the baseline logic or mixed-signal CMOS processes, in order to reduce the cell area or to handle the high voltage needed to program or erase the NVM cells.

As an example, the commonly used SST SuperFlash[1] embedded memory technology uses 8–10 extra masks layers from the baseline CMOS process to manufacture the memory cell and the high-voltage programming circuitry. This could add up to 50 % to the cost of fabricating a wafer.

The added processing cost, combined with lower chip yields, can hinder the use of eNVMs in designs where only a small amount of NVM is needed. It is thus very desirable to develop NVM in standard CMOS logic or mixed-signal processes, see for example (Wang & Ma, 2008) and (Hsu et al., 2014). This type of eNVM is called logic NVM (LNVM).

As is the theme for this book and emphasized repeatedly, eNVM that can be implemented using standard CMOS process is desirable as it represents the lowest

[1] http://www.sst.com/technology/SuperFlash-Overview.

cost solution in terms of wafer processing cost. Since there is no free lunch in life, LNVMs have some drawbacks. The most important one is that LNVMs are usually larger in area than eNVMs that use modified processes.

All engineering designs involve trade-off. Given a market need, the choice of whether to use LNVM or eNVM with modified process involves many trade-offs as well. The following illustrates a trade-off analysis from a cost point of view.

To compare the cost of two options, one using LNVM with low wafer cost but large memory cell area and the second using a modified NVM process with smaller memory cell area but higher wafer cost, we assume the following:

1. A is the area of the non-NVM part of the chip.
2. The NVM block (so-called an IP) occupies an additional area, $N \times c + B$, where N is the total number of NVM bit and c is the bit cell area, B is the overhead area including the charge pump, row and column decoder, and controller.
3. A process cost adder of $x\,\%$ for the modified embedded NVM process. The baseline is the logic process, i.e., $x = 0$ for LNVM.

In the following analysis, we assume the area $A = 1{,}000{,}000\,\lambda^2$, where λ is the half pitch of the technology. Two embedded NVM processes with $x = 20\,\%$ and $50\,\%$, respectively, are considered. Two logic NVMs, one with a unit cell area that is 10 times and the second that is 50 times that of the modified NVM processes are also included. We use $B = 100{,}000\,\lambda^2$ for the LNVM cases while for the modified NVM process, we use $B = 50{,}000\,\lambda^2$.

The reason for the smaller B and memory cell area for the modified NVM processes is because the modified processes often allow optimization of both the memory cell and the peripheral circuits. For example, reduced high-voltage design rules allow smaller memory cells and smaller area for the high-voltage control transistors.

In Fig. 8.2 we plot the relative areas of the four cases. As expected, for smaller embedded NVM blocks, LNVM solutions have the cost advantage. The cost curves cross over at larger N (number of bits of NVM). For example, the cost of LNVM with $c = 10$ is lower than that of the modified NVM with a process cost adder of $20\,\%$ for $N \leq$ ~24 kb. That is to say, if a design requires an NVM IP of ~24 kb or less, it makes sense to use the LNVM solution. Above this size, it is more economical to use the modified NVM process.

Of course, the cost is often is not the only concern. Other factors to consider include the ease of design and availability of models for each of the memory options, the speed of read and write of each memory options, the reliability and scalability of each memory for future processes.

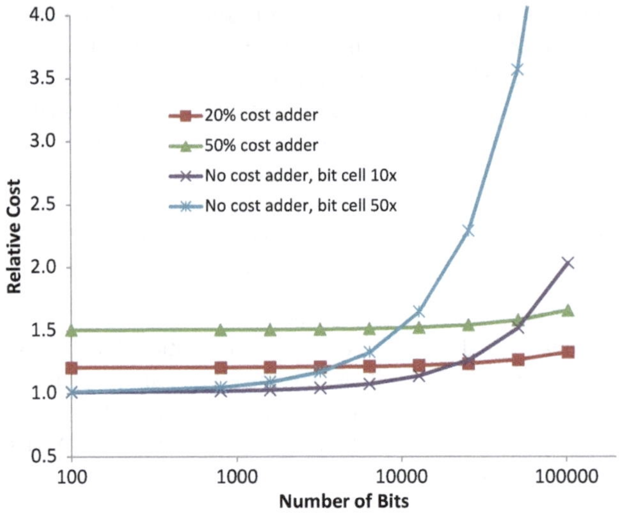

Fig. 8.2 Relative cost of embedding NVMs using logic and modified NVM processes

8.3 Programming and Erase Methods

Before discussing the details of LNVM memory cells and array architecture, we first review the methods for programming memory cells. We have discussed in Chaps. 2 and 5 some of the basic principles that can be used to programming NVM cells, including tunneling and hot carrier injection (HCI).

HCI is the key programming method found in a large number of embedded flash memories, especially in NOR type of flash memories. The advantage is its fast programming speed. Most commonly used memory cells use nFET and channel hot electron for programming the floating gate.

Another programming method is to use band-to-band tunneling (BTBT). One advantage of BTBT is that it may be utilized to inject both electrons for programming and holes for erase of the floating gate. BTBT is usually more energy efficient than the HCI method but the BTBT tunneling process can also be more damaging to the tunneling oxide, resulting in lower reliability for the NVM.

Fowler–Nordheim (FN) tunneling is the workhorse found in NAND flash memories and is becoming more prevalent in embedded NVMs as well. The key advantage of FN tunneling is its low power consumption, as such it is especially useful in low power applications. The disadvantages include the requirement of erasure for a whole block/sector and slower programming speed. The program voltage is significantly higher than that of HCI, but only on a high-impedance node with negligible current. HCI needs a high drain or bulk voltage (>3.1 V) with simultaneously high current.

Table 8.2 Comparisons of programming methods

Method	Use	Speed	Major advantage	Major issues
FN tunneling	Both program and erase	Medium ~ milliseconds	Lowest power	Slow and block erase. High voltage
Hot electron injection	Program	Fast ~ microseconds	Byte programmable	High power
Hot hole injection	Erase	Fast	Enabling prog/erase	Limited endurance
Band-to-band tunneling	Program and erase	Fast	Lower power than HCI	Reliability

Table 8.2 compares the characteristics of the leading programming and erase methods. In general, Fowler–Nordheim tunneling has been gaining popularity due to its extreme low power consumption vs other methods. On the other hand, HCI does have the advantage of faster programming and the ability to program each cells individually with lower program voltage. FN tunneling programming is usually 10–100 times slower than HCI programming methods. For FN tunneling, since the high voltage is applied to multiple cells at the same time, write disturb tends to be a bigger issue and judicious array design is needed to minimize the disturb effect on other cells.

By using different types of polysilicon, FN tunneling current may be changed to suit the overall design requirement. In Fig. 8.3 we illustrate the FN tunneling current density from two different devices, normal pFET with p+ gate and pFET with n+ polysilicon gate. The FN tunneling can be measured under two bias conditions: inversion and accumulation. Both devices are made with same tunneling oxide, using the 3.3 V I/O transistors from a logic CMOS process offered by a foundry (Wang, Ma, Horch, & Paulsen, 2007).

The difference in observed current density can be explained by the realization that tunneling, for the same oxide thickness, is related to the V_{ox}. As we discussed in Chap. 2, Eq. 2.17

$$V_{ox} = V_{GB} - V_{fb} - V_{B},$$

where V_{fb} is the flatband voltage and V_{B} is the voltage drop across the well. For the case of accumulation, $V_{B} = 0$. For inversion, $V_{B} \sim 2\phi_n \sim 1$ V for MOSCAP (Taur & Ning, 1998).

Table 8.1 listed the examples of V_{fb} of pFET for both n+ polysilicon gate and p+ polysilicon gate. Note that for a n+ gate/n-well MOS structure, the V_{fb} is close to zero, $V_{fb} \sim 0.15$ V, while $t{\cdot}V_{fb}$ of the p+ gate/n-well MOS structure is $V_{fb} \sim 0.95$ V, the difference is equal to ~ the bandgap voltage V_{BG}. This explains the ~1 V difference FN tunneling current in the inversion case.

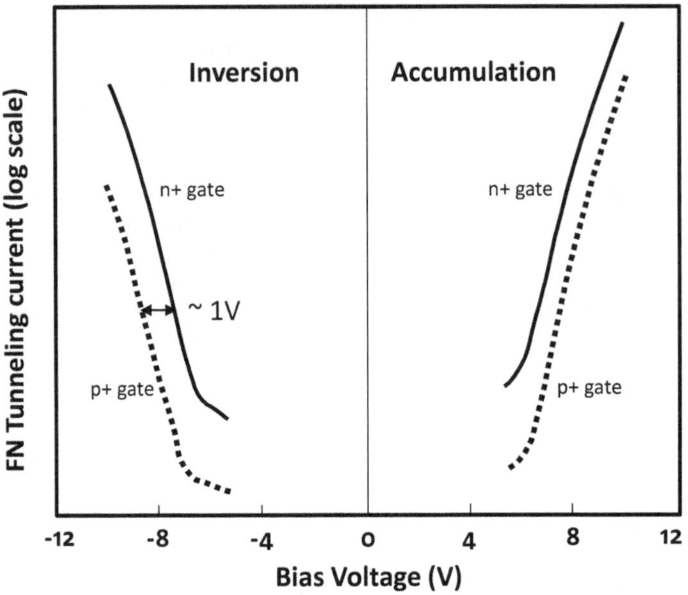

Fig. 8.3 F–N tunneling of pFET with p+ and n+ polysilicon gate under positive and negative bias conditions. The oxide thickness is about 7 nm

For FN tunneling, the charge change on the floating gate can be computed from the current density,

$$\Delta Q_{\mathrm{fg}} = \int I_{\mathrm{FN}}(t)dt = WL\int J_{\mathrm{FN}}(t)dt \tag{8.1}$$

W and L are the width and length of the tunneling junction, respectively.

The FN tunneling current density J_{FN} given by Eq. 2.21 is determined by the electric field in the oxide,

$$E_{\mathrm{ox}} = V_{\mathrm{ox}}/d, \tag{8.2}$$

where d is the thickness of the tunnel oxide. According to Eq. 5.5, V_{ox} is a function of the charge on the floating gate as well. So Eq. 8.1 is an integral equation that cannot be solved analytically but can be solved numerically to simulate the tunneling process.

It is interesting to note the effect of source and drain biases on the FN tunneling when the tunneling structure is a MOSFET, as opposed to a MOSCAP. The effect is different when the FET is biased in inversion and when the FET is biased in accumulation.

For simplicity we assume that the source and drain are biased at the same voltage. In the case of inversion, as illustrated in Fig. 8.4a, the channel is inverted

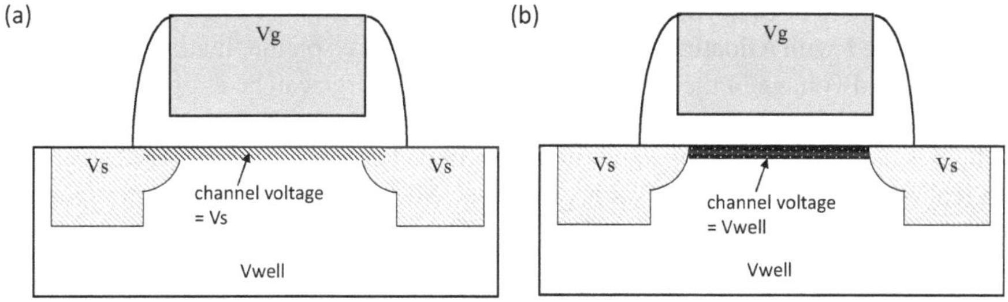

Fig. 8.4 Effect of source–drain biases on FN tunneling current for (**a**) inversion and (**b**) accumulation

and is at the same voltage as source–drain. The FN tunneling current is determined by the voltage difference, $V_g - V_s$. The well bias has no effect on FN tunneling, as it is screened out by the inversion layer.

For accumulation, shown in Fig. 8.4b, there are two regions from which tunneling can occur. In the accumulated channel region, the channel is shorted to the well and the tunneling is determined by the voltage difference $V_g - V_{well}$ and the area of the channel. The source and drain regions are biased by a separate voltage, V_s. The tunneling is determined by the voltage difference $V_g - V_s$ and the gate to source–drain overlap area. The total tunneling current of the whole MOSFET is the sum of these two tunneling current contributions.

This fact may be used in memory array design to reduce disturb (Pesavento & Hyde, 2013).

As stated in Eq. 5.6, the change of the amount of charge on the floating gate will manifest as a change of the threshold voltage of the floating-gate transistor. Appropriate sensing circuit can be connected to the floating-gate device to sense the information stored on the floating gate.

8.4 Memory Cell Design Considerations

Depending on the number of bits, typically around 60–80 % of an embedded NVM's area is occupied by the memory cells. The rest are taken by the memory controller, the high-voltage system, sense amplifiers, and row and column decoders. The performance, such as programming speed, read speed, and reliability of a given memory is also dominated by the property of the memory cell. Thus, the memory cell design is perhaps the most critical part of the NVM design from both cost and performance points of view.

Within the constraints of logic CMOS processes, the main device we can use to design the memory cell is the floating-gate transistor discussed in Chap. 5. Other options include fuse and antifuse, which can be used to build a one-time-programmable (OTP) memory cell. One may also use the nitride spacers found on MOSFETs as storage device, described in more details in Chap. 10.

As described in Chap. 5, for floating-gate-based NVM, the basic storage element is a MOSFET with a floating gate. The amount of charge on the floating gate affects the threshold voltage of the transistor and the store charge can be assigned a "1" or a "0" depending on the threshold voltage. The floating gate can store multiple bits, which is commonly used in stand-alone multilevel cell (MLC) NAND flash chips. In reality, in embedded NVMs it is rare to see MLC, mainly due to the complexity of programming and reading of MLC cells. The requirement of parametric yield in MLC is also much higher than single-level cells (SLC).

Recall that there are two types of FETs in a typical 0.25 μm and below CMOS process, the core and the I/O FETs. Because the core FETs usually have a gate oxide that is too thin to retain the charge on the floating gate, only I/O FETs have been used in making LNVMs. For the LNVM IPs described in this chapter and in Chap. 10, the floating-gate transistors are built using the 3.3 V I/O transistors with ~7 nm gate oxide. In Chap. 10 we will show that it is very challenging but possible to use the 5 nm oxide of the 2.5 V I/O transistor as the floating-gate transistor.

Having decided on the type of floating-gate devices, the other factors for the design of a memory cell are as follows:

1. FET type: n- or p-type floating-gate transistors;
2. Programming and erase methods;
3. Single or differential design.

Conventional flash memories use n-type of floating-gate FET as the storage device. This is mainly due to historical reasons as nFET is usually faster than pFET for reading operations. nFET-based logic dominated earlier IC designs. In addition, HCI is more efficient for nFET, as the tunnel oxide has a higher barrier for holes. As a result, EEPROM and NOR type flash by nFET using HCI programming dominated earlier floating-gate memories.

On the other hand, pFET-based floating-gate memory has found acceptance in the LNVM arena. One reason for using pFET is the use of the n-well of the pFET as the control gate node in Fig. 8.5. In Chap. 10, we will discuss several LNVMs using pFETs that have seen commercial deployment.

In order to add or remove (erase) charge on the floating gate another terminal is typically needed to capacitively couple to the floating gate with the terminal, as shown in Fig. 8.5. In NAND flash memory, this capacitor is provided by the ONO layer with the second polysilicon layer as the control terminal.

For logic CMOS process where only one layer of polysilicon is available, the capacitor can be implemented using either

1. A MOSCAP or
2. Another capacitor such as a poly–poly capacitor or MIM capacitor.

In Chap. 5, we introduced the concept of coupling ratio that determines how much voltage is needed on the control terminal,

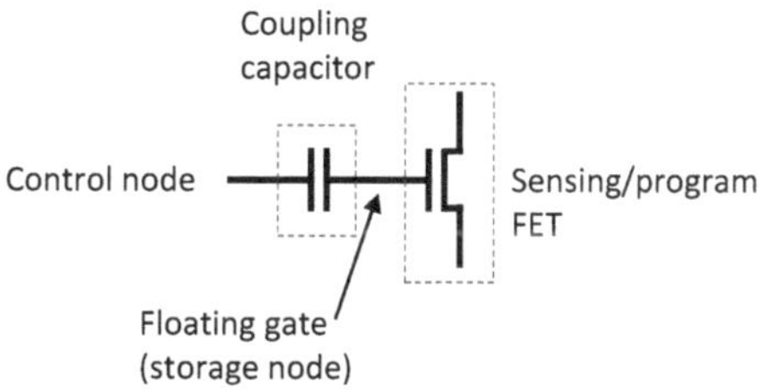

Fig. 8.5 Control terminal and capacitor for controlling the floating gate

$$\alpha_{CG} = \frac{C_{CG}}{C_{CG} + C_{FET}}, \tag{8.3}$$

where C_{FET} is the capacitance of the FET and C_{CG} is the capacitance of the capacitor in Fig. 8.5.

The importance of the coupling coefficient can be seen from rearrangement of the Eq. 5.5,

$$V_{cg} = \frac{1}{\alpha_{CG}} V_{ox} + V_w + \frac{Q_{fg}}{C_{CG}}. \tag{8.4}$$

The control gate voltage is inversely related to the coupling coefficient. Since the required V_{ox} is usually fixed by tunnel oxide thickness and the required programming speed, the control gate voltage is a sensitive function of the coupling coefficient.

As an example, we chose a 65 nm logic process and use its 3.3 V I/O transistor as the floating-gate device. Our NVM requires an erase speed of 10 ms. This translates to a minimum $V_{ox} = 7$ V for the desired programming speed. Also assume $V_w = 0$. If $\alpha_{CG} = 0.7$, we obtain from Eq. 8.4 a minimum $V_{cg} = 10$ V. If $\alpha_{CG} = 0.5$, the minimum V_{cg} becomes 14 V. As we have seen in Chap. 4, the conventional LDMOS may be able to go as high as 15 V but without much margin. It is much harder to handle a high voltage that is much higher than 10 V using the LDMOS. Another problem is the control gate leakage, which is of the opposite direction of the channel injection. When $\alpha_{CG} < 0.5$, C_{CG} will drop more voltage than C_{FET}, which can cause leakage charge injection into the floating node through C_{CG} that can lead to memory window reduction, if not total malfunction.

Thus, we need the coupling coefficient to be as close to one as possible. On the other hand, the coupling coefficient has a dominant effect on the size of the memory cells. So the choice of the coupling coefficient is a trade-off between the high-voltage constraint and the desire to reduce the memory cell size.

For example, to reach the desired program and erase speed, or to lower the required voltage, typically a coupling ratio greater than 0.6 is required. This translates into large C_{CG} or small tunneling/programming junctions.

For the 3.3 V transistors, the minimal size is typically $L = 0.3$ μm and $W = 0.3$ μm, the capacitance of the tunneling junction is then about 0.4 fF.

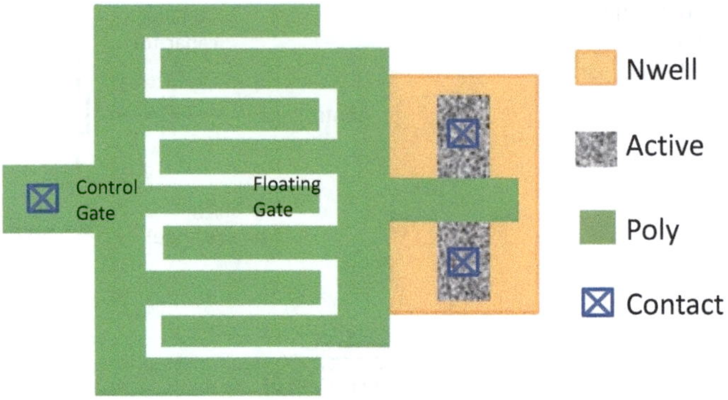

Fig. 8.6 Interdigitated coupling between polysilicon, adapted from Na et al. (2007)

Considering the parasitic capacitances, the capacitance of the control capacitor needs to be ~4 fF or higher to achieve a coupling ratio of 0.8 or higher.

Unfortunately, in standard CMOS or RF/mixed signal processes, there is no backend capacitor that can satisfy the requirements of: (1) higher capacitance per area and (2) low leakage, $< \sim 10^{-23}$ A as required for data retention.

For this reason, most current LNVMs use a MOSFET or MOSCAP as the coupling capacitor in the memory cell of Fig. 8.5. As a consequence, LNVM floating-gate memory cells usually have at least two transistors (2T cells) and often three transistors (3T cells). As we will see in detail later in this section, 2T cells are common for memory using FN erase and HCI programming method. 3T cell is more common for low power memories using FN tunneling for both programming and erase.

Instead of using MOS capacitor in the floating-gate memory cell, it is possible to use poly to poly coupling and replace the MOS capacitor by a poly finger cap (Na, Kim, & Kim, 2007). An example layout is shown in Fig. 8.6. In this case since the height of the poly and the minimum spacing between the poly is fixed, the coupling capacitance can only be varied by the length. For older technologies, this results in a large cell due to the long poly segment that is necessary to achieve an acceptable coupling ratio.

The areas of LNVM memory cells that are capable of reprogramming are usually much larger than the flash memory cells and much larger than the size of a typical FET. For example, in the single poly NVM in a 0.18 μm logic poly processes (Na et al., 2007), the memory cell has an area of over 50 μm^2, while the NAND flash cell is 0.13 μm^2, and a typical FET is about 0.2 μm^2.

For a more advanced, e.g., 65 nm, technology, the polysilicon spacing is 150 nm. The interpoly capacitance is then

$$C = \frac{\epsilon A}{d} = 8.854 \times 10^{-12} \frac{kA}{d} (\mathrm{F/m})$$

For silicon dioxide, the relative dielectric constant $k = 3.8$, for polysilicon height of 0.2 μm and spacing of $d = 0.15$ μm, we get $C = 0.045$ fF per micron length of polysilicon. To get a 5 fF capacitance, one needs poly length of about 100 μm. With interdigitated design as shown in the following figure, the total length of polysilicon roughly is halved. Still the control capacitor occupies an area of roughly 20 μm^2.

As technology scales down, the polysilicon spacing continues to reduce while the polysilicon height barely scales. The area of the coupling capacitor is then much reduced. For example for 20 nm process, the polysilicon spacing is around 30 nm. The capacitance per unit length is increased by 5× with the corresponding reduction in the length of the poly and the total cell area.

However, the dielectrics between the poly fingers are deposited oxide, which is generally more leaky than thermally grown oxide. As a result, the leakage between the polysilicon fingers could pose the ultimate limit to the inter-poly spacing and to the size of the memory cell.

Despite of the large area, the concept of lateral coupling has been used to develop a single poly NVM for LCD drivers, where high voltage is usually required (Park et al. 2015) for analog trimming and code storage applications using a 0.13-μ m BCDMOS process. In this design the so-called select gate lateral coupling (SGLC) cell has a very small unit cell size of 2.82 μm^2, which is comparable to the corresponding size of SRAM. The SGLC cell is fabricated using a combination of only 1.5-V and 5-V transistor-related processes for channel HCI programming. The SGLC cell exhibits a high programming speed of 100 μs and is over-erase-free, which is suitable for a NOR array structure.

In addition to using the poly–poly capacitor, the MIM capacitor can also be used. High-performance single polysilicon electrically erasable programmable read-only memories (EEPROMs) with stacked metal–insulator–metal capacitor as a control gate have been proposed in a 0.25 μm single poly logic process (Na & Kim, 2006). The cell size of the proposed MIM control gate structure is similar to that of the n-well control gate devices. The effective control gate coupling ratio of the proposed EEPROM cell was higher than that of cells with n-well control gate because of the absence of depletion capacitance in the n-well region.

To summarize, the process of designing an embedded NVM memory cell is as follows:

1. Choose the process node based on the cost and the requirement of the system;
2. Choose the floating-gate transistor, normally the 3.3 V I/O transistor;
3. Specify the V_{ox} range based on the required program and erase speed for the NVM IP;
4. Set the maximum programming voltage, V_{pp}, based on the reliability of the high-voltage transistors;
5. Set the memory window from the reliability requirement;
6. Obtain the coupling ratio, e.g., from Eq. 8.4;
7. Size the devices in the memory cell;
8. Optimize the cell layout.

8.5 Memory Arrays

A nonvolatile memory array is architecturally similar to other memory arrays like SRAM and DRAM, where a row and column arrangement is almost universally used. The main challenge and a unique aspect of NVM arrays is the requirement that the memory cell needs to operate at a high voltage, usually much higher than either the I/O or core supplies. As indicated in the last section, most floating-gate-based logic NVMs need a voltage typically on the order of 10 V to program the memory cells. In the case of fuse or antifuse-based OTP memories often higher current is also needed to program the fuse.

A high-voltage (HV) system, including the HV generator, is needed to generate the high voltage necessary for these operations. In addition, the row and/or column decoders need to be able to handle the high voltage. This can be a serious constraint since the high-voltage transistors, e.g., those described in Chap. 4, can be large when designed in the logic CMOS processes.

Another important aspect of the NVM array is the disturb issue where, when one row of the memory is being read or programmed, the other rows may be affected. We will discuss this aspect with some examples in Chap. 10.

Figure 8.7 shows the major blocks of a typical nonvolatile memory.

The controller circuit may include a finite state machine which contains the algorithms for memory operations such as erase, program, and reading. Charge pump provides the high voltage, generated from the supply of core or IO voltage.

Cell array is the core of the memory array comprising of the memory cells arranged in rows and columns. The memory cells in a column usually share a bit

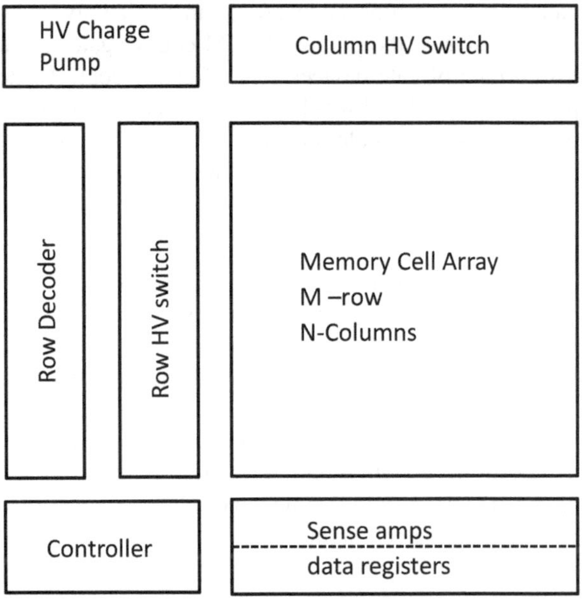

Fig. 8.7 Architecture of an embedded NVM array

line, while the memory cells in a row share the word line. The bit lines are connected to a row of sense amplifiers. The sensed results are usually stored in a data register before being sent to the NVM controller. The cell array usually is the largest block of the NVM in terms of area. HV row and column decoders border the cell array and usually have to be pitch matched to the row and columns.

Note that not all circuit blocks are necessarily present in all NVMs. For example, for one-time programmable memory or ROM memory, the memory may only be programmed at wafer testing. In this case, a HV supply from the tester can be used to program the memory. No high-voltage charge pump is then needed on the chip.

8.5.1 High-Voltage Generators

High-voltage generator is the circuit that brings the supply voltage, e.g., 3.3 V, to the programming voltage V_{pp}, e.g., 12 V. The classical HV generator is a voltage multiplier or a charge pump, with the Dickson voltage multiplier (Dickson, 1976) or its variants being the most commonly used. Figure 8.8 shows the implementation of Dickson charge pump using diode-connected nFETs. The capacitors can be implemented using MOSCAPs. Figure 8.9 shows the timing diagram and the ramping of voltages at different stages of the charge pump.

Starting with a clock signal CLK at low, the first diode-connected FET D_1 turns on and charge the capacitor C_1 so the voltage at the node V_1 becomes

$$V_1 = V_{in} - V_t \tag{8.5}$$

Note at this period, $\overline{\text{CLK}}$ is high and diode D_2 is off.

At the next phase, CLK turns high, the voltage at node V_1 is capacitively coupled high to:

$$V_1 = \frac{C_1}{C_1 + C_{node}} V_{clk} + (V_{in} - V_t) \tag{8.6}$$

where C_{node} is the rest of the capacitance, not including capacitor C_1, at the node V_1.

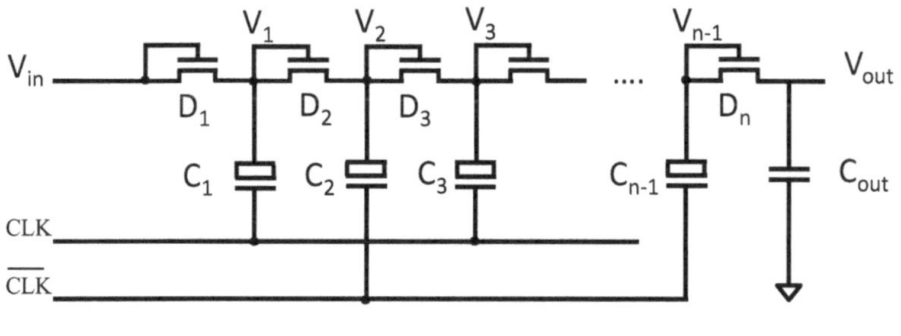

Fig. 8.8 A Dickson charge pump using CMOS FETs

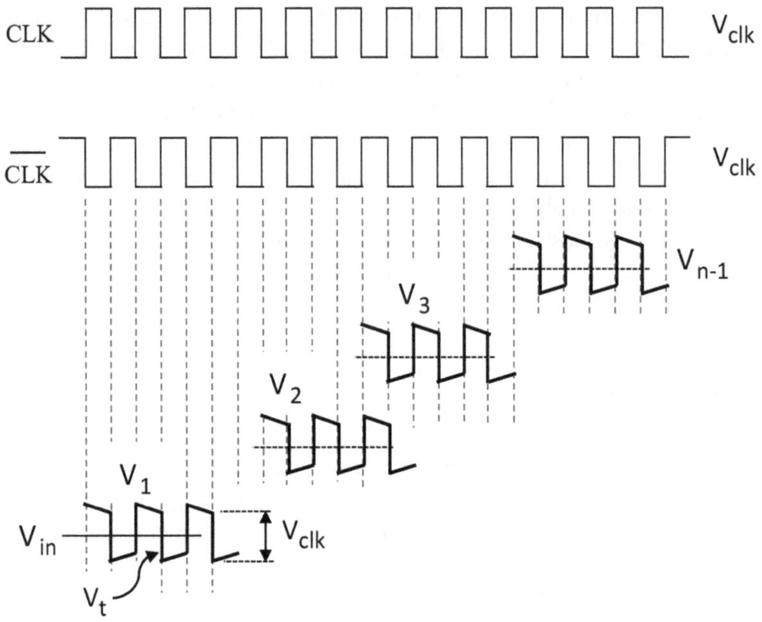

Fig. 8.9 Timing pattern of charge pump circuit

We assume the capacitor C_1 is much larger than the rest of capacitance associated with node V_1, so the coupling ratio is close to one.

At this moment, the capacitor at node at V_2 is coupled to the low clock $\overline{\mathrm{CLK}}$. The diode D_2 conducts and charges second capacitor. The voltage is given by $V_2 = V_1 - V_t$. Or,

$$V_2 = V_{\mathrm{in}} + \frac{C_2}{C_2 + C_{\mathrm{node}}} V_{\mathrm{clk}} - 2V_t. \tag{8.7}$$

After stage n, the output voltage becomes

$$V_{n+1} = V_{\mathrm{in}} + n\left(\frac{C}{C + C_{\mathrm{node}}} V_{\mathrm{clk}} - V_t\right) - V_t \tag{8.8}$$

We have assumed here V_{clk} is larger than V_t so that the pump can work correctly. Also for simplicity, we assumed $C_1 = C_2 = \cdots C_{n-1}$ and the node capacitance are identical at each node. Notice that for very small V_{clk}, as in power scavenging circuits, V_t will need to be close to 0. Including a load directly connected to the output reduces the output voltage by $I_{\mathrm{out}}/(C \times f_{\mathrm{clk}})$, where f_{clk} is the frequency of the clock, further reduces the output voltage to

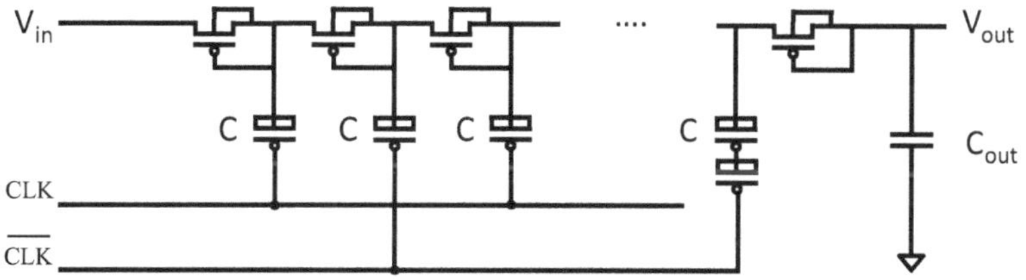

Fig. 8.10 Dickson charge pump built entirely with pFET

$$V_{out} = V_{in} + n\left(\frac{C}{C + C_{node}}V_{clk} - V_t - \frac{I_{out}}{(C + C_{node})f_{clk}}\right) - V_t \qquad (8.9)$$

Besides the parasitic capacitance which reduces the charge pump efficiency, other factors, including the body effect which changes the threshold voltage of the diode connected FETs and the leakage of diodes and capacitors, can also reduce the efficiency of the charge pump. As the node voltage increases, even though the voltage drop across the diode is nearly constant given by V_t, the absolute voltage to any substrate may cause high leakage and limits the efficiency. At the extreme, the voltage may be high enough to cause breakdown to limit the highest voltage this circuit can achieve.

The original Dickson charge pump of Fig. 8.8 is based on nFETs. The voltage across each diode-connected FET is only the low-voltage V_{clk}. So I/O nFET can be used in this design without worrying about breakdown. However, at higher stages the absolute voltage at the nodes is moving higher. This high voltage cannot be supported by the source and drain junctions of even the 3.3 V I/O nFET.

To design a charge pump capable of going up to 12 V or more, a pFET-based charge pump is often used. Figure 8.10 shows a basic version showing all the terminals of the devices used in the design. In this design all the voltage nodes are connected to an n-well which, as we have seen in Chap. 4, can sustain voltages of up to 15 V in the CMOS process. The source to n-well/drain voltage is at most V_{dd}, which is within the design spec of all the transistors.

For the charge pump used for LNVM, the input voltage is the 3.3 V V_{dd} and the maximum output voltage needs to be about 10 V. The drive current requirement depends on the size of the array, including how many cells will be erased at once. For FN tunneling-based NVM cells, I_{out} is usually on the order of 1 μA.

For later stages, the node voltage, e.g., V_{n-1}, could exceed the breakdown of the gate oxide of the 3.3 V I/O transistor. The capacitor can be replaced by two series connected MOSCAPs with double the area, as shown in last stage of the charge pump in Fig. 8.10.

8.5.2 *Row and Column Decoders*

The memory cells in the array need to be addressed by the row and column decoders. Depending on the memory architecture, one or both of these decoders need to be able to handle high voltages. For design simplicity, the architecture shown in Fig. 8.8 separates the row and column decoders into standard decoders that run on the core supply voltage and the corresponding HV part that utilize high-voltage switches to control the voltage generated by the HV generator. In almost all cases, the logic part of the decoder can use core transistors while the memory arrays have to deal with HV. A level shifter is necessary to bridge these two parts of the decoder.

The conventional row and column decoders are extensively discussed in many texts on memories. For column decoder, there are two types of approaches, AND-decoder based and tree based. AND-based decoder usually has on the order of $N \times 2^N$ transistors, while binary tree-based decoder has less transistors but is slower because of the series of pass transistors. Usually a combination of the two approaches is used for various address bits.

The decoders presented earlier drive the HV switches. A HV switch is capable of switching, on/off or vice versa, a HV signal (>5 V) from the HV generators in response to a control signal at the usual logic supply voltage from the decoders.

As we have seen in previous chapters, the n-well in a CMOS process can usually handle voltages as high as 15 V and is used often in HV circuits. The LDMOS of Chap. 4, where the drain of an nFET is replaced by n-well, is a good example. Another method to handle high voltage is cascading, i.e., arranging in series, of several pFET transistors. By cascading, the voltage drop across individual source—to drain will be more tolerable. Both of these structures are evident in the following example of a HV switch, Fig. 8.11 (Bernard & Diorio, 2006). The symbol for the LDMOS is on the right.

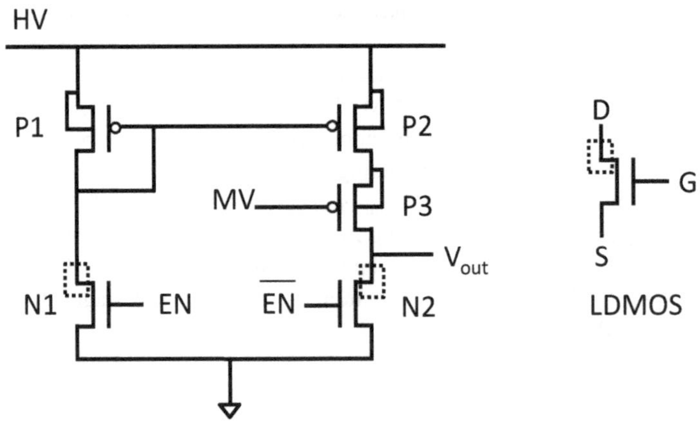

Fig. 8.11 A simple HV switch using conventional pFET and HV LDMOS nFET. The symbol for the LDMOS is on the right with its n-well drain indicated by the *dashed square*

The principle of operation of the earlier HV switch is as follows. When the node EN is set (=V_{dd}), LDMOS N1 is on. The second LDMOS, N2, is off. This brought the gate of both pFETs, P1 and P2, to low, turning on both. Intermediate voltage MV is set at an intermediate voltage, between V_{dd} and the HV, e.g., HV/2. PMOS P3 is on under this condition. The output voltage, V_{out}, is then equal to HV. Note that the high voltage HV (~10 V) is only applied to the n-wells of the pFET as well as the n-wells of the LDMOS N1 and N2. So there is no reliability problem.

To turn the switch off, EN is set at low. nFET N2 is then turned on which brings V_{out} to low. P1, P2, and P3 are all off at this state. The purpose of the cascode P3 and MV is to reduce the voltage drop from HV to V_{out} which is presently at ground. Without P3, the circuit looks like part of a simple differential amplifier (but work in gain saturation), with the danger of very high V_{ds} in P2 when V_{out} is low.

An issue with the earlier design is that when the HV switch is turned on (EN = V_{dd}), there is a conducting path from HV through P1 (which is on), and N1, to ground, causing a parasitic leakage that burns power not used by V_{out}. A solution is to replace nFET N1 by a capacitor, either an MIM or an MOSCAP. This way, the leakage path is turned off and there will be no static current.

Another variant of the earlier design is shown in Fig. 8.12. It is again driven by the high-voltage HV and an intermediate-voltage MV ~ HV/2.

This design consumes no static power when it is not switching. Figure 8.13 shows an example of the timing signals for operating this switch.

In Fig. 8.12, N1, N2, N7, and N8 are LDMOS transistors. Again it is worth noting that only the drains of these LDMOSes are subject to high voltage. This is why the LDMOSes in Sect. 4.3.2 are used while the symmetric LDMOS, where both the source and the drain use n-wells, is not needed. The symmetric LDMOS

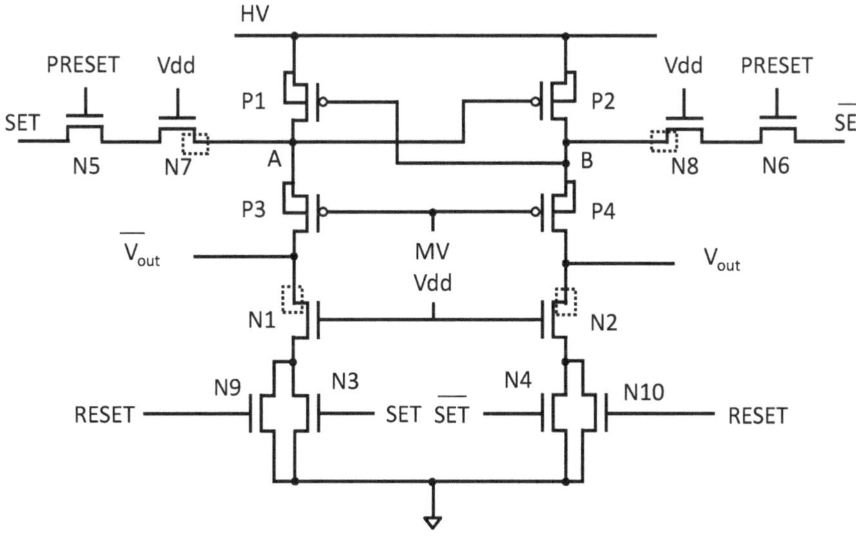

Fig. 8.12 A resettable HV switch, adapted from Pesavento (2009)

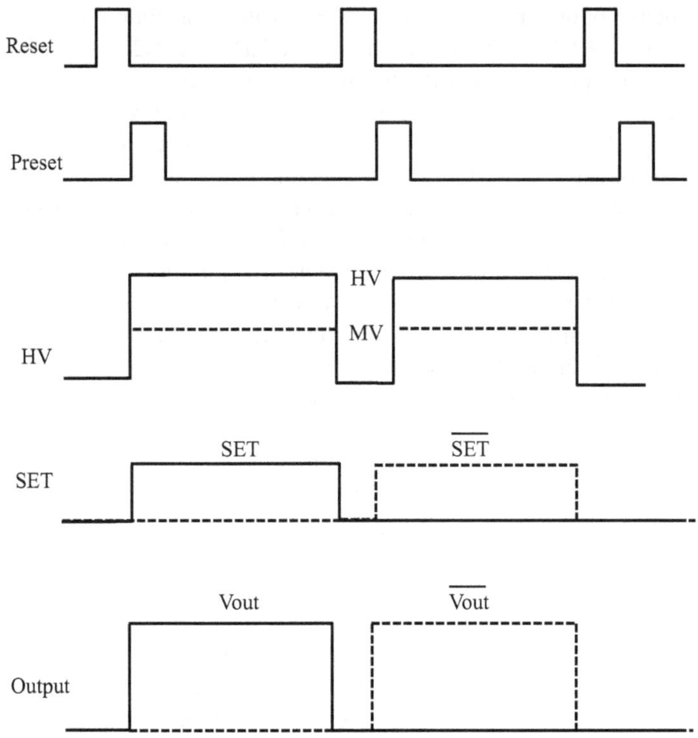

Fig. 8.13 Signals to control the HV switch in Fig. 8.12

can be used as pass transistors for HV. But since it is very large due to the use of two n-wells, they are not used unless there is no design around it.

Some added functionalities include SET, RESET, and PRESET control signals and associated transistors in this fully functional HV switch. LDMOS transistors N7 and N8 are used to isolate the conventional FET, e.g., N5 and N6 from the high voltage. With the PRESET control signal applied to the gates of N5 or N6, nodes labeled A or B are charged or discharged before the HV operations to avoid erroneous transient operations, e.g., transistors N3 and P1 turning on simultaneously when ramping the HV. This could happen due to the capacitive coupling effect of their gates to the ground.

8.5.3 Sense Amplifiers

The row and column structure means many memory cells share a column (the bit line). To speed up the read process, dedicated sense amplifiers are often used. The normal sensing process is similar to the sensing in SRAM or DRAM. The selected bit lines are first precharged to a predetermined voltage. Then the selected row is

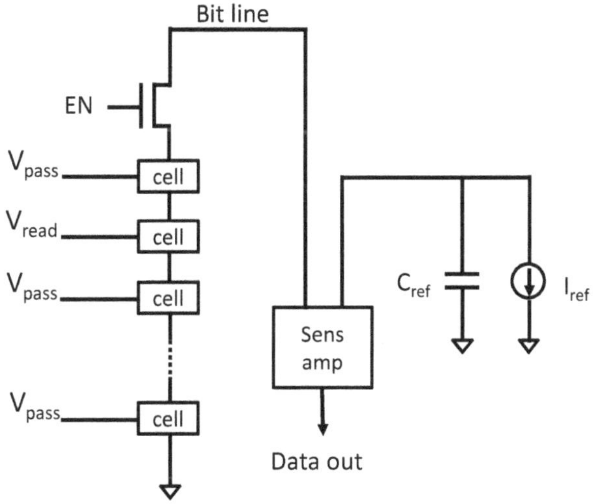

Fig. 8.14 Sensing of NAND memory. Only one column of the memory is shown. In this case V_{pass} is set to turn on the cells that are not selected

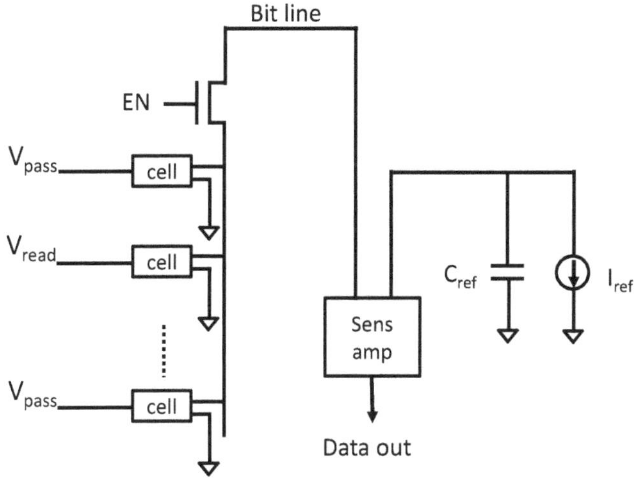

Fig. 8.15 Sensing of NOR memory. Only one column of the memory is shown. In the NOR case the V_{pass}, e.g., V_{pass}= GND for n-type of floating-gate cells, is set to turn off the cells so that selected cell can be read

enabled. All the unselected rows are "by-passed." The method to bypass the unselected rows are different for NAND type of architecture vs NOR type of architecture.

Figures 8.14 and 8.15 show the circuits for sensing of NAND and NOR memory architecture, respectively. In the NAND architecture the memory cells on the same bit line are arranged in series. To bypass the unselected rows, a high word-line voltage is applied to turn on the floating-gate transistors to allow the bit line current

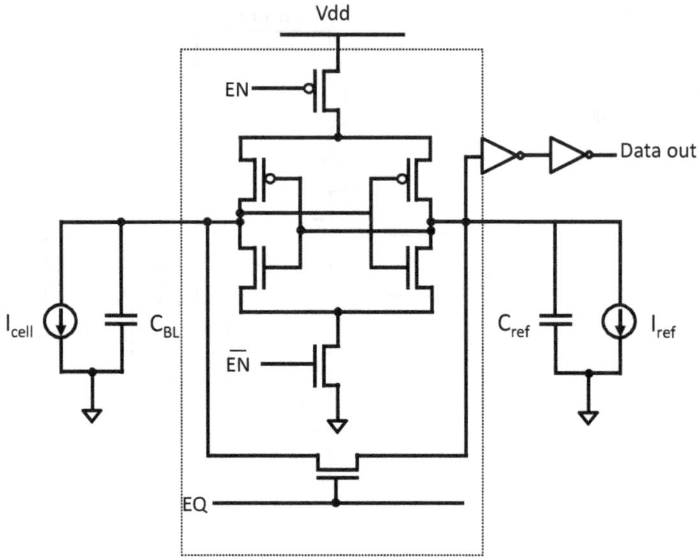

Fig. 8.16 The current-mode sense amplifier

to flow through. The magnitude of the current is determined by the floating-gate transistor on the selected row.

For NOR, the unselected word-line transistors are turned off and the unselected rows do not contribute to the discharging of the bit lines. Most LNVMs discussed in this book are NOR type.

Examples of sense amp designs are shown in Figs. 8.16 and 8.17.

The sense amplifier shown in Fig. 8.16 works on sensing the current, in particular, the magnitude of the current is sensed by the rate of discharging bit line and the capacitor on the reference side. This is the same circuit for an SRAM-based latch circuit.

Figure 8.17 is an example of a voltage-mode sense amplifier. The bit line is modeled by a bit-line resistance while there is a reference resistance in the sense amplifier. The two resistance forms a voltage divider which drives the inverters. Ideally the reference resistance should be tunable to adjust for the process variations.

For example, for a programmed cell, the equivalent bit-line resistance is 10 KΩ while an erased cell the bit-line resistance is 100 KΩ. The reference resistance is chosen at 55 KΩ. The input to the inverters in the sense amplifier is $0.84V_{dd}$ for the programmed cell, which will output a logic high signal at the data out node. For the erased cell, the input to the inverters is $55/155 = 0.35V_{dd}$, giving a logic low to the sense amp output.

Figure 8.18 is a differential voltage-mode sense amplifier. This amplifier is popular because it is easy to use and reliable. The differential nature of the design makes it resistant to supply noise due to its ability to reject common mode voltages.

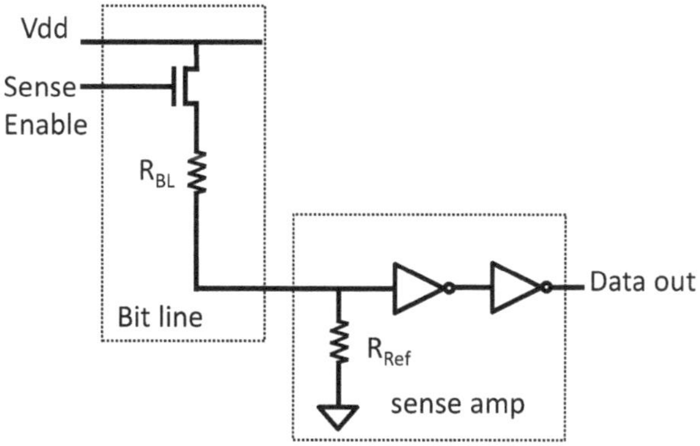

Fig. 8.17 Voltage-based sense amplifier

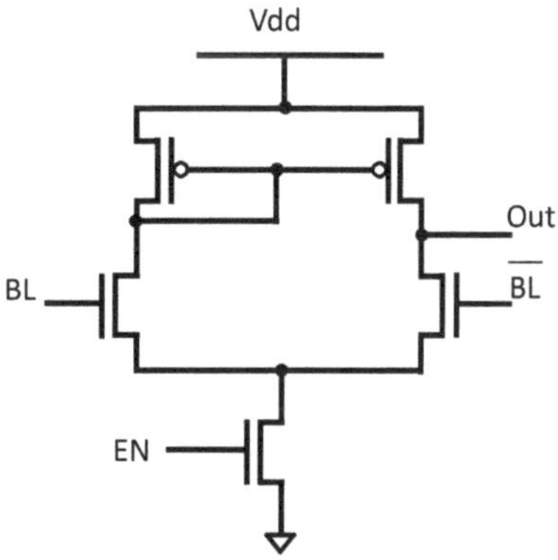

Fig. 8.18 Differential voltage mode sense amplifier

In this design, the two pFETs form a current mirror, and have identical size, as are the two nFETs.

The sense amplifier may be enabled at the same time as the word line during a read cycle, making timing requirements for this amplifier very easy. The bit line and the reference are precharged to equal voltage. A read cycle begins when the signal EN is turned on to enable current to flow through the current source at the bottom. The difference in the bit line and the reference voltage is then sensed.

One of the drawbacks of this amplifier is that the biasing current source must remain on throughout the entire read cycle. This constant current draw causes the device to consume a significant amount of power throughout the entire read cycle.

In general, the current sensing sense amplifiers are faster and more tolerant to mismatches. On the other hand, the voltage mode sense amplifiers consume less power but are typically slower.

8.6 Memory Reliability

Reliability is one of the most important metrics of NVM. There are several aspects to the NVM reliability, including:

Data retention—the ability to retain the data during power off is the key difference of an NVM from volatile memories such as DRAM and SRAM. Usually 10-year retention under the specified operating temperature is required. Although for archive applications up to 50 or 100 years may be necessary.

Endurance—The number of cycles a multi-times programmable (MTP) NVM can be programmed and erased is another distinguishing metric from SRAM and DRAM, which normally have unlimited read and write endurance. The limited number of endurance, especially for floating gate or charge trapping based NVMs, is a direct result of damage that can be done to the storage medium from the high-voltage programming and erase operations.

Disturb—There are two types of disturb. Write or program disturb is the disturb of storage state of unselected cells when writing or programming of selected cells. Read disturb is the disturb of both selected and unselected cells during the read process. When there is enough disturb, the state of the cells can change, causing a bit to flip. Both disturb mechanisms become worse with thinner tunnel oxide and when more memory cells are packed together.

The reliability of an NVM can be improved by process optimization and by design optimization. The unique aspect of LNVM is that the process is fixed and only design can be changed to optimize the reliability for LNVM. In the rest of this chapter and Chap. 10, we will discuss some of the methods used for improving the reliability of LNVM.

8.6.1 Data Retention

Data retention of a NVM can be classified into intrinsic or extrinsic retention. Intrinsic retention is the retention property of a memory cell that is undamaged from subsequent operations of the memory. It is mostly due to the leakage of insulators surrounding the floating gate, such as through the undamaged tunnel oxide or the oxide surround the floating gate.

Extrinsic data retention is defined as the additional leakage caused by the operation of the memory cells after usage, such as after programming/erase cycles. In particular, program and erase (P/E) cycles cause the most damage to the tunnel oxide (often through a mid-oxide trap generation) and are the dominant factor affecting the data retention characteristics of floating-gate NVMs.

As an example, a 16 kb test array is designed to evaluate the intrinsic reliability of LNVMs (Ma et al., 2007). The memory cells were programmed to have a range of initial cell currents.

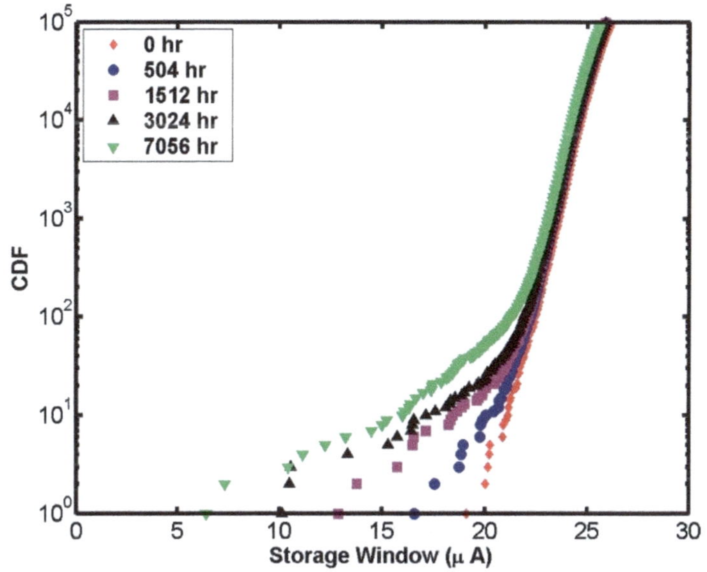

Fig. 8.19 Cumulate distribution of nearly 16 Kb memories as a function of storage window for different bake times at 85 °C. These memories have been cycled 1K times prior to bake (Ma et al., 2007)

Figure 8.19 shows the cell current distribution of a memory array after extended baking at 85 °C up to 2 years, and after 1k program and erase cycles. In Fig. 8.19, we can see clearly a main distribution, where there is little shift in the cell current, and a tail distribution, where the current moves fairly rapidly. The main distribution is attributed to the intrinsic retention while the tail bits are attributed to the extrinsic retention due to cells that have been damaged from the program and erase process.

In a more detailed intrinsic retention study (Ma et al., 2004), it is found that the cell in the 16 kb array and found the following empirical relation,

$$I_c \propto V_{fg}^2 \tag{8.10}$$

From this relationship, it can be easily shown that $\Delta V_{fg} \propto V_{fg}$ and $\Delta I_c \propto I_c$ observed in the 16 kb array are equivalent.

From the linear relationships we can derive an empirical model that is capable of predicting the cell current after any bake period. If we assume the leakage is a first-order process, i.e., one dominant barrier without an intermediate state, the leakage process can be described as a RC discharge process, with the cell current decaying exponentially according to

$$I_c(t) = I_c(0)e^{-t/\tau} \tag{8.11}$$

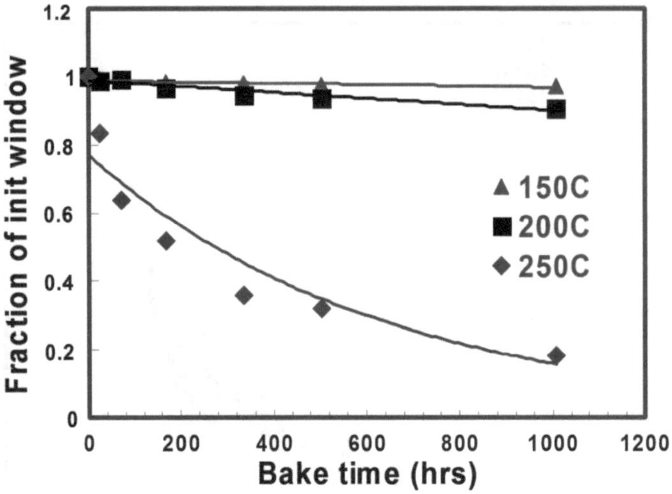

Fig. 8.20 Average cell current normalized to their prebake values vs bake time. Data points were taken after bake at the indicated temperatures and solid lines are fits using the model in Eq. ***8.12***

with $\tau = RC_{\text{total}}$, where C_{total} is the total capacitance of the floating gate and R is a resistance of the leakage path from the floating gate to the ground.

The model described by Eq. 8.11 was used to analyze the retention data of the 16 kb array. Equation 8.11 describes the data well after either short or long periods of bake, but not both. To model both short and long bakes together, a second exponential term was added to give the final analytical model for the cell current as follows:

$$I_c(t) = I_c(0)\left(\alpha e^{-t/\tau_1} + (1-\alpha)e^{-t/\tau_2}\right). \tag{8.12}$$

This two-time-constant model suggests two independent leakage paths and is used to fit the bake results. Figure 8.20 shows some measured data and associated fits at different temperatures. The long and short time constants differ by about a factor of 100, with $\tau_1 \sim 10$ h and $\tau_2 \sim 1000$ h at 250 °C. The constant α is found to be near constant at about 0.2 for all splits and at all temperatures. It measures the portion of the cell current decay that is due to the faster leakage path.

The time constants at different temperatures obtained from the curve fits have been plotted in an Arrhenius plot (Ma et al., 2004). Good linear relationship is observed for both time constants, indicating that the time constants both follow the Arrhenius relationship:

$$\tau = \tau_0 e^{-E_a/kT} \tag{8.13}$$

The two modes of leakage have different activation energies: the fast leaking mode has activation energy of about 1.0 eV while the slow leak mode is about 1.7 eV. The

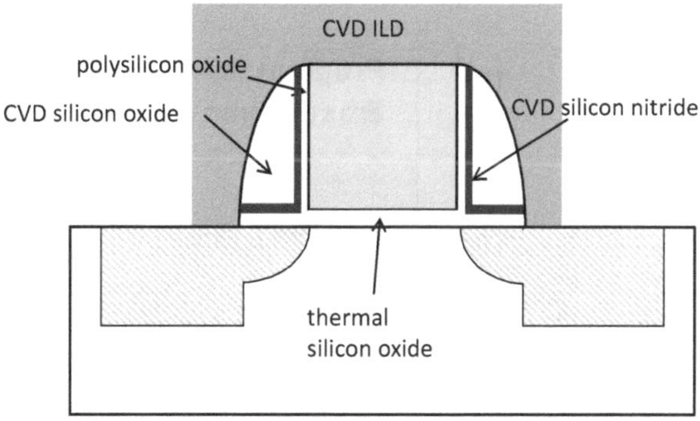

Fig. 8.21 Different types of oxide surrounding the floating gate can contribute to the data retention

activation energy and the Arrhenius relations can be used to extract the time constant at the chip operating temperature.

The retention behavior discussed earlier can be explained by charge redistribution in the insulators surround the floating gate, for example as in (Liao et al., 2013). This is one of the intrinsic characteristics of LNVM. As we discussed in Chap. 5, the charge retention is strongly dependent on the insulators surrounding the floating gate. Some of the insulators are shown in Fig. 8.21.

The bottom of the polysilicon floating gate is the gate oxide. Only 6.2 nm thick in the earlier example, this oxide is thin but generally is very high quality and does not have much mobile charge. On the other hand, the rest of insulator surrounding the polysilicon gate is usually deposited oxide or composite SiO_2/Si_3N_4. On the side of the polysilicon is a thin polysilicon oxide which is usually thermally grown. But since it is grown out of polysilicon with many grain boundaries and typically of 10 nm or less, it can leak. On the other side of this layer are the silicon oxynitride spacers. The top of the polysilicon gate is the interlayer dielectrics (ILD) which is usually deposited with high density plasma **chemical vapor deposition** (CVD) process. Both ILD and nitride can have a lot of hydrogen ions and can also easily trap charges. The behavior observed earlier can be explained by the charge redistribution in these trappy insulator layers. The good news is that it does not appear to affect the intrinsic data retention too much.

The insulators surrounding the polysilicon gate in most logic CMOS processes are not developed with NVM data retention in mind. They are usually not optimized to be the best insulators. Some CMOS processes do not have the appropriate insulators and floating-gate-based LNVM may not be feasible in those processes.

Extrinsic retention is related to stress-induced defects in the oxide. These defects can cause trap-assisted tunneling which will result in leakage current at even very low gate voltage. This **stress-induced leakage current** (SILC) can cause floating gates to leak charge and reduce the lifetime of the floating-gate NVMs. As the trap generation is a stochastic process, it usually affects a small percentage of the bit array.

Figure 8.22 shows direct measurements of the change in tunneling current of a 7 nm oxide as a function of the amount of tunneled charges (Wang et al., 2005), as

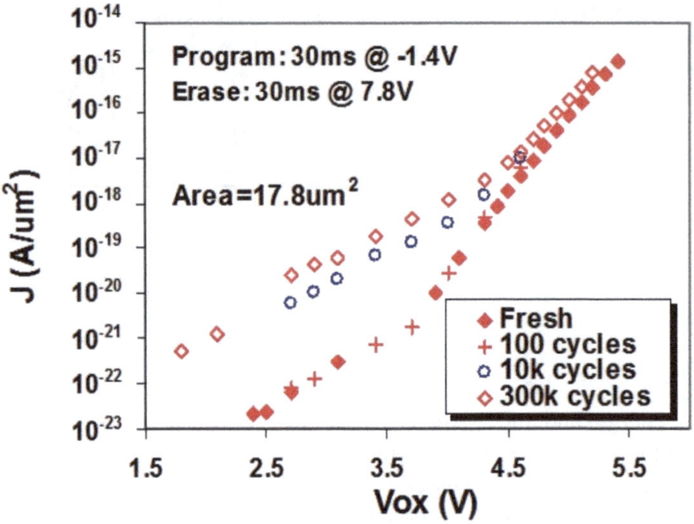

Fig. 8.22 SILC current after cycling stress, from Wang et al. (2005)

measured by the number of the program and erase cycles. As can be seen in the figure, the tunneling current increased appreciable in the low-voltage region. This is the voltage region for the gate voltage during the storage and disturb conditions.

SILC has been studied extensively as it sets the limit to EEPROM tunnel oxide scaling. For example, see the following studies for details including quantitative modeling (Moazzami & Hu, 1992; Olivo, Nguyen, & Ricco, 1988). From these and other studies, it is concluded that the stress-induced leakage current is the main reason that the tunneling oxide thickness limit is about 7 nm. The 7 nm limit can also be illustrated qualitatively using the trap-assisted model in Fig. 8.23.

In the most simplistic model, for a tunnel oxide of 7 nm or thicker, the SILC due to one trap involves two direct tunneling steps, as shown by the path labeled "1" in Fig. 8.23. Each of the tunneling leg needs to go through a distance of ~3.5 nm. Recall from quantum mechanics that direct tunneling current is exponentially dependent on the thickness of the barrier. Order of magnitude estimates using measured results (Lee, King, & Hu, 1999) for direct tunneling through a silicon oxide show that for ~3 nm oxide, the direct tunneling current of less than 10^{-13} A at a bias of -1 V. For two-step tunneling, the overall tunneling current is the probability product of the two one-step tunneling and should be less than 10^{-23} A. This is low enough to provide 10 years of retention, as suggested in Sect. 5.2.

The next dominant tunneling mechanism involves the presence of two or more traps, e.g., those of paths labeled (2) and (3) in Fig. 8.23. However, those leakage paths will occur with a much smaller probability as it needs the presence of two traps almost lined up. Of course for an oxide that has gone through extensive erase cycles, there will be many traps and the earlier situation is more likely. This is why

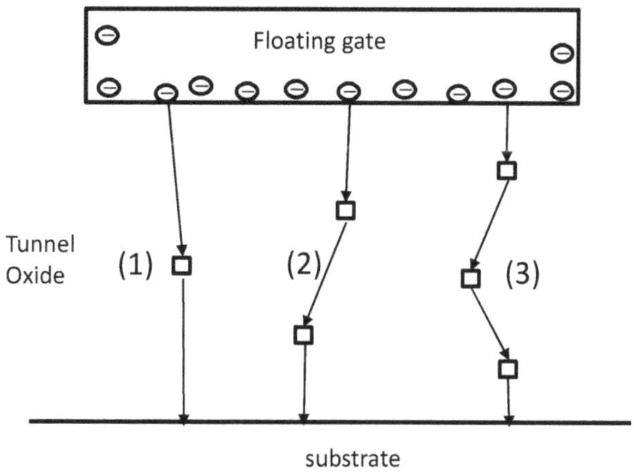

Fig. 8.23 Trap-assisted tunneling from floating gate to substrate. The traps are indicated by the squares

there usually are limits to the number of program and erase cycles, even for thicker oxide with high initial quality.

On the other hand, if the tunnel oxide thickness is thinner than 7 nm, then SILC involving only one trap becomes significant. This is a much more easily reached condition and explains why the oxide thickness limit is about 7 nm for conventional EEPROM.

There are many attempts to reduce SILC, including process and design. In the process front, more robust oxides such as nitrogen or fluorine doping may reduce but not eliminate SILC. In the design front, SILC may be reduced by using slower ramp on the tunneling voltage. Still the 7 nm limit remains for general floating-gate NVM.

One conclusion one can draw from the band diagram of the MOS structure, e.g., Fig. 2.8, is that because of the barrier height differences, the electrons on p+ floating gate has a higher barrier that they need to overcome to leak to the substrate. That is, p+ gate-based floating-gate NVM may provide a better retention than n+ polysilicon floating gate.

From experimental point of view, p+ poly has also been shown to have much lower SILC than n+ poly floating gate (Houtsma et al., 1999). This may be attributed to the observation that SILC is mainly due to traps located at about 2.6 eV from the top of the valence band in the oxide, thus much closer to that of the silicon conduction band edge. It has also been observed that the SILC for negative gate voltage is lower than for positive V_g for p+ floating gate. This point illustrates the importance in choosing the right charge storage level in the floating gate to minimize the effect of SILC.

On the other hand, it is well known that the defect creation in oxide is exponentially dependent on the applied voltage. Since, as shown in Fig. 8.3, it takes about 1 V higher voltage for the p+ poly PMOS device than n+ poly PMOS device, the tunneling from p+ poly floating gate may increase the degradation of the oxide. As such, the endurance performance of a p+ poly floating gate must be carefully monitored. It is expected that the p+ polysilicon gate NVM may have worse program/erase endurance performance (Ma et al., 2007). In general, P+ poly is well suited for moderate cycle count-embedded NVM applications requiring scalability to thinner oxides.

It is possible to do bandgap engineering and design the memory cells that optimize both endurance and data retention (Wang et al., 2007) for pFET-based LNVM cells.

Finally, the unique characteristic of extrinsic retention is that the stress-induced traps are point defects that are randomly located. The failure of one floating gate is uncorrelated to the failure probability of next floating gate, no matter how close or far these two floating-gate transistors are located. One may take advantage of this fact to design redundancy into memory cells or employ error correction codes to enable the use of thin tunneling oxide.

In Sect. 10.3 we will show that it is possible to develop small memories using 5 nm oxide available from the 2.5 V I/O transistors. The key here is the use of built-in redundancy of differential cells, discussed later, that reduce the failure probability to acceptable level for a small memory, e.g., <1 kb.

8.6.2 Program/Erase Endurance

In Chap. 2 we discussed gate oxide reliability. One of the parameters commonly used to measure oxide quality is the charge to breakdown (Q_{bd}). Q_{bd} is measured by stressing the oxide with a constant current. The current density multiplied by the time for the oxide to breakdown is the Q_{bd}.

Given that a tunnel oxide has a finite charge to breakdown, we expect that there will be a limit to the number of program and erase cycles that a floating-gate memory cell can perform before the oxide becomes unusable, e.g., does not hold any charge. The limit can be estimated by the following relation, $N_{\mathrm{limit}} \sim Q_{\mathrm{bd}}/\Delta Q$, where ΔQ is the amount of charge transported through the tunnel oxide during each program and erase cycle. In reality, long before the oxide breaks down as measured by Q_{bd}, the oxide quality will start to be degraded, the SILC appears, and the floating gate will start to lose charge. Therefore, we expect the real endurance is much lower than N_{limit}.

Still it is an instructive exercise to estimate the limit using Q_{bd}. An estimation is presented in Table 8.3.

Thus, there can be an upper limit of about 50 million program and erase cycles before the tunnel oxide breaks down.

This limit far exceeds the best endurance for floating-gate EEPROM of about 1 million cycles. More typically the endurance spec for embedded floating-gate EEPROM is only 1000–100,000.

This is not surprising. One effect that contributes to the reduced endurance is due to charge trap generation in the tunnel oxide. The trapping will be severe enough to affect the programming/erase behavior beyond what the program can handle. With

Table 8.3 Estimate of the program/erase endurance limit

Quantity	Typical value	Unit	Notes
ΔV	2	Volt	Floating-gate voltage change between erased and programmed state
C_{tot}	1×10^{-15}	Farad	Total capacitance of floating gate
$\Delta Q = C_{\mathrm{tot}}\ \Delta V$	2×10^{-15}	Coulomb	Charge transfer per tunnel event
A_{tun}	1×10^{-9}	cm^2	Size of the tunneling junction—a design parameter
Δt	2	millisecond	Average tunnel time = erase time
$\Delta Q/A_{\mathrm{tun}}/\Delta t$	1×10^{-3}	$\mathrm{A/cm}^2$	Average tunnel current density
Q_{bd}	100	$\mathrm{C/cm}^2$	Q_{bd} at $J = 0.01\ \mathrm{A/cm}^2$
$Q_{\mathrm{bd}}/(\Delta Q/A_{\mathrm{tun}})$	5×10^7		Endurance, number of program/erase cycles at oxide breakdown

more endurance cycles, it might take higher and higher programming voltage or longer and longer programming time to reach the original memory window.

Similarly as we noted earlier, oxide breaks down when there is a sizable conducting path (aka percolation path) in the oxide, which will cause more significant current to flow and heat the oxide along the conducting path. The self-heating can soon start a self-destruction process and breaks the oxide. Long before the current starts to be sizable to cause local heating, SILC will appear. Even a minute amount of SILC will cause data retention errors.

8.6.3 Disturb

Disturb refers to unintentionally affecting the storage state of cells in a memory array. There are two types of disturb. Write or program disturb is the disturb of content of unselected cells when writing or programming of selected cells. Read disturb is the disturb of both selected and unselected cells during read process. Both disturb mechanisms become worse with thinner oxide and when more memory cells are packed together.

Both disturb mechanisms are strongly dependent on the array architecture and are best described using examples. We will discuss both disturbs in Sect. 10.3.

Read disturb is becoming an almost universal problems for essentially all memories, including DRAM, SRAM, and flash memory due to the continued scaling and denser packing of cells.

For example in DRAMs, the continued reading of one DRAM row will cause increased error rate in adjacent rows. This is due to the word-line coupling effect where leakage in the word-line transistor can reduce the effectiveness of the isolation between the storage capacitor and the bit-line voltage swings.

For NAND flash memory, as the cells gets smaller and more densely packed, inter-floating-gate coupling becomes significant. Similar to Eq. 5.1, the floating gate voltage is not only a function of the charge on its own floating gate but increasingly dominated by the charges on the neighboring floating gates.

8.6.4 Improving Data Retention Reliability

As discussed in Sect. 8.6.1, due to the use of thinner tunneling oxide, the extrinsic data retention is a serious limitation to the reliability of logic NVM.

One of the methods for improving the reliability of flash memory is the error correction codes (ECC) (Gregori, Carini, & Torelli, 2003). On-chip ECC has gained widespread use in all newer flash memories. However, implementing ECC usually incur high overhead, including both the redundant bits and the coding and decoding circuits. For these reasons embedded NVMs with small bit counts rarely use the traditional ECC techniques.

A more easily adopted method is by adding direct hardware redundancy, especially in the form of differential memory cells (e.g., Lindhorst et al., 2005) in LNVMs. Differential memory cells have many other advantages which will be discussed in more details in Chap. 10. Here we focus on the inherent improvement in data retention. In Fig. 8.24 we illustrate distribution of threshold voltage for differential cells as well as conventional, single-ended, memory cells.

As we know, the digital state stored in a floating-gate memory cell is classified to a "0" or a "1" state by the threshold voltage of the floating-gate transistor. When the threshold voltages of a group of cells are plotted, the distributions can look like the two curves in Fig. 8.24. In a single-ended memory cell, a read threshold divides the "0" from the "1" state. When one of the distributions, either the threshold voltage of a "0" state transistor moves up or a "1" state transistor moves down, encroaches and exceeds the threshold, a data retention failure occurs.

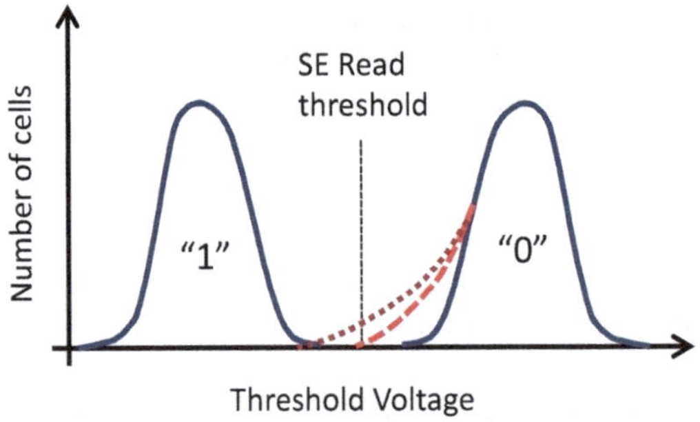

Fig. 8.24 Cell threshold voltage (or cell current) distribution for "0" or "1" state. The SE read threshold is the threshold where single-ended cell may fail when a cell's threshold crosses, as illustrated by the red dashed line. For differential cell, the cells only begin to fail when the threshold voltage of one FG transistor crosses that of the corresponding FG transistor in the same differential cell

In place of a fixed read threshold, the differential cell uses two cells to self-reference each other. Only when the threshold or current of "1" side of the cell exceeds that of the "0" side does a failure occur. As illustrated in Fig. 8.24, the failure modes of a differential cell can be as follows:

1. One side of the cell leaks and the V_t or current moves to cross the other side.
2. Both sides move and cross each other.

For the first failure mode, the memory window essentially doubles that of the single-ended cell. Since the failure probability is strongly dependent on the window size, including exponential dependence, the failure probability of a differential cell will be much reduced from that of the single-ended cell.

The failure probability of the second failure mode can be estimated as the following. Since for a differential cell to fail, both floating gates will have to fail. We may argue that the differential cell failure probability can be written as $f_{\mathrm{diff}} = f_1 \times f_0$, where f_1 and f_0 are the failure rate of the floating gate on the "1" and "0" side. For simplicity assuming the single floating-gate failure probability is $f = f_1 = f_0$, then for an N-bit array, we can calculate its failure probability to be

$$F = 1 - (1 - f_1 f_0)^N$$

Another commonly used reliability metric is *FIT* (failure-in-time) rate. One *FIT* is one device failure in one billion device hours. For a memory with a failure probability of F, the FIT of this memory after 10 years if operation is then,

$$\mathrm{FIT} = \frac{10^9}{10 \times 365 \times 24} F.$$

Figure 8.25 shows the dramatic improvement in FIT for a 16 Kb array with single ended and differential cells (Ma et al., 2008). For a fixed FIT rate, e.g., 10, we can see that differential cell can tolerate almost four orders of magnitude worse single cell failure rate.

Also shown in Fig. 8.25 is the dependence of product failure rate for memories of different size as a function of single bit failure rate. This dependence illustrates the well-known fact that smaller memory is more tolerant of the single bit errors than a larger memory. From this figure, we can see that with $f \sim 10^{-3}$ to 10^{-4}, a 16 Kb memory array may be built with a reliability target of 10 FIT. On the other hand, it is nearly impossible to build a memory with 1 Gb density, even using differential cells, and ECC coding techniques will need to be used for large arrays.

The differential cell method stores a bit in two cells, and hence it has a relatively large area overhead. To offset that overhead, differential cells can be generalized to the rank modulation (RM) scheme (Jiang et al., 2009), where the two cell method is a special case. We will refer the readers to the research papers for the RM implementation.

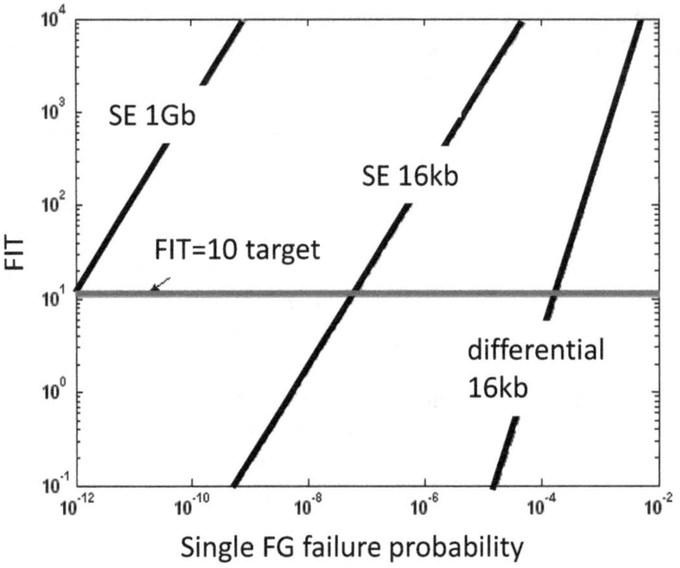

Fig. 8.25 Reliability (in FIT) after 10 years of operation of 16 Kb and 1 Gb arrays vs single FG failure rate for single-ended (SE) and differential cells, assuming a constant gate oxide area for each floating gate

8.7 An Application of LNVM

One typical applications of LNVM is for storing the trimming bits in analog and mixed-signal ICs. The need for trimming capacitor, resistor, and even inductor values are common in precision analog circuits. For example, capacitor trimming is often used in precision oscillators and other timing circuits. Precision resistors are important in digital-to-analog converters (DAC) based on resistor ladders.

NVM can be used to store the trimming values either in digital or analog form. Multiple times of programmability is often necessary as the trimming values may need to be calibrated because of environment change, such as temperature, or as the devices age.

The following figures show a few examples of capacitor and resistor trimming using NVM. Figure 8.26 shows the schematic for a trimmable capacitor. It is composed of an array of capacitors C_i, of equal or unequal values, in series with select transistors. The gates of the select transistors are controlled by NVM bits, from LSB to MSB. The total capacitance is then given by

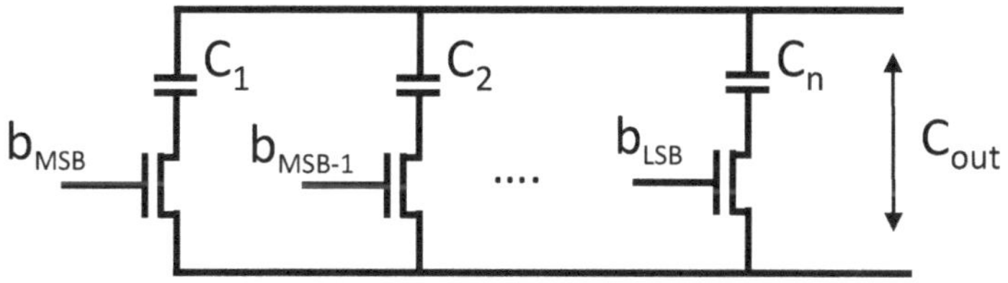

Fig. 8.26 A trimmable capacitor array with select MOSFETs controlled by NVM bits

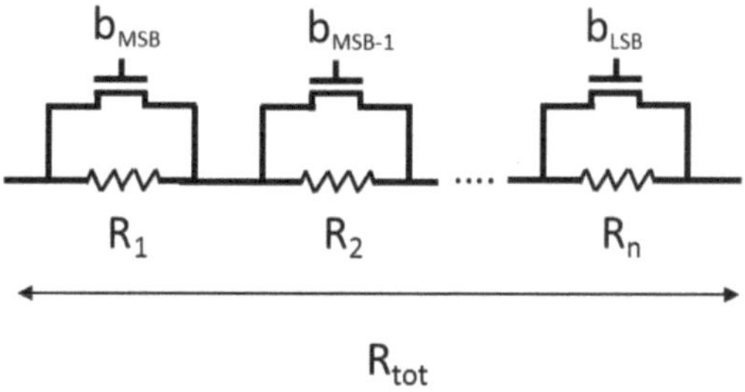

Fig. 8.27 A trimmable resistor with bypass FETs controlled by NVM trim bits

$$C_{\text{out}} = \sum_{i=\text{LSB}}^{\text{MSB}} b_i C_i \tag{8.14}$$

Note that an additional penalty will come from the quality factor (Q) for the capacitor, defined by the energy stored in the capacitor over the energy dissipated by the access transistor over a full cycle.

A trimmable resistor is shown in Fig. 8.27. Assuming that the resistor R_i is much larger than the on-resistance of the FETs, the total resistance is given by

$$R_{\text{tot}} = \sum_{i=\text{LSB}}^{\text{MSB}} (1 - b_i) R_i \tag{8.15}$$

The precision of the earlier circuits is a direct function of the number of bits used. For example, for a four-bit design, LSB = 0, MSB = 3, the value of the capacitor or resistor values would be in the ratio of 8:1, 4:1, 2:1, and 1:1 to a base value, to give 16 output levels.

Series version of the trimmable capacitor or the parallel version of the trimmable resistors can be also constructed. Section 11.2 will discuss the analog and digital trimming circuits using floating-gate transistors.

Note that in general the memory size needed for storing the trim bits is fairly small. For example, for an oscillator chip, a number of bits might be used to center the nominal frequency; a number of bits may be used to store the temperature coefficients to compensate for the temperature drifts. And finally a number of configuration bits may be used to set output, i.e., final frequency and voltage levels.

Typical size may be around 100 bits for an analog chip. For this type of memory density, as discussed in Sect. 8.2 LNVM provides the best cost solution.

The NVM bits can be both OTP and MTP type. In Chap. 9 we show some of the OTP memory while in Chap. 10 we show some MTP memory designs.

Patent Litigations Involving Flash Memory Because of the rapid growth of flash memory and its large commercial market, there are many patent litigations involving flash memory.

Sandisk, due to its early start in this area and leading role in developing multilevel cells (MLC), is the beneficiary of many such litigations. Samsung was a late comer to flash memory. As a result, it had to pay early flash memory developers like Toshiba and Sandisk a lot of royalties to use their flash technology.

For example, according to an *EE Times* report (Ojo, 2002), SanDisk settled a costly legal battle with Samsung Electronics Co. Ltd., in 2002 and renewed a cross-licensing and royalty-generating agreement. They had a previous 5-year contract between SanDisk and Samsung.

Samsung sued SanDisk in 2002 alleging that the company violated four patents held by Samsung. In turn, SanDisk countersued that the Samsung patents, which revolve around NAND flash memory technology, were unenforceable and breached the 5-year contract between the two companies.

At the same time, Sandisk is still locked in litigation with many other memory companies: Memorex Products Inc.; Micron Technology Inc.

According to *Wall Street Journal* (Clark, 2009), "analysts estimated that South Korea's Samsung has been paying hundreds of millions of dollars a year to SanDisk in patent royalties, though the amounts have gone down lately as fierce competition has caused flash-memory prices to plunge. Edwin Mok, an analyst at Needham & Co., puts Samsung's payments at about $350 million in 2008 and $250 million in 2009. Based on the companies' remarks, he estimates that Samsung would pay SanDisk around $125 million a year starting in 2010."

The main dispute appeared to be related to the TLC and QLC technology that store three and four bits, respectively, in one NAND floating-gate transistor.

References

Bernard, F., & Diorio, C. (2006). Patent No. 7145370. USA.

Clark, D. (2009, May 28). *Sandisk, Samsung forge patent deal*. Retrieved from Wall Street Journal http://www.wsj.com/articles/SB124343824782558893

De Vries, A., & Ma, Y. (2007). A logical approach to NVM integration in SOC design. *Electronic Design News*, 73–80.

Dickson, J. (1976). On-chip high-voltage generation in MNOS integrated circuits using an improved voltage multiplier technique. *IEEE Journal of Solid State Circuits, 11*, 374–378.

Gregori, S., Carini, A. K., & Torelli, G. (2003). On-chip error correcting techniques for new-generation flash memories. *Proceedings of the IEEE, 91*(4), 602–616.

Houtsma, V., Holleman, J., Salm, C., de Haan, I. R., Schmitz, J., Widdershoven, F., et al. (1999). Minority carrier tunneling and stress-induced leakage current for p+ gate MOS capacitors with poly-Si and poly-SiGe gate material. In *Technical Digest of IEDM* (pp. 457–460). IEEE.
Hsu, C., Lin, Y., Yang, E., & Shen, R. (2014). *Logic non-volatile memory: The NVM solutions from eMemory*. Singapore: World Scientific.
Jiang, A., Mateescu, R., Schwartz, M., & Bruck, J. (2009). Rank modulation for flash memories. *IEEE Transactions on Information Theory, 55*(6), 2659–2673.
Lee, W., King, T., & Hu, C. (1999). Evidence of hole direct tunneling current through ultrathin gate oxide using P+ poly-SiGe gate. *IEEE Electron Device Letters, 20*, 268–270.
Liao, Y., Tsai, L., Leu, L., Lee, Y.-H., Wang, W., & Wu, K. (2013). Investigation of data retention window closure on logic embedded non-volatile memory. In *2013 I.E. International Reliability Physics Symposium (IRPS)* (pp. MY.7.1–7.4). IEEE.
Lindhorst, C. A., Diorio, C. J., Gilliland, T. N., Pesavento, A., Srinivas, S., Ma, Y., et al. (2005). *Differential floating gate nonvolatile memories*. US Patent 6,950,342.
Ma, Y., Deng, R., Nguyen, H., Wang, B., Pesavento, A., Niset, M., et al. (2008). Floating-gate nonvolatile memory with ultrathin 5-nm tunnel oxide. *IEEE Transactions on Electron Devices, 55*, 3476–3481.
Ma, Y., Deng, R., Wang, B., Horch, A., & Paulsen, R. (2007). *Data retention reliability of p+ poly floating gate memories in logic CMOS processes*. Cannes, France: ICMTD.
Ma, Y., Gilliland, T., Wang, B., Paulsen, R. P., Wang, C., Nguyen, H., et al. (2004). Reliability of pFET EEPROM with 70-A tunnel oxide manufactured in generic logic CMOS processes. *IEEE Transactions on Device and Materials Reliability, 4*, 353–356.
Moazzami, R., & Hu, C. (1992). Stress-induced current in thin silicon dioxide films. In *Technical Digest, International Electron Devices Meeting* (pp. 139–142). San Francisco: IEEE.
Na, K.-Y., & Kim, Y.-S. (2006). High-performance single polysilicon EEPROM with stacked MIM capacitor. *IEEE Electron Device Letters, 27*, 294–296.
Na, K.-Y., Kim, Y.-S., & Kim, Y.-S. (2007). Novel single polysilicon EEPROM cell with a polyfinger capacitor. *IEEE Electron Device Letters, 28*, 1047–1049.
Ojo, B. (2002, August 22). *SanDisk, Samsung settle legal battle*. Retrieved from EE Times http://www.eetimes.com/document.asp?doc_id=1133551
Olivo, P., Nguyen, T., & Ricco, B. (1988). High-field-induced degradation in ultra-thin SiO_2 films. *IEEE Transactions on Electron Devices, 35*(12), 2259–2267.
Park, S., Choi, K., Kim, N., Kimg, J., Kwon, Y., Ko, K., et al. (2015). Single-Poly Embedded NVM Solution for Analog Trimming and Code Storage Applications. Proceedings of IEEE International Memory Workshop. Monterey: IEEE..
Pesavento, A. (2009). *Reduced area high voltage switch for NVM*. US Patent 7,580,311.
Pesavento, A., & Hyde, J. (2013). *PFET nonvolatile memory*. US Patent 8,416,630.
Taur, Y., & Ning, T. H. (1998). *Fundamentals of modern VLSI devices*. Cambridge: Cambridge University Press.
Wang, B., & Ma, Y. (2008). Opportunities and challenges in multi-times-programmable floating-gate logic non-volatile memories. In *2008 International Memory Workshop* (pp. 120–125). IEEE.
Wang, B., Ma, Y., Horch, A., & Paulsen, R. (2007). Reliability of a 90 nm embedded multi-time programmable logic NVM cells using work-function engineered tunneling device. In *45th Annual IEEE International Reliability Physics Symposium* (pp. 662–663). IEEE.
Wang, B., Ma, Y., Paulsen, R., Diorio, C., & Humes, T. (2005). Measurement of ultralow gate tunneling currents using floating-gate integrator technique. *IEEE Electron Device Letters, 26* (5), 329–331.
Wang, B., Nguyen, H., Ma, Y., & Paulsen, R. (2007). Highly reliable 90-nm logic multitime programmable NVM cells using novel work-function-engineered tunneling devices. *IEEE Transactions on Electron Devices, 54*, 2526–2530.

Chapter 9
One-Time Programmable Memories in Logic Processes

Abstract In this chapter, we focus on the One-Time Programmable (OTP) embedded NVM using basic logic CMOS processes. We review poly fuse, antifuse, and floating-gate-based OTP memory cell and arrays.

Keywords Nonvolatile memory • ROM, OTP • One-time programmable • MTP • Multiple-times programmable • Logic nonvolatile memory • Fuse • Antifuse • Reliability • Data retention

9.1 Introduction to OTP Memory

Read only memory (ROM) is used in many applications. ROM is essentially a one-time programmable (OTP) memory that can be programmed either during manufacturing or in the field only once. The original floating-gate memory, EPROM, invented by Kahng and Sze is one type of OTP that can be programmed in the field.

In the era of EEPROM and flash memory that can be programmed up to one million times, it is interesting to note that OTP continues to find many applications. The main reason for using OTP memory is its low cost (think DVD vs. hard drive) and for many applications, one time programmability is sufficient.

In standard logic CMOS processes, there are three types of memory element for OTP memory: floating-gate transistor, fuse, and antifuse.

We have already discussed the use of floating-gate transistor for memory and will discuss it again in the context of OTP in Sect. 9.4. Fuses in ICs can be made of metal or polysilicon. Metal fuses that can be cut by laser have been widely used for, e.g., redundancy repair of DRAMs. Electrically programmable polysilicon fuses are more recent and have found wide uses for chip ID, redundancy, trimming, and calibration.

Antifuse, as the name suggests, is the opposite of a fuse. In normal conditions, antifuse is in an "open" or high resistance state. In the programmed state, antifuse turns into a low resistance state. As is the case of fuse, antifuse can only be programmable once and can be used to build OTP memory. Antifuses have been used in many of the same applications as fuses, such as chip ID, as configuration bits for analog ICs and for redundancy repair in DRAMs, and in FPGAs.

Y. Ma, E. Kan, *Non-logic Devices in Logic Processes*,
DOI 10.1007/978-3-319-48339-9_9

The recent development of two-terminal resistive RAM (ReRAM) can be considered as fuse or antifuse with multitimes program and erase capabilities. Special metal oxide and contact metal materials are required for the reversible fusing operation and are not yet available in standard CMOS processes. We will not treat ReRAM in this book.

9.2 Polysilicon Fuse

Electrically programmable polysilicon fuse, or poly eFuse, was developed by IBM in the early 2000s (Tonti et al., 2003). The big improvement of poly fuses over laser-cut metal fuses is its field programmability. Poly fuses have been used for storing the chip identification and part numbers (Rohrer et al., 2005).

Poly fuse uses silicided polysilicon as the fuse and the mechanism of electromigration for programming. By passing high current through a silicided polysilicon fuse structure, the metal in the silicide will migrate and cause break in the poly silicon line. This dramatically increases the resistance across the fuse which can be detected with a sensing circuit.

Figure 9.1 shows a circuit for an OTP memory cell based on a poly fuse that can be arranged in row and column structure (e.g. Kim et al., 2011). Separate read word line (RWL) and write word line (WWL) transistors are used as well as separate bit line (BL) for read and program line (PL) for programming.

Table 9.1 presents the operations of this memory cell. Note that usually high current, but not high voltage, is needed to cause electromigration and to program a poly fuse. As such the programming voltage V_{pp} may be the core or the I/O voltage. But high current, ~100 μA to 10 mA, is needed to cause electromigration and program the fuse in 100 μs or less.

The mechanism of programming can be estimated by the following method. Electromigration follows the so-called Black equation,

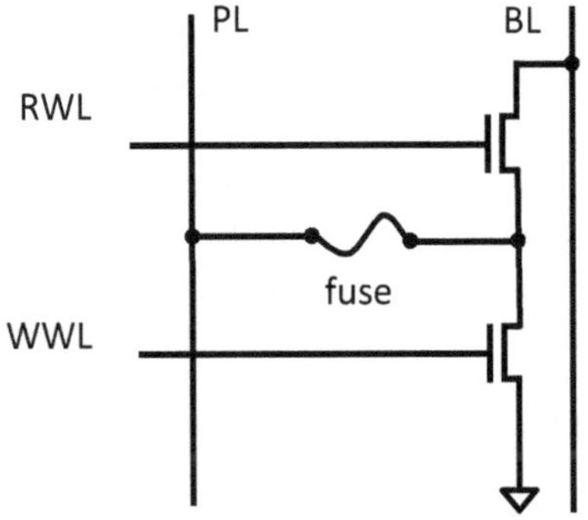

Fig. 9.1 An OTP memory cell based on a fuse

Table 9.1 Operation condition for the fuse OTP cell

	WWL	RWL	BL	PL
Program	V_{dd}	0	Float	V_{pp}
Read	0	V_{dd}	Sense amp	0

$$\mathrm{MTTF} = AJ^{-n}\mathrm{e}^{-E_a/kT} \tag{9.1}$$

where MTTF is mean time to failure, J is the current density, T is temperature, E_a is an activation energy of about 0.9 eV, A is a constant, k is Boltzmann's constant, and n is a model parameter that ranges from 1 to 2.

Assuming a poly cross section of about 1 μm^2, and a current of 10 mA, and temperature from self-heating of around 800 °C, MTTF is on the order of 1 μs. To guarantee the blow of the fuse, we may need to extend the time to 10 μs.

Poly fuse is attractive since it uses only polysilicon and, since no high voltage devices are needed, can be made in logic CMOS process. Many foundries offer poly fuse in their most advanced processes. The overall area of poly fuse is large due to the need of large transistors to drive high current and additional area to isolate other circuits from the heat generated during programming. Thus, poly fuse is used mostly in low density applications.

In the more advanced processes, polysilicon gate has been replaced by metal or silicide gate. Therefore, polysilicon is no longer available as a fuse element. Still, it is possible to use metal gate as the fuse in a manufacturable process as demonstrated by (Kulkarni et al., 2015). The cell uses one pMOS transistor and one metal fuse resistor (1T1R), with a cell area of 16.4 μm^2 in a 22 nm metal gate tri-gate process with high-k gate dielectrics. The fuse element is a metal1–via1–metal2 stack and again uses electromigration mechanism to program. This is a big leap over the silicided polysilicon fuse as the metal fuse can now be stacked on top of the select or access transistors. In addition, the heat generated is better isolated than polysilicon fuse, because poly is closer to the silicon substrate which is a good thermal conductor and hence is harder to isolate. The new metal fuse therefore is more area efficient as well as providing the possibility for further scaling in 3D.

Using short bit lines to reduce IR drop, a low programming voltage V_{pp} of 1.6 V is enough to reliably blow the fuses of an 80b (5 rows × 16 columns) array. Because of the low programming voltage, a voltage doubler working on a 1 V supply rail can be designed on chip and used to program the fuses in 50 μs. The fuse can be arranged in an array shown in Fig. 9.2.

Poly or metal fuses have some drawbacks. The main problem is that high current is needed to blow the fuse, usually on the order of 10 mA. Thus, the poly fuses usually have to be programmable sequentially bit by bit, resulting in a slow programming speed. High current means the power consumption is high during programming. Finally, even though the fuse itself is small in area, the associated select transistors and programming circuits have to be large to be able to drive the large current. Furthermore, because of the high temperature during programming, isolation area around the fuse is also necessary, further to enlarge the cell size.

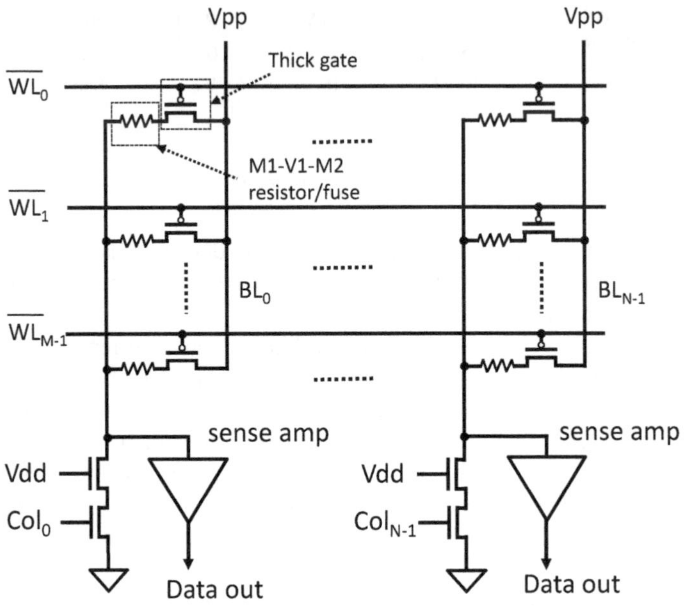

Fig. 9.2 A fuse array made of 1T1R cells. The metal fuse uses the metal1–via1–metal2 stack

In the following section, the antifuse-based OTP is discussed that usually required much less current and therefore more power efficient.

9.3 Antifuse-Based OTP

An antifuse is a capacitor before programming and a resistor after programming. There are many ways to implement antifuse in logic CMOS processes. The most common is to use a MOSCAP. The unprogrammed state is the normal MOSCAP with the gate oxide intact, providing the insulating or high resistance path between the two nodes of the MOSCAP. The antifuse can be programmed by breaking the gate oxide, whereas the two nodes are shorted. Antifuse can also be implemented using a metal–insulator–metal (MIM) structure. However, in a pure CMOS process, no suitable MIM structure exists that can be broken with reasonable low voltage, i.e., less than 10 V. Such a structure may exist in mixed-signal process where thin-oxide MIM capacitors are available.

As the programmed antifuse is always on, there is a need for a select transistor in an antifuse memory cell. So a basic antifuse cell is usually a 1T1C device. Since MOSCAP and MOSFET are often used interchangeably in this context, this type of antifuse cells may also be called a 2T cell.

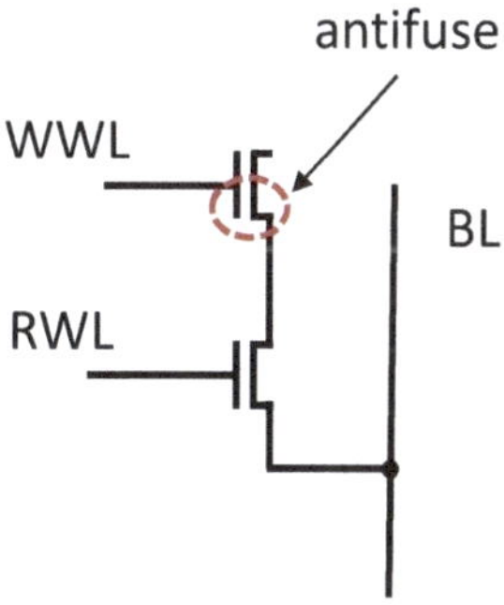

Fig. 9.3 An antifuse cell based on one transistor and one antifuse

Table 9.2 Operation condition for the antifuse based OTP cell of Fig. 9.3

	WWL	RWL	BL
Program	V_{pp}	V_{dd}	0
Read	0	V_{dd}	Sense

A 2T antifuse cell schematic is shown in Fig. 9.3. The gate of the select transistor is connected to the read word-line RWL and the gate of the antifuse is connected to the write word line (WWL). The operations of the cell are listed in Table 9.2.

To program the antifuse, WWL is brought to a high voltage, V_{pp}. The BL is set at a low voltage. When the RWL is enabled, there will be a high voltage across the gate to source in the antifuse transistor and this may break the gate oxide. To read, BL is charged, then WWL is set to ground, and RWL is enabled. If the antifuse is programmed, BL will start to discharge through the broken oxide through WWL and be detected by the sense amplifier connected to the BL.

Note that the program transistor (WWL) only needs to be a half MOSFET, with no use for the drain, which is most often left without an external contact. So the memory cell is basically a 1.5T cell and can be very small.

When V_{pp} is at WWL, once the antifuse has broken down, V_{pp} will be at the drain of the RWL transistor. To ensure that the RWL transistor does not break during programming, the cell may use transistors with different gate oxides, for example, a thin FET as the antifuse and a thick FET as the select transistor (Wang & Yu, 2003). The thick oxide transistor can handle the high voltage that is needed to break the gate oxide of the thin FET. This transistor can be implemented with the I/O transistor in a CMOS process. The thin oxide can be implemented with the core transistor in the same process. For example, for the 65 nm process, the core oxide thickness is about 1.2 nm with breakdown voltage of about 3 V.

To reduce the cell area, Kurjanowicz (2008) combined the two transistors in a 2T antifuse cell into a single "split channel" transistor where, the channel has one part with thin gate oxide and the other part with thick gate oxide, as shown in Fig. 9.4. The fabrication of this particular cell was discussed in Sect. 3.1.2.

It appears that similar concept was proposed in the patent application of Peng et al. (2006) but was not in the claims. This simple omission may have caused a

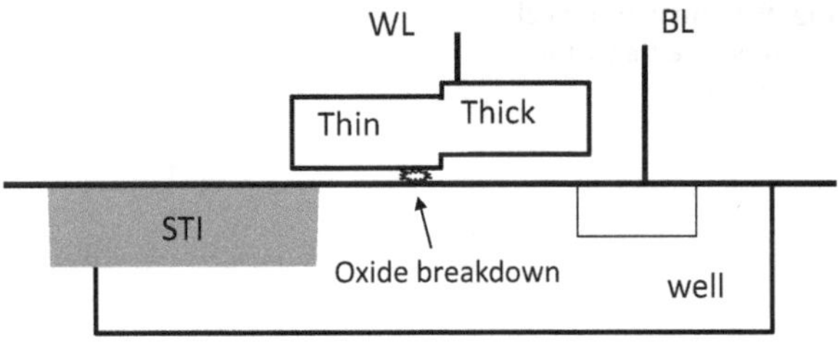

Fig. 9.4 Sidense 1T hybrid cell

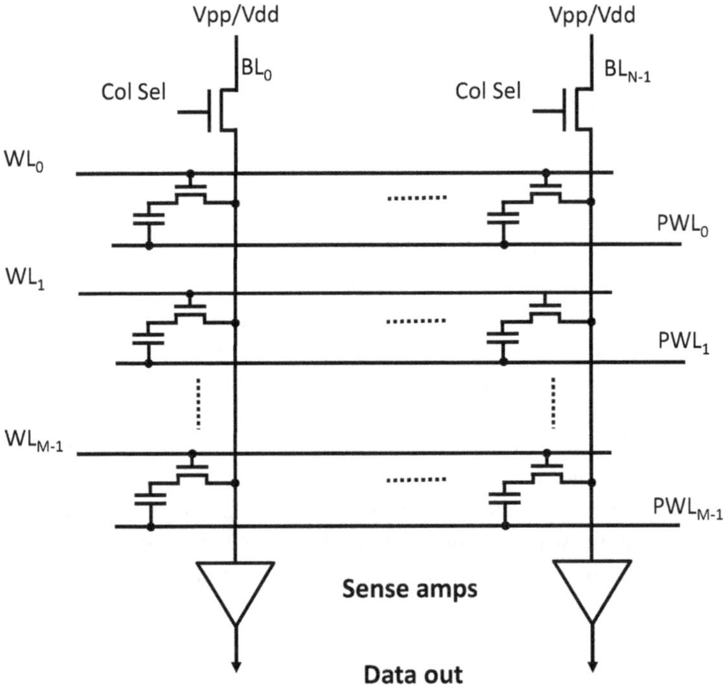

Fig. 9.5 An M row by N (column) 1T1C OTP array

10-year litigation fight between two companies (see the notes at the end of this chapter) with millions dollars of fees paid to the attorneys on both sides.

Using the 2T memory cell, a commercial OTP memory array is shown in Fig. 9.5 (Peng et al., 2006). Word line (WL), programming word line (PWL), and bit line (BL) are the signals in the row and column structure.

The programming and read operations are as follows. During programming of a cell on the selected row and selected column, $V_{pp}/2$ is applied to the selected WL and V_{pp} is applied to the selected PWL. BL is kept at 0 or V_{dd}, depending on the intended data. This will break the thin gate oxide of the capacitor. Since the WL transistors use thick gate oxide they will remain intact.

To avoid disturbing other unselected cells, unselected WL and PWL lines are kept at ground while unselected BL are biased at $V_{pp}/2$ or V_{dd}.

Fifty microsecond pulses are typically needed to program the antifuses. For a 1 Mb array at 50 μs per byte, the total programming time is less than 10 s, which is slow for most applications. Fortunately since this is an OTP device, the programming can be done automatically during testing. To ensure that the oxide can be broken, the cell programming current needs to be larger than several tens of μA.

The 1T split channel cell can be arranged in a crossbar array with the cell at the intersection of the word line and bit line. The drain is left unconnected. This results in a very compact array, with corresponding smaller peripheral circuits. The trade-off is that the hybrid transistor is not a standard CMOS transistor and not a self-aligned device. The boundary between the thin and the thick FET region needs to be well controlled.

One such crossbar array is shown in Fig. 9.6 (Kurjanowicz & Smith, 2012). This array only has one set of bit lines and one set of word lines.

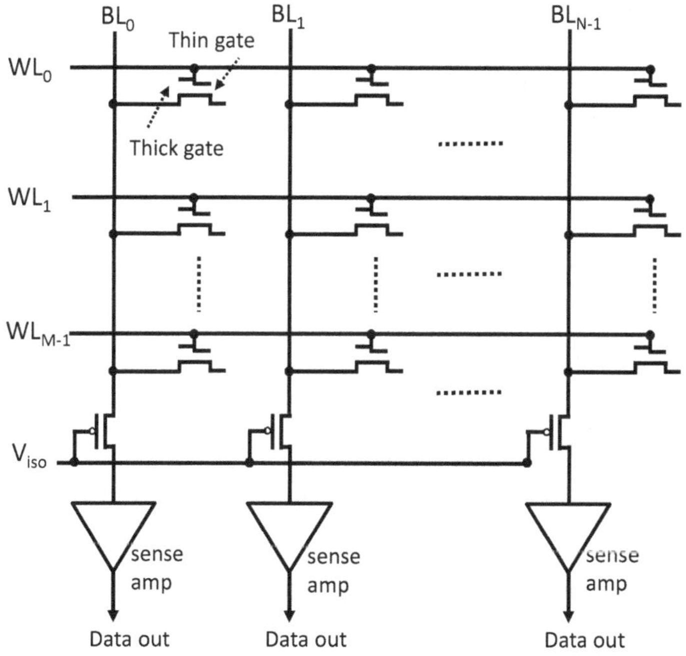

Fig. 9.6 Cross point OTP array based on split channel antifuse cell

Table 9.3 Operations of OTP memory array of Fig. 9.6

	WL	BL	Unselected WL	Unselected BL
Program	$-V_{pp}$	V_{dd} for “1” 0 for “0”	Float	Float
Read	0	V_{dd}	V_{dd}	Float

The operation condition of this memory array is presented in Table 9.3. During programming, a WL is brought to $-V_{pp}$, for the antifuse that is to be programmed the corresponding BL is bring to V_{dd}. This creates voltage drop of $V_{dd}+V_{pp}$ across the oxide and causes the oxide to rupture. The unselected BL is allowed to float, resulting in the unselected cells to have a much lower voltage, i.e., not to be broken.

During read, the BL is charged to V_{dd}. WL is first charged to V_{dd} and then switch to ground, which will allow programmed cell to discharge the BL if the oxide has been broken.

One complication of the cross point-connected array is that once an antifuse is programmed, the corresponding word line and bit line are shorted. An HV isolation transistor, enabled by the signal V_{iso}, is inserted to isolate the BL to the sense amp. In a typical array, several bit lines may share a single sense amplifier. In that case, additional column select transistors can be deployed.

The previous antifuse OTP memories mostly use n-type select FETs and capacitors. The select transistor can also be a pFET. The drawback is that pFET will have its own n-well and this can result in a larger cell. However, since a whole row of cells may share the same n-well, the penalty is not very high. For more advanced process the thin FETs have high-k gate dielectrics. An antifuse array with pFET select transistor and high k gate dielectric as the antifuse has been demonstrated by Kulkarni et al. (2012). Their cell is a 1T1C structure using a 1.8 V I/O pMOS transistor as the select transistor. The antifuse capacitor is an inversion n-MOSCAP structure discussed in Chap. 3, with N+-metal gate and a high-k dielectrics in a n-well. The array is shown in Fig. 9.7 and the operating conditions are summarized in Table 9.4.

The cell area is only about 1 μm^2 in a 32 nm metal gate process with high k gate dielectrics. The programming voltage is about 4.5 V with the programming accomplished in about 200 μs. The n-well surrounding the capacitor allows the containment of the postbreakdown current, which is routed to the peripheral programming and sensing circuits.

For this antifuse cell design, it was reported that breakdown at the LDD region can have 20× lower resistance than breakdown in the channel region (Kulkarni et al., 2012). This nonuniform breakdown and resultant resistance distribution have to be taken into consideration when designing sensing circuits.

Cascode transistors are extensively used in the peripheral circuits, including the sense amplifiers, level shifters, and pulldown circuits so they can sustain the high voltage during programming.

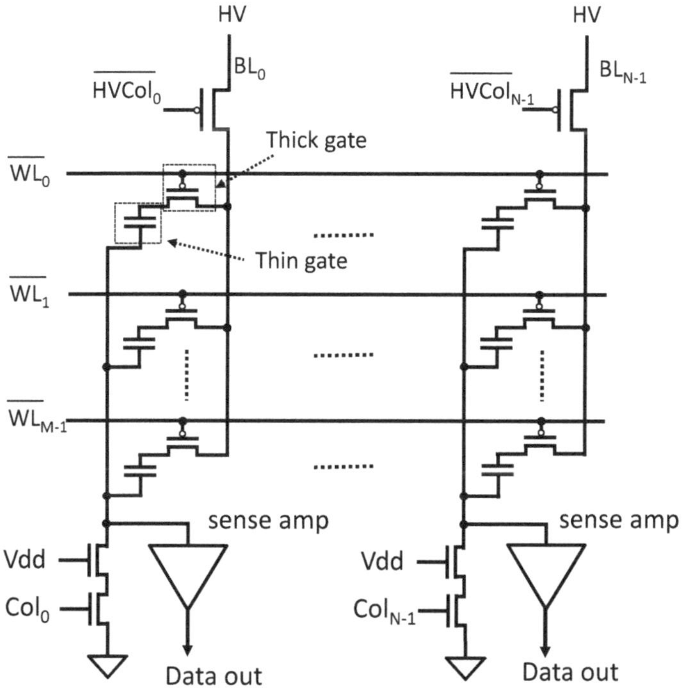

Fig. 9.7 An OTP memory array using 1T1C (pFET select transistor and n-MOSCAP in n-well) antifuse cells

Table 9.4 Operations of memory array of Fig. 9.7

	HV	Col	HVCol_bar	Unselected HVCol_bar	WL_bar	Unselected WL_bar	V_{dd}
Program	4.5	1.05	0	4.5	0	4.5	1.05
Sense	1.5	1.05	1.5	–	0	1.5	1.05

9.4 Floating-Gate OTP Memory

Another way to design an OTP memory is to use the floating-gate transistor we discussed in Chaps. 5 and 8. This practice dated back to the original PROM as UV erased PROM is the first floating-gate OTP memory, and it is still being used quite extensively.

A more recent implementation in logic CMOS process is the pFET-based OTP memory called NeoBit (Lee et al., 2006). NeoBit is composed of two p-MOSFET connected in serial, one as a floating-gate transistor, and the other as a select transistor. The schematic for the cell is the same as that in Fig. 5.3 and the cell

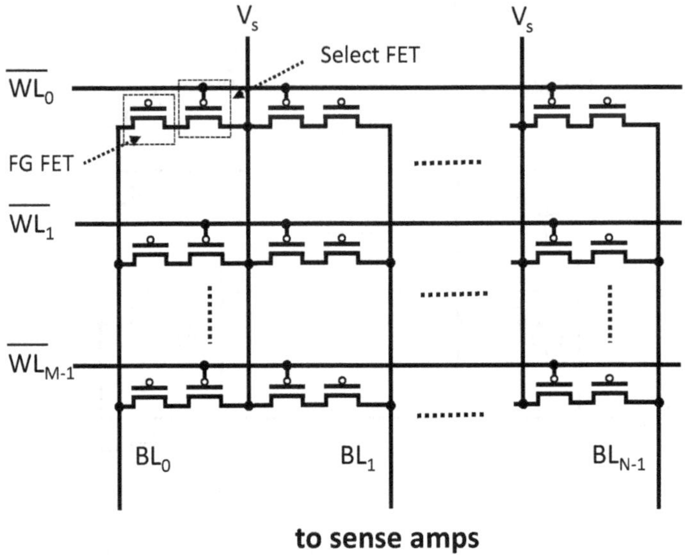

Fig. 9.8 An OTP memory array based on 2T pFET cell

Table 9.5 Operation condition for an UV erase EPROM cell

	WL	V_s	BL	NW
Erase	UV			
Program	0	V_{pp}	0	V_{pp}
Read	0	V_{dd}	Sense	V_{dd}

array is shown in Fig. 9.8. The source of the select transistor is biased to the source line voltage V_s. The drain of the floating-gate PMOS transistor is connected to the bit line voltage. The n-well of these two PMOS transistors is biased to an n-well voltage. Note that all pFET transistors in the memory array can be placed in one n-well. Thus, this OTP memory can be packed fairly densely.

Because of the data retention requirement, the floating-gate transistor needs to be a 3.3 V transistor with 7 nm oxide. The select transistor could be a core transistor but it may still need to sustain operation at V_{pp} of up to 6 V. For this reason and for simplicity in layout, both transistors may be implemented using the p-type I/O transistor.

The operating conditions are presented in Table 9.5.

The floating-gate transistor can be erased coming out of the fab by a UV erase step. NeoBit uses channel hot hole-induced HCI into the floating gate (CHEI) to program the bit cell, which is discussed in Sect. 2.4.4. Recall that the peak gate injection current is, as shown in Fig. 2.23, $V_g \sim 1/3$ to $½V_{ds}$. This condition can be achieved by self-bias for pFET as illustrated later.

Under the erased condition, there is no charge on the floating gate. The pFET is at the cutoff region when both the n-well and the source are biased at V_{dd} and the drain is biased to 0 V. This can be understood from the capacitances of the pFET discussed in Sect. 2.4.2. In this situation,

$$C_{gb} = WLC_{ox}$$
$$C_{gd} = C_{gdo} = C_{ov}$$
$$C_{gs} = C_{gso} = C_{ov}$$
$$C_{tot} = WLC_{ox} + 2C_{ov}$$

The parasitic capacitances, C_{gov}, C_{gdo}, and C_{gso}, are typically small but can cumulatively account for ~10–20 % of the total capacitance. The coupling ratio of gate to well can then be as high as 0.9. Thus, the floating-gate voltage, according to Eq. 9.2,

$$V_{fg} = -\frac{Q_{fg}}{C_{tot}} + \alpha_s V_s + \alpha_d V_d + \alpha_{cg} V_{cg} + \alpha_w V_w \ldots = -\frac{Q_{fg}}{C_{tot}} + \sum \alpha_i V_i \qquad (9.2)$$

is close to V_s and V_{nw}, or 0.9 V_{dd}. Since $V_{sg} \sim 0.1 \times V_{dd} < V_{tp}$, for $V_{dd} = 3$ V, the pFET is off.

During programming, the source and n-well are biased at a higher voltage, $V_{pp} \sim 2V_{dd}$. Under this condition, V_{sg} may be close to $0.2 \times V_{dd}$ or 0.6 V and the transistor will be biased in the saturation region.

The self-biasing and self-limiting characteristics of pFET are what makes floating-gate pFET-based OTP memory possible. As electrons are injected into the floating gate, the threshold voltage is reduced. Electrons stored in the floating gate turn on the channel of the p-type FG-MOSFET, increasing the channel current. More electrons are then injected onto the floating gate, enough to overcome the increase in the negative gate voltage.

For n-FET, the gate injection current will be too small as the peak gate current is approximate $V_{gs} \sim V_{ds}$. In addition, as the electron injection initiates, the stored electrons increase the threshold voltage of the MOSFET and further reduce the channel current. The MOSFET is gradually turned off and stops the injection of more electrons on to the floating gate. Therefore, pFET in Fig. 9.8 is a more frequent choice.

Fuse and antifuse OTP have many advantages over floating-gate-based OTP memory. The main attraction is that both technologies can be more easily adapted to new process technology, while the floating-gate-based OTP memory needs the 3.3 V transistor and feasible pFET self-biasing by the capacitive ratios. They generally have good manufacturability for mass production and yield and can meet the design for manufacturing (DFM) rules. However, they usually consumer a lot more current and can be slow to program.

On the other hand, floating-gate-based memory can be UV erased and also easily adapted to make MTP memories with electrical erase, as we will discuss in the next chapter.

A Patent Litigation Case in Antifuse OTP Technology Patent wars are a common place ever since the patent laws are in practice for more than 100 years. More recent patent litigations are fought over billions of dollars in damages between highly visible companies, like Apple vs Samsung.

An interesting battle between two little known companies, Kilopass and Sidense, over a relatively obscure antifuse technology is an interesting case study on many fronts[1]:

Kilopass is a California-based semiconductor company. Sidense is a Canadian company that is competing with Kilopass in the embedded nonvolatile memory (NVM) market. Kilopass has a longer history and holds many earlier patents utilizing antifuse for OTP memory.

In May 2010, Kilopass initiated a patent infringement suit against Sidense, alleging that Sidense infringed three of Kilopass patents related to programmable memory cells. Claim 1 of the representative patent, US 6,940,751 (filed in January 2004 and granted in September 2005), is one of the most contentious and listed as follows:

> "1. A programmable memory cell useful in a memory array having column bitlines and row wordlines, the memory cell comprising:
> a transistor having a gate,
> a gate dielectric between the gate and over a substrate,
> and first and second doped semiconductor regions formed in said substrate adjacent said gate and in a spaced apart relationship to define a channel region there between and under said gate;
> and where in the second doped semiconductor region of the transistor is connected to one of said row wordlines."

In and around 2005, Kilopass learned through patent publication that Sidense has a published patent application on a 1-T antifuse memory cell that looked similar to Kilopass patent. So Kilopass engaged a patent law firm to assess whether technology described in the Sidense application infringed Kilopass' patents. Kilopass was advised that, if Sidense's actual 1T-Fuse product was the same as that described in the international patent application, Sidense's product did in fact infringe Kilopass's patents. However, if that assumption was wrong and Sidense had eliminated any particular claimed element, most notably, the first doped region, and replaced it with a shallow trench isolation region, then Sidense 'would not infringe [the] claims literally.'

Indeed, Kilopass later learned that the Sidense 'area of interest' had indeed been replaced, and that the redesign had avoided infringement of Kilopass's patents or 'at least make our case much tougher.' Another firm retained by Kilopass confirmed the earlier advice it had received—although Kilopass had reasonable arguments that Sidense's field oxide region was equivalent to the doped region in Claim 1, arguing literal infringement 'may be difficult.'

[1] *Kilopass Technology, Inc v Sidense Corporation* No 2013-1193 (Fed Cir Dec 26, 2013), Court of Appeals for the Federal Circuit, USA. See also, http://jiplp.blogspot.com/2014/03/patent-litigants-beware-awards-of.html.

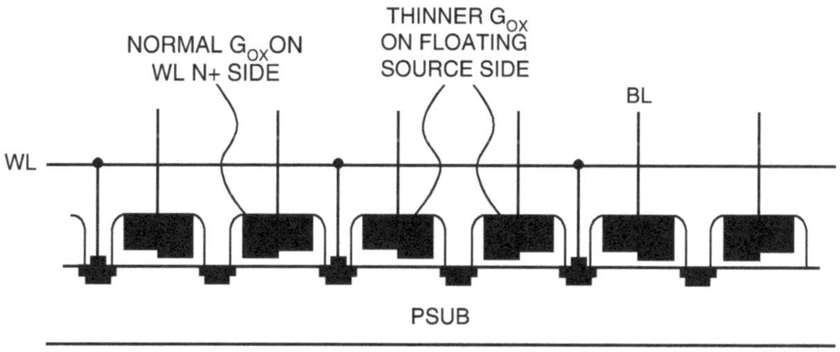

Fig. 9.9 Figure 16 of the Kilopass patent 6,940,751 showing the 1.5T thick/thin oxide hybrid OTP cell (from figure 16 of US Patent #6,940,751)

At trial in the US District Court in Northern California, the court found that Sidense's product did not infringe any of Kilopass's patents. The court's decision was based primarily on the difference in the 'area of interest' and the claimed region. The decision was affirmed by the US Court of Appeals for the Federal Circuits (CAFC) which handles patent appeals for all patent cases.

The difference can be seen Fig. 9.9 where Kilopass has discussed the 1.5T cell of Fig. 9.4. Note that in Kilopass' implementation, there is a doped region in the floating source side, while Sidense (Fig. 9.4) does not have the doped source but sits on the oxide of a STI region.

So it is worth noting the importance to claim all the possible implementations of an idea and use broad terms in patent applications. According to one of the Kilopass inventors, they actually thought about the Sidense approach but forgot to cover this situation in the patent application. Indeed, according to the court document (Court of Appeals for the Federal Circuit, Dec 26, 2013),

> "In 2005, Kilopass's founder and an inventor on all three of Kilopass's patents, Jack Peng, reviewed an international patent application submitted by Sidense that was directed to protecting Sidense's competing 1T-Fuse memory cell. Peng believed that the 1T-Fuse was similar to Kilopass's patented cells, except that Sidense used a split gate implementation. Peng contacted a patent attorney at the law firm Perkins Coie to discuss potential infringement. In an email to the Perkins attorney, Peng explained that '[Kilopass] did not file [a] dedicated patent for this split gate implementation' and that 'we should [have] . . . a long time ago even though we were very busy.' According to Peng, it was not a priority to Kilopass at that time because Sidense's 'split gate [memory cell] is not self-aligned, so their practical cell size will be larger than [Kilopass's] 1.5T cell.'
>
> To add insult to injury, following the District Court's ruling, Sidense filed a motion in the District Court seeking an award of over $4.2 million in attorneys' fees." (Court of Appeals for the Federal Circuit, Dec 26, 2013).

The ruling was partly based on some unfavorable behavior and document trails on Kilopass' part.

1. One is "Kilopass's 'gamesmanship'" in the proceedings. The lower court had found that Kilopass had made contrary construction arguments to the USPTO's Board of Patent Appeals and Interferences.
2. The court also criticized Kilopass for attempting to advance previously undisclosed arguments on the doctrine of equivalents without the court's permission. This, the court held, was 'particularly inappropriate in light of evidence that Kilopass has known for many years that Sidense does not literally infringe its patents.'

(Court of Appeals for the Federal Circuit, Dec 26, 2013).

In the end, the court ordered Kilopass to pay for Sidense' attorney's fee of over $4 M.

References

Court of Appeals for the Federal Circuit. (2013, December 26). *Kilopass Technology, Inc v Sidense Corporation No 2013-1193*. Washington, DC: Court of Appeals for the Federal Circuit.

Kim, J., Kim, D.-H., Jin, L., Ha, P.-B., & Kim, Y.-H. (2011). Design of 1-Kb eFuse OTP memory IP with reliability considered. *Journal of Semiconductor Technology and Science, 11*, 88–94.

Kulkarni, S., Chen, Z., Srinivasan, B., Pedersen, B., Bhattacharya, U., & Zhang, K. (2015). Low-voltage metal-fuse technology featuring a 1.6V-programmable 1T1R bit cell with an integrated 1V charge pump in 22nm tri-gate process. In *VLSI Technology Symposium* (pp. 174–175). IEEE.

Kulkarni, S., Pae, S., Chen, Z., Hafez, W., Pedersen, B., Rahman, A., et al. (2012). A 32nm high-k and metal-gate anti-fuse array featuring a 1.01μm^2 1T1C bit cell. In *VLSI Technology Symposium* (pp. 79–80). IEEE.

Kurjanowicz, W. (2008). *Split-channel antifuse array architecture*. US Patent 7,402,855.

Kurjanowicz, W., & Smith, S. (2012). *High speed OTP sensing scheme*. US Patent 8,130,532.

Lee, H., Woo, S., Chen, H., Shen, R., Wang, C., Hsia, L., et al. (2006). NeoFlash—True logic single poly flash memory technology. In *IEEE Non-volatile Semiconductor Memory Workshop* (pp. 15–16). Monterey, CA: IEEE.

Peng, J., Rosendale, G., Fliesler, M., Fong, D., Wang, J., Ng, C., et al. (2006). A novel embedded OTP NVM using standard foundry CMOS logic technology. In *IEEE Non-volatile Semiconductor Memory Workshop* (pp. 24–26). Monterey: IEEE.

Rohrer, N., Lichtenau, C., Sandon, P. A., Kartschoke, P., Cohen, E., Canada, M. G., et al. (2005). A 64-bit microprocessor in 130-nm and 90-nm technologies with power management features. *IEEE Journal of Solid State Circuits, 40*, 19–27.

Tonti, W., Fifield, J., Higgins, J., Guthrie, W., Berry, W., & Narayan, C. (2003). *2003 I.E. International Integrated Reliability Workshop Final Report* (pp. 36–40). IEEE.

Wang, S.-Y., & Yu, T.-L. (2003). *Low programming voltage anti-fuse structure*. US Patent 6,580,145.

Chapter 10
Multiple-Times Programmable Logic Nonvolatile Memory

Abstract In this chapter, we focus on the Multiple-Times Programmable (MTP) embedded NVM using basic logic CMOS processes. Several types of floating-gate-based NVMs are discussed, including medium density eNVM based on two-transistor (2T) memory cell, and low density but also low-power three-transistor (3T) memory cell. In the last section, we review the charge trapping-based NVM cell design.

Keywords Nonvolatile memory • Logic nonvolatile memory(LNVM) • NVM • Embedded NVM • Floating-gate transistor • Memory cell • Memory array • Reliability • Data retention • Endurance • Disturb • Tunneling • Fowler–Nordheim tunneling • HCI • Hot carrier injection • Coupling ratio • Threshold voltage • Program • Erase • High-voltage switch • Sense amplifiers • Row and column decoders • Readout buffer • Charge pump • High-voltage generator • Dickson charge pump • One-time programmable • MTP • Multiple-times programmable

10.1 Introduction

One-time programmable nonvolatile memory (NVM) is very useful and, since the invention of PROM, has found wide range of applications in integrated circuits. However, many more applications demand reprogrammable nonvolatile memory. For example, the trimming of an analog circuit, shown in Sect. 8.7, may change as the device ages, or when it is used in a different environment from where the initial calibration was done. It would then be necessary to reprogram the trim bits to reflect the change. In many cases, a few times of programmability may be enough. This can be achieved by using a few banks of OTP memory—one bank of the OTP memory is used for each programming round. This practice, however, may not be cost effective and often not practical for many applications. In addition, if the cost differential is small, customers would always prefer to have a many-times programmable memory "just in case" additional programming rounds are needed.

One of the biggest markets for multiple-times programmable (MTP) NVMs is microcontrollers where the NVM is used to store the firmware code and also computed data. For microcontrollers often megabytes NVM is needed to store

Y. Ma, E. Kan, *Non-logic Devices in Logic Processes*,
DOI 10.1007/978-3-319-48339-9_10

large program codes. As the use of microcontrollers is prolific, demand for low cost MTP NVM becomes an important requirement for logic CMOS processes.

As discussed in Sect. 8.2, the main option for MTP NVM in logic processes is to use floating-gate transistors as the storage element. In Sects. 10.2–10.4 we will present several commercial MTP NVM based on the floating-gate approach. In Sect. 10.5 we will discuss a charge trapping memory where the spacer of a MOSFET is used to store data.

10.2 A Medium Density Logic NVM

It is easy to perform single functions of adding or removing charge from single polysilicon devices. It is much more challenging to be able to both add and remove charges from the same floating gate without large area penalty. The programming can be done using hot carrier injection of a single pFET as discussed in the OTP memory of Sect. 9.4. To enable both erase and programming, we need to add one more capacitor to couple to a control gate signal for erase, as shown by the basic memory in Fig. 10.1a. The memory cell comprises two standard pFETs. A small pFET with source, drain, and n-well tied together is used as the tunneling junction, and a larger pFET serves as the readout and programming transistor.

The cells are erased by biasing the tunneling junction to a high positive voltage (~10 V), thereby removing electrons from the floating gate by Fowler–Nordheim tunneling.

The cells are programmed by bringing the drain node of the readout transistor to about −3 V, and using impact-ionization-generated hot-electron injection to add electrons to the floating gate. Because the drain has the most negative voltage, the holes generated in the impact-ionization process will be collected by the drain. This process reduces hole injection into the tunnel oxide, which has been recognized as a serious oxide-degradation mechanism.

Unfortunately, overprogramming is problematic for the pFET-based cell, where the floating gate becomes so negatively charged that the FET is always on, irrespective of the bias on the control gate. To form arrays a select transistor is

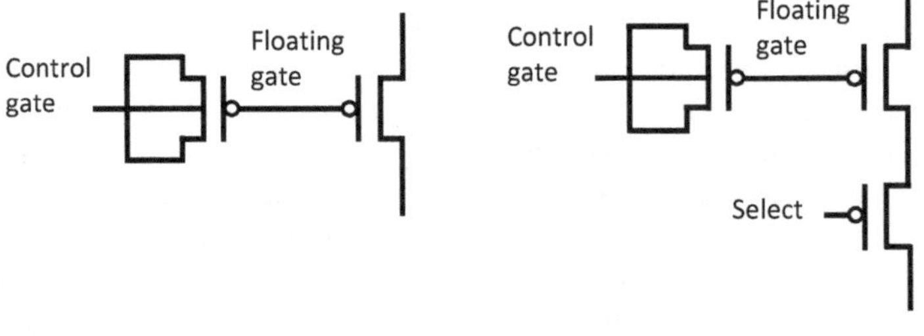

Fig. 10.1 2T and 3T floating-gate memory cells for MTP LNVM. The select transistor on the 3T cell is added to solve the over programming problem

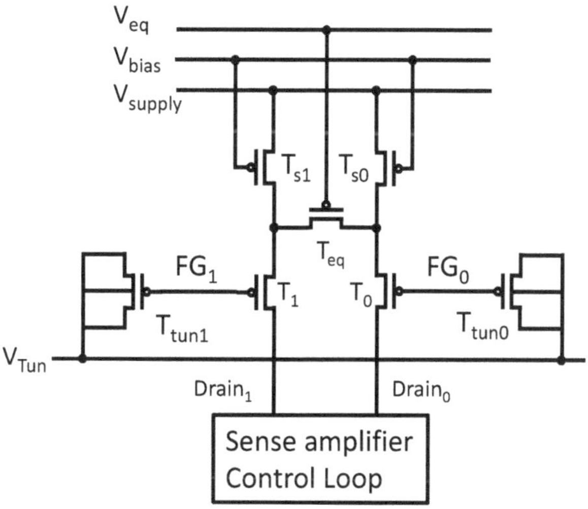

Fig. 10.2 AEON memory cell and operations, after (Pesavento et al., 2004)

added to isolate the read-out transistor, thereby avoiding read/write disturb problems. This brings the memory cell to a three-transistor (3T) cell.

One of the commercially available memories based on the earlier architecture is the AEON MTP memory, originally designed by (Pesavento et al., 2004). The design has since been ported to many logic processes and been used in numerous chips.

To make the memory robust and reliable for even automotive applications, a differential architecture was adopted. The resulting pFET-based memory cell is shown in Fig. 10.2 (Pesavento et al., 2004). The memory cell in Fig. 10.2 uses a differential approach to increase storage window, as discussed in Sect. 8.5. The differential cell design has other benefits that include easy sense amp design and low power read.

Note that for the differential pair, there are four possible states and only two are used to define a binary bit with the complementary states. The states of a differential pair and the assignments are as follows:

FG_1 on, FG_0 off = ‘1’
FG_1 off, FG_0 on = ‘0’
FG_1 off, FG_0 off—this is an undefined state and is used as an initial state for programming “1” or “0” later on. This can be potentially used as a state to denote the uninitialized state, as in the encoding scheme in asynchronous circuits.
FG_1 on, FG_0 on = this is another undefined state and not used in normal operation. This can be potentially used as a state to denote exception.

The operations of the memory cell in Fig. 10.2 are listed in Table 10.1. The memory cell first has to be initialized by FN tunneling to remove electrons from both floating gates. This is accomplished by ground all nodes except the V_{tun}, and

Table 10.1 Operations for the 3T MTP memory cell of Fig. 10.2

Signal	Initialization	Program "1"	Program "0"	Read
V_{supply}	GND	V_{dd}	V_{dd}	V_{dd}
V_{tun}	Ramping to V_{pp}	GND	GND	GND
V_{eq}	GND	V_{dd}	V_{dd}	GND pulse
V_{bias}	GND	Biased to program level	Biased to program level	Biased to read level
Drain_1	GND	V_{dd}	Pulse to neg voltage for injection	
Drain_0	GND	Pulse to negative voltage for injection	V_{dd}	

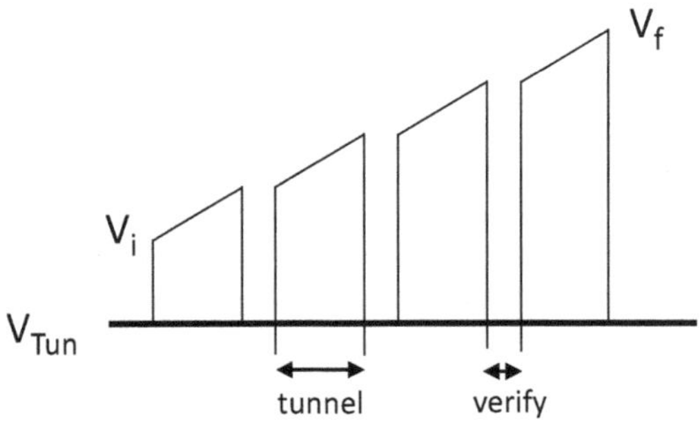

Fig. 10.3 Tunneling pulses for initialization of the memory cell

ramp up the V_{tun} signal for a fixed period of time. The floating gates are then positively charged and both transistors T_0 and T_1 will be off.

The memory cell is programmed using injection through either T_0 (programming a "0") or T_1 (for programming a "1").

There is a major problem with using pFETs as the floating-gate transistors. During initialization phase, some floating gate can be overerased. In this case, the floating gate is overly positively charged that when the cell is biased for injection, the transistor will remain turned off, resulting in a stuck-at bit. The overerase is usually due to mismatch between tunneling junctions in an array. There are large differences between the tunneling rate of the slow and fast tunneling junctions. Then during tunneling, when the FG with the slow tunneling junctions reach the erase target, the fast floating gate with the fast tunneling junction may be overerased.

To overcome the overerase problem, a controlled loop (CLOOP) is implemented in the memory cell of Fig. 10.2 for program/erase and verify. During erase, the tunneling voltage V_{tun} will ramp in pulsed fashion, e.g., Fig. 10.3. Between pulses, a verify step is performed where CLOOP monitors the state of the two floating gates in each memory cell, by checking the current through transistors T_0 and T_1. When the current reaches a target value for certain cells, e.g., in the fast tunneling cell, the CLOOP will stop injection for those cells and continue for other cells until the erase is properly done for all the floating gates (Lindhorst et al., 2005).

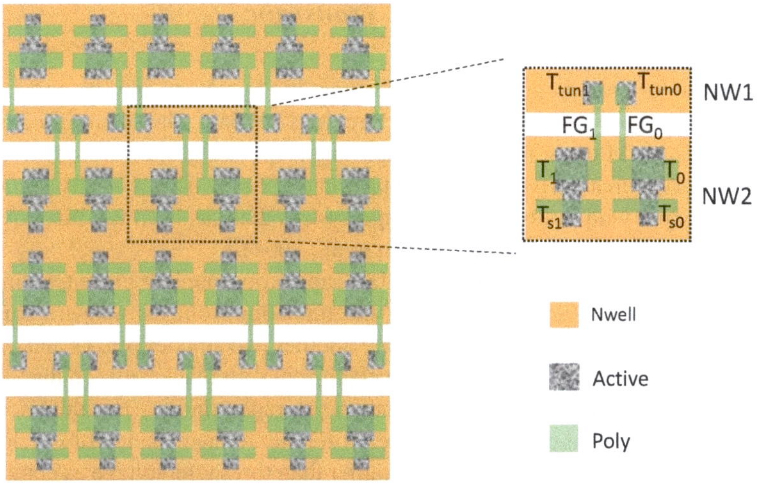

Fig. 10.4 Example layout of the AEON memory cell and cell array

The earlier operation and schematics shown in Fig. 10.2 also provide some guidance to the layout of the cell array. The tunneling junctions of transistor T_{c1} and T_{c0} can be placed in the same n-well, while all other transistors can be placed in a second, independent n-well. The first n-well needs to be able to tolerate the highest tunneling voltage and is required to initialize (remove the electrons) from the floating gate, which can typically be about 10 V for 7.0 nm oxide of the 3.3 V CMOS transistors.

On the other hand, the second n-well will only need to bring to the usual V_{dd}, i.e., 3.3 V. So the layout of the memory array will be stripes of Nw1 separated by stripes of Nw2, as illustrated in Fig. 10.4.

10.3 Low Density, Low Power LNVM

The eNVM of last section uses FN tunneling to erase and hot carrier injection for programming. One of the drawbacks is that the programming consumes a lot of power as HCI is not a very efficient method. Tens or hundreds of μA is needed to provide reasonable gate injection current for each cell to achieve acceptable target programming speed. One low power solution is to use FN tunneling for both program and erase.

10.3.1 Logic NVM Using 3.3 V Transistor

FN tunneling for both program and erase is possible by adding another transistor to the 2T floating-gate memory cell in the previous section. The memory cell for a commercial LNVM using FN/FN programming and erase is shown in Fig. 10.5, as shown in Pesavento and Hyde (2013) and Ma et al. (2008). This is a fully functional memory product with built-in high-voltage system that has been used in, e.g., low power and low cost RFID chips (Glidden et al., 2004).

Devices T_{C1} and T_{C0}, acting as the control capacitors coupling to the two bit lines are large. T_{t1} and T_{t0} are connected to the tunneling voltages and are small to achieve large coupling ratios. T_1 and T_0 are another two smaller pFETs and control the cell currents to sense amplifiers. The cell current difference between the "1" and "0" side floating gates, $i_1 - i_0$, determines the storage window. For this memory, there is no independent control gate. Therefore, the voltage change of the floating gate is exhibited in memory cell current rather than threshold voltage as commonly done in traditional floating-gate flash memory.

Table 10.2 details the programming and erase operations for such a memory cell (Pesavento & Hyde, 2013). In contrast to the differential memory cell of Fig. 10.2, there is no initialization step.

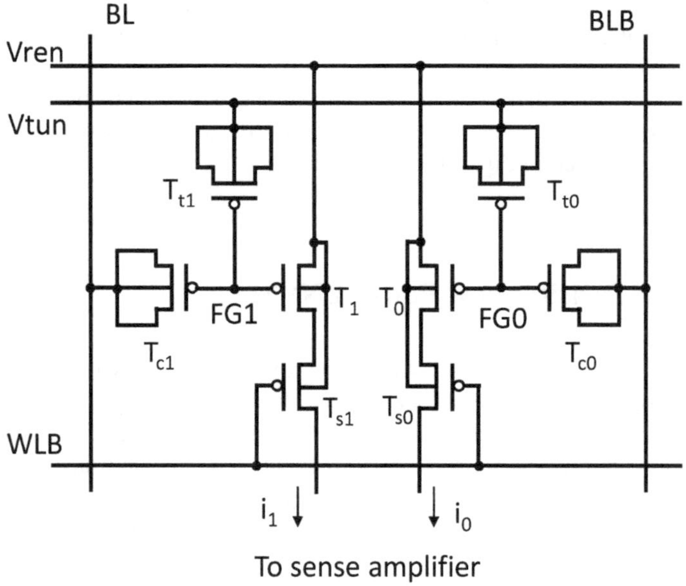

Fig. 10.5 A differential FG memory cell built with standard pFETs. After programming the FG1 ("1") or FG0 ("0") side are oppositely charged

Table 10.2 Operation condition of a differential memory cell of Fig. 10.5

	V_{tun}	V_{ren}	WLB	BL	BLB
Write 1	V_{pp}	0	$V_{pp}/2$	V_{pp}	0
Write 0	V_{pp}	0	$V_{pp}/2$	0	V_{pp}
Unselected row	$V_{pp}/2$	$V_{pp}/2$	$V_{pp}/2$	X	X
Read	0	V_{dd}	0	0	0
Unselected row for read	0	0	V_{dd}	0	0

The programming voltage, V_{pp}, is determined by the desired programming time. The cell program time is typically 5–10 ms through multiple program/verify cycles. For FN tunneling, recalled from Sect. 2.5.2, the electric field needs to be on the order of 10 MV/cm to get meaningful gate current, which is set by the desired programming speed of ~1–10 ms. With the use of 3.3 V I/O transistors for the memory cells, V_{pp} is determined to be about ~10 V, a voltage that is less than the n-well breakdown voltage of a typical logic CMOS process.

To program a "1" to the memory cell, the bit-line BL is set to V_{pp} (~10 V), and the complementary bit-line BLB is set to 0. V_{tun} is also set to V_{pp}. Due to capacitive coupling (the large transistors T_{C1} and T_{CO}), the floating gate FG1 is pulled to a high voltage, e.g., to about 7–8 V, while FG0 remains low. With $V_{ren} = 0$, this condition results in a 7 V across the gate to the channel of the T1 transistor and ~10 MV/cm across the 7 nm oxide. Electrons will tunnel from the n-well of T1 to floating-gate FG1. On the contrary, floating gate FG0 will have low voltage due to the large coupling capacitance of T_{CO}. Since V_{t0} is at V_{pp}, electrons will be tunneled from FG0 to T_{t0}. As a result of the dual tunneling action, FG1 will be negatively charged while FG0 will be positively charged.

When BL and BLB are flipped to program a "0" to the cell, FG0 will be negatively charged while FG1 will be positively charged.

During read, both bit lines as well as V_{tun} remain low. V_{ren} is brought to a read V_{dd} and the select transistors T_{s1} and T_{s0} are activated by bringing WLB to low. If the cell is programmed to a "1," because of the negative charge on FG1 and positive charge on FG0, the cell current i_1 will be larger than cell current i_0 and be sensed by the sense amplifier.

For the unselected rows, a half voltage, $V_{pp}/2$, e.g., 5 V is applied to all the relevant nodes such as V_{tun} and V_{ren}. This couples the floating gate to close to 5 V. Thus even though BL = 10 V, the voltage across the 7 nm oxide is reduced to ~ 10 – 5 = 5 V. There would be negligible tunneling current on the unselected cells.

The layout of the earlier memory cell is very different from that of Fig. 10.4. While the 3T cell in the previous section only needs two n-wells, you can see from the cell and the operation table that all three pFETs associated with the same floating gate, i.e., T_1, T_{c1}, T_{t1} for FG1, are independently biased and need their own wells. The corresponding cell size will be much larger.

Some sharing of the n-wells are possible between the two floating gates. In particular, we note that T_{c1} and T_{c0} share the same tunneling voltage and can be placed in the same well. Similarly T_1, T_0, T_{s1}, and T_{s0} only go up to $V_{pp}/2$ and can share the same well. Fig. 10.6 shows an example of layout (Pevavento & Langlinais, 2010).

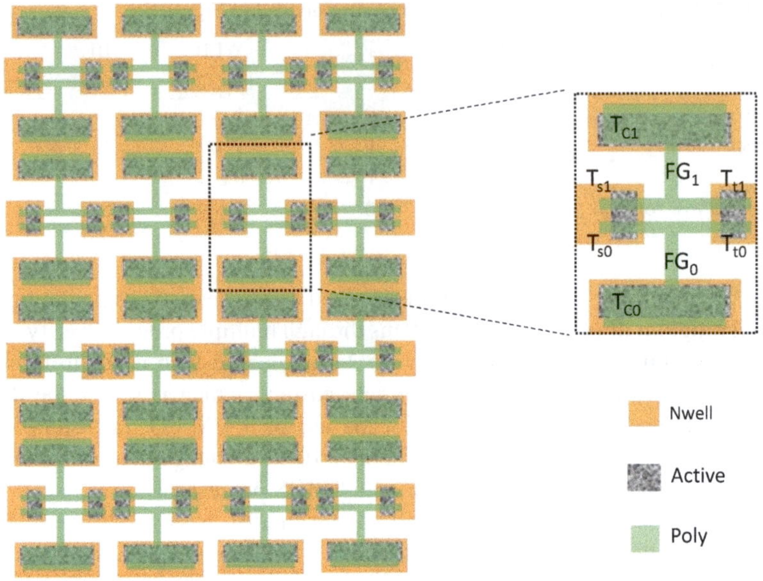

Fig. 10.6 Array and cell layout of memory in Fig. 10.5, adapted from Pevavento and Langlinais (2010)

The advantage of this memory cell is the much less power consumption. The main drawback is its large cell size. As a result, a number of RFID chips have used this type of memory cells. As we will discuss in Sect. 11.2, passive RFID chips do not have its own power source, but rely on harvesting power from the readers to power the chip. As a result RFID chips demand the lowest power consumption for both programming and erase. Also since usually only a small amount of NVMs, e.g., <2 kb, are needed for RFID, the large memory cell size does not affect the final cost of the chip significantly.

10.3.2 Disturb

As discussed in Sect. 8.6, whenever memory cells are placed in an array, various disturb mechanisms will occur. One of the most serious disturb mechanisms is the program disturb—when programming one row of cells, data in the other unselected rows could be affected. This could result in reduced retention for the data in other rows or even data errors. In the worst-case scenario, one particular row could be written repeatedly while other rows have static data. As a result, all the rest of the rows are disturbed and subjected to failures. This scenario happens in any type of EEPROM as well, where it is often countered by load balance, i.e., we try to use all rows in an evenly manner to distribute the stress, which are more commonly called

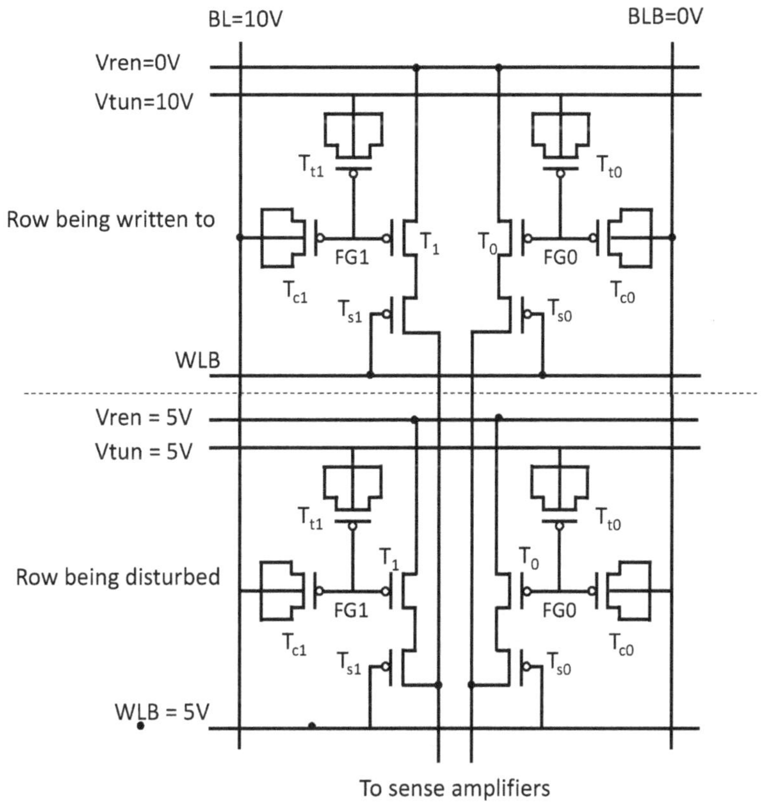

Fig. 10.7 Bit array and the bias condition for program disturb

wear leveling in NAND flash memory technology. This would minimize the program disturb and lengthen the endurance.

As an example let us consider the program disturb pertaining to the array presented in the previous section. In reality, significant program disturb still occurs on the unselected rows in Fig. 10.7 for the following typical case. Assume that the memory cell stored a "1." Then the voltage of FG1 may be close to $V_{fg1} = -1$ V. With $T_{c1} = 10$ V, assuming a coupling coefficient of 0.2 and T_{t1} and T_1 are at 5 V, and a coupling coefficient close to 0.7, then the floating gate voltage is close to

$$V_{fg} = -1.0 + 0.7 \times 5 + 10 \times 0.2 = 4.5\,\text{V}.$$

Thus, the voltage across the T_{c1} to FG junction is 5.5 V. Thus, according to Fig. 8.3, the FN tunneling currents at the normal tunneling voltage of $V_{ox} = 8$ V vs 5.5 V differ by a factor of only ~10,000. This means that the write disturb immunity is only about 10,000, i.e., the cell in the unselected rows can be wrongly programmed if the neighboring rows have been programmed for 10,000 times.

Note that under the worst condition, one of the rows of the memory may hold permanent data while other rows may be repeatedly programmed and erased. In this case, the total number of write cycles is reduced. For example, for a memory with 16 rows, if one of the rows contains "permanent data," the other 15 rows can only be written on the average 10,000/15 = 666 times.

Read disturb is another issue that can be serious. During read, for a floating gate that is programmed to be $V_{fg} = 1$, when $V_{ren} = V_{dd} = 2.5$ V the floating gate may be at a voltage,

$$V_{fg} = 1.0 + 0.2 \times 2.5 = 1.5\,\text{V},$$

assuming a coupling ratio of 0.2.

This might look benign and will not normally cause any disturb until we note that under the worse-case condition, a memory will be repeatedly read over 10 years. Even with a duty cycle of 10 % on time, the total biased time can be as high as

$$10 \times 365 \times 25 \times 60 \times 60 \times 0.1 = 3 \times 10^8\,\text{s}$$

In other words, we have to assure that there will be over eight orders of magnitude difference in tunneling current between the program condition and the read condition.

The problem will get worse if the cell is slightly damaged during the program/erase cycles and have SILC. As we have seen SILC is characterized by the increased leakage current at low voltages, which would be closer to the bias under the read condition.

10.3.3 LNVM with 5 nm Tunnel Oxide

The LNVMs discussed in previous sections all used 7 nm oxide from the 3.3 V I/O transistor as the tunnel oxide. It is well known that the tunnel oxide in floating gate (FG)-based NVMs cannot be scaled below ~7 nm, a practical limit due to stress-induced leakage current (SILC), which will cause larger bit error rates or more "tail" bits after the program/erase cycles. These conclusions were drawn from traditional high density EEPROM and flash memory studies. Various studies have estimated that the intrinsic scaling limit of the tunnel oxide thickness is about 5–6 nm (Runnion, Gladstone, Scott, Dumin, & Lie, 1996).

Since many advanced technologies only offer a 2.5 V I/O transistor option, where the available gate oxide thickness is only ~5 nm, it is of great interest to investigate the scaling of logic NVM solutions to thinner tunnel oxides. As we discussed in the reliability in Sect. 8.6.4, using the thinner tunnel oxides for small bit count, embedded applications in advanced processes may be possible.

Indeed it has been shown that ~5 nm tunnel oxide can be used in LNVMs in several baseline foundry logic CMOS processes (from 0.25 μm to 65 nm) without any process modification (Ma et al., 2008). The oxide thickness studied was between 4.8 and 5.3 nm. The memory cell and array architecture are the same as those in the previous section. Good intrinsic retention with gate oxides as thin as 4.8 nm is reported. There were many leaking bits observed and they were able to be overcome thanks to the use of differential memory cells. If the chip allows the implementation of error correction code, the error rate would be further improved.

As we have mentioned in Chap. 8, the intrinsic retention property measures the data retention quality of the dielectrics surrounding the poly silicon floating gate, including gate oxide, spacer, and the interlayer dielectrics (ILD) and are typically similar among all the cells in the array. While other reliability issues such as program disturb can be designed around, intrinsic retention is a prerequisite for any NVM design.

The activation energies extracted for all four processes using the method detailed in Ma et al. (2004) are also similar to those found for the 7 nm oxide. We thus conclude that from the intrinsic retention point of view, 5 nm from the 2.5 V I/O gate oxide is not a tunnel oxide thickness limit for floating-gate memory cells in the processes investigated.

Figure 10.8 shows cumulative distribution function (CDF) of the cell current (both i_1 and i_0) as well as the memory window (as measured by $i_1 - i_0$) for the 0.18 μm memory. We note that significant number of fast leaking or "tail" bits are observed and the number increased with increased bake time. Bake temperature dependence of the cell current is also investigated for bake temperature of 25, 85,

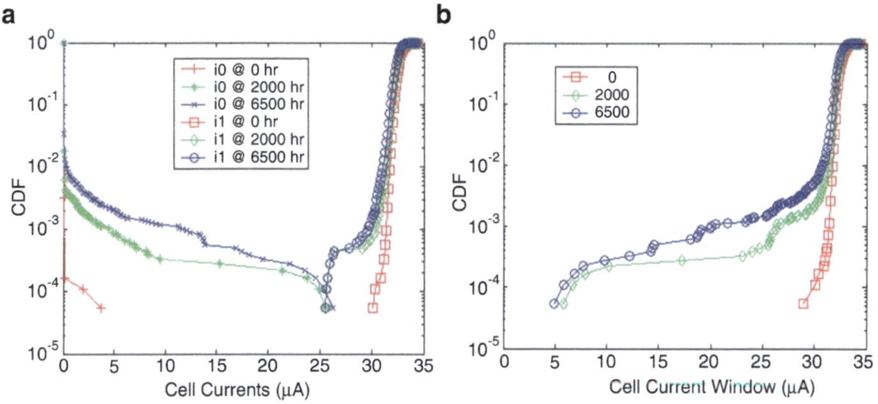

Fig. 10.8 Cumulative cell current distribution functions (CDFs) of the i_1 and i_0 sides of a differential cell (**a**) and the cell current window ($i_1 - i_0$) (**b**). Bake temperature is 85 °C after the cells have been cycled 100K times and the bake time was over 6500 h. Before bake, the cells appear to stop leaking at about 25–27 μA, a range determined by transistor mismatch. Cell window ($i_1 - i_0$) distribution showing that the memory window remains even after long bake, i.e., no cell with both i_1 and i_0 leaked to the neutral point, even though some floating gates have lost almost all their charges. The *arrow* indicates the failure point due to sense amp sensitivity

and 150 °C. It was found that while the main distribution shifts more at 150 °C, corresponding to the intrinsic retention shift discussed in the previous section, the 25 °C and 85 °C temperature bake produced larger tail bit populations, with 85 °C bake temperature produced the highest number of tail bits. The fewer tail bits at 150 °C can be attributable to the annealing of the SILC defects.

As the abundance of tail bits suggests, precautions must be taken to mitigate tail bit reliability problems for 5 nm NVM solutions. Since the tail bits are understood to be caused by localized defects, the failure rate of two separate floating gates is independent of each other. The use of the differential cell greatly increased the reliability, as discussed in Sect. 8.6.4.

This point is illustrated in Fig. 10.8a. Figure 10.8a shows the cell currents for both the 'on' (i_1) and 'off' (i_0) sides of a differential memory cell from a chip that has gone through over 100K write cycles. The cell currents appear to stop moving at about 25–27 μA, which we interpret as corresponding to the floating gate having reached the charge neutral point. Even though the i_1 and i_0 cell current distributions overlap, the cell current window ($i_1 - i_0$) distribution showing that the memory window remains in Fig 10.8b. This indicates that no differential cell has both of its FG1 and FG0 floating gates leaked to the neutral point, demonstrated the advantage of the differential cell.

We further note that the cell current distributions of Fig. 10.8a suggest that for this memory the failure rate of the FG1, f_1, is approximately equal to that of FG0, f_0, Thus, for the differential cell we can argue the overall cell failure rate is approximately the square of the single floating-gate failure rate, $f_{\text{diff}} = f_1 \times f_0 \sim f^2$, dramatically increasing the reliability of memory based on the differential cell architecture. Figure 10.8b shows the cumulative distribution of the cell current window. It confirms that the window for the differential cell remains, despite the fact that some of the floating gates have lost almost all charge.

Note that we do not expect $f_1 \sim f_0$ to be true in general, due to the fact that the charge on the two floating-gate sides of the differential cell is usually programmed to have opposite sign. This gives memory cell designers another tool to optimize the reliability of the cell by tweaking the cell design to minimizing the product, using the general failure probability, $f_{\text{diff}} = f_1 \times f_0$.

In summary, we can see that the reliability is much improved using differential cells. This is the key to the success of developing thin tunnel oxide-based floating-gate memory—use of redundancy such as those contained in the differential cell to lessen the impact of oxide defect associated with thin tunnel oxide.

10.4 Multiple-Times Programmable NVM Latch

A variant of the floating-gate nonvolatile memory is the rewritable NVM latch. This is a combination of the ideas from the OTP fuse in the previous chapter and the multiple-times programmability of the floating-gate memory.

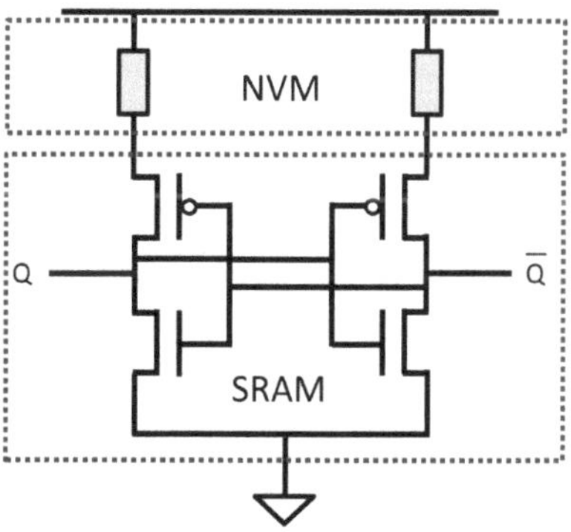

Fig. 10.9 nvSRAM cell comprised of an SRAM cell with NVM backup cells

As we discussed before, electronic fuse or antifuse has been extensively used in ICs. One of the main applications is the memory repair where redundant rows of memory replace defect rows during testing. Fuses are also frequently used as configuration memory to store such information as the trimming bits of analog circuits. For these applications it is useful to be able to reprogram the fuses to account for, e.g., device aging or environmental changes, such as temperature changes. SRAM cells backed by NVMs can be used for this purpose. For this type of application, the NVM latch is also called a programmable fuse. For this approach, there will be a setup time after power up as the data stored in the NVM will need to be read and then loaded into the SRAMs. This process can take anywhere from milliseconds to seconds.

It is sometimes desirable to be able to load the configuration information upon power up and without delay. Similarly it is equally useful, and essential in some cases, to be able to back up the data in SRAMs in NVMs in the case of power failure. One way to do this is to use the so-called shadow SRAM or nonvolatile SRAM (nvSRAM or NVRAM). Figure 10.9 shows a typical NVSRAM in a differential geometry. The earliest nvSRAM uses SONOS memory as the NVM backup cell, see for example Sharma (1993).

SONOS NVM needs extra process steps. Similar approach can be based on the logic NVM cell discussed in Chaps. 8 and 10. For example, Fig. 10.10 shows a combination of the FN tunneling logic NVM cells with a SRAM cell (Diorio et al., 2007). Since many sense amplifiers use an SRAM-based circuits, the circuit in Fig. 10.10 can also be alternatively viewed as a LNVM with its dedicated sense amp.

By adding a number of additional circuit elements, a more robust design is obtained and shown in Fig. 10.11. In this design, output buffers are added to reduce errors induced by load noises. If the loads on the two outputs are unbalanced due to mismatches, e.g., one has a much higher capacitive load than the other, then during the power up or reset, this node may rise much slower than normal and slower than the second nodes. The latch may then latch into an incorrect state.

Reset circuit makes it possible to initialize the latches. There are several situations where the initiation could be warranted: before the power up to ensure

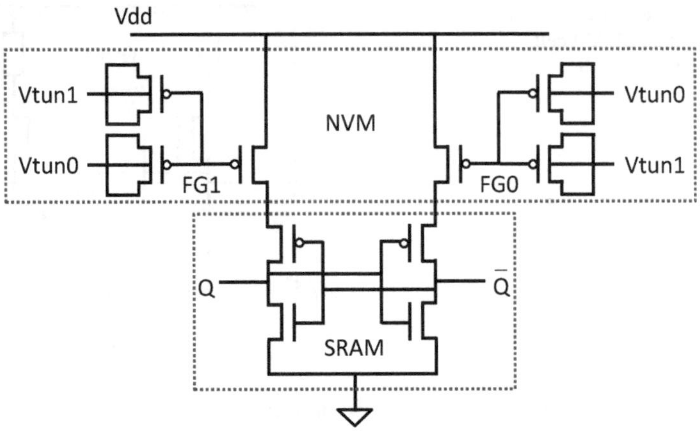

Fig. 10.10 nvSRAM cell with logic NVM cells as backup

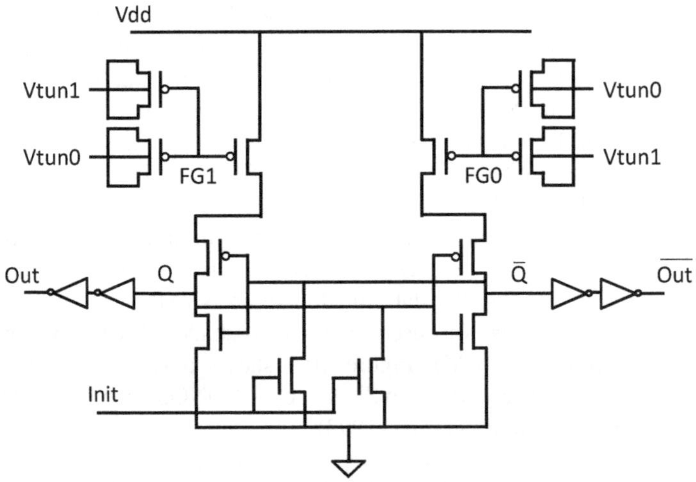

Fig. 10.11 A reprogrammable fuse with output buffers and initialization

that both nodes are totally discharged or during power up to hold these two nodes at ground to ensure that power supply is stable, and lastly after the power is on to correct for errors such as soft errors caused by cosmic rays.

10.5 Charge-Trapping Memory

Previous examples of MTP NVM are all floating gate based. As we have stated many times, for floating-gate transistor-based memory, the tunnel oxide thickness is limited to about 7 nm due to data retention requirement. This requirement restricts to the processes that can be used for LNVM to those with 3.3 V I/O transistor. We also saw in the previous section that 5 nm oxide, available with the 2.5 V I/O transistors, can be also used for LNVM but only at very small densities. As semiconductors continue to scale to smaller dimensions, the trend is to use lower and lower I/O voltages. Thus for better scalability, it is desirable to investigate logic NVM solutions that do not rely on the use of the oxide from 3.3 V or 2.5 I/O transistors.

It is well known that charge-trapping (CT) memory such as SONOS-based NVM can tolerate thinner oxide. The reason is that the trapped electrons are generally localized in the isolated traps. If one of the traps leaks, other traps may not and the memory cell as a whole will still work. An innovative use of the charge-trapping concept is by the use of silicon nitride in the spacer of the MOSFET. Embedded nitride-spacer-based logic NVMs fabricated with a pure CMOS logic compatible processes have been demonstrated. For example, a saw-like self-recovery Self-Aligned Nitride (SAN) memory cell is proposed and fabricated in 28 nm high-k metal gate (HKMG) CMOS process for high-density logic NVM applications (Sung et al., 2014). Nitride spacer-based charge-trapping NVM can push the scaling limit of the oxide to 3 nm.

The cross section of the memory cell is illustrated in Fig. 10.12. In the SAN cell, two transistors, both nFETs in this case, are placed close together. One of the transistors serves as a select transistor and the other as the program transistor. The nitride spacers from these two adjacent FETs merge and serve as the storage medium.

As illustrated in Chap. 2, the nitride spacer is separated by the gate oxide to the silicon substrate. Thus, the gate oxide is the tunnel oxide for this type of SONOS memory.

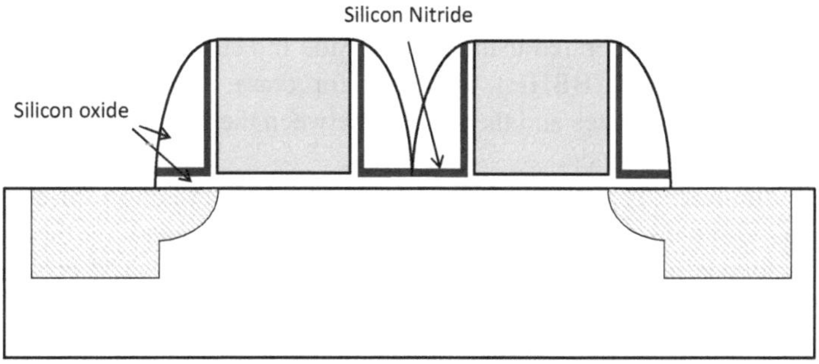

Fig. 10.12 A 2-transistor memory cell using nitride in sidewall spacer for charge storage

Table 10.3 Operating conditions for memory cell in Fig. 10.13

	PG	SG	BL	SL
Erase	−0.5	0	0	4.5
Program	3	0.7	0	3
Read	0.4	1	1	0

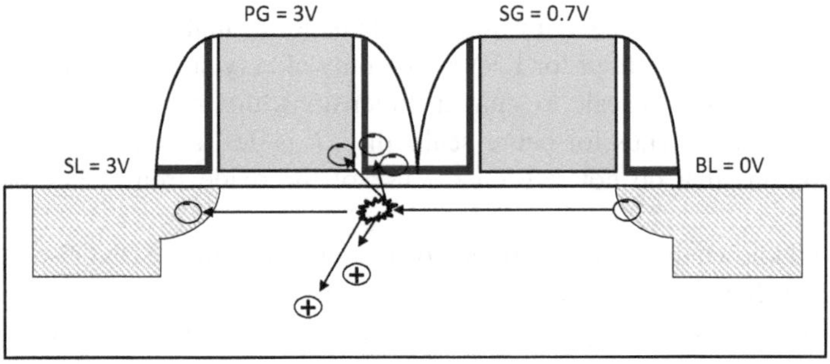

Fig. 10.13 Programming bias condition for the SAN cell

Electrons or holes can be tunneled or injected into the spacer storage region, on the side of the program transistor, and affects the on/off characteristics of the PG transistor.

To be clear, Fig. 10.12 is not exactly two transistors in series. In particular, the source of the program transistor and the drain of the select transistor are eliminated. This can be done without modifying the process but rather through a modification of the masks for the LDD and source/drain implants. In this case we will not draw the source of the program transistor and the drain of the select transistor in the layout. Recall the logic operations in Table 3.1, the corresponding LDD implants will not be done in the area of between the two FETs. Based on 28 nm HCMOS logic design rules, a very small cell size of 0.0535 μm^2 can be achieved and fabricated.

The operations of the SAN cell are presented in Table 10.3 and illustrated in Fig. 10.13. The cell is operated using Source-Side Injection (SSI) for program and band-to-band hot holes (BBHH) tunneling for erase. The program and erase voltages depend on the sizes and the spacing between the two FETs, which directly depends on technology node.

Because the memory cell uses band to-band hot hole injection for erase which tends to be more damaging than electron injection or FN tunneling, the endurance of this memory is not great. It appears to be able to generate 100 program erase cycles (Sung et al., 2014).

It is worth noting that the silicon nitride spacer is not being used after 40 nm process for the pMOSFETs. Due to the use of stress film to enhance the mobility of the MOSFETs, SiGe stress film is used in the source–drain of pFET to apply compressive stress to enhance the hole mobility of pFET while silicon nitride film is used to apply tensile stress to enhance the mobility of electrons in the nMOS. Hence only nMOS-based SAN is possible, as have been demonstrated in 28 nm process (Sung et al., 2014).

References

Diorio, C., Bernard, F., Humes, T., & Pesavento, A. (2007). *Rewritable electronic fuses*. US Patent 7,242,614.

Glidden, R., Bockorick, C., Cooper, S., Diorio, C., Dressler, D., Gutnik, V., et al. (2004). Design of ultra-low-cost UHF RFID tags for supply chain applications. *IEEE Communications Magazine, 42*(8), 140–151.

Lindhorst, C. A., Diorio, C. J., Gilliland, T. N., Pesavento, A., Srinivas, S., Ma, Y., et al. (2005). *Differential floating gate nonvolatile memories*. US Patent 6,950,342.

Ma, Y., Deng, R., Nguyen, H., Wang, B., Pesavento, A., Niset, M., et al. (2008). Floating-gate nonvolatile memory with ultrathin 5-nm tunnel oxide. *IEEE Transactions on Electron Devices, 55*, 3476–3481.

Ma, Y., Gilliland, T., Wang, B., Paulsen, R. P., Wang, C., Nguyen, H., et al. (2004). Reliability of pFET EEPROM with 70A tunnel oxide manufactured in generic logic CMOS processes. *IEEE Transactions on Device and Materials Reliability, 4*, 353–356.

Pesavento, A., Gilliand, T., Lindhorst, C., Srinivas, S., Bernard, F., Salazar, S., et al. (2004). Embedded non-volatile memory in logic CMOS. In *Non-volatile Semiconductor Memory Workshop* (p. 49). Monterey: IEEE.

Pesavento, A., & Hyde, J. (2013). *PFET nonvolatile memory*. US Patent 8,416,630.

Pevavento, A., & Langlinais, J. (2010). *Radio frequency tag including configurable single-bit/dual-bits memory*. US PTO. US Patent 7,796,450.

Runnion, E., Gladstone, S., Scott, R., Dumin, D. J., & Lie L. (1996). Limitation on oxide thicknesses in flash EEPROM applications. In *Proceedings of International Reliability Physics Symposium* (pp. 93–99). IEEE.

Sharma, U. (1993). A novel technology for megabit density, low power, high speed NVRAMs. In *VLSI Symposium* (p. 53). Japan.

Sung, Y., Lin, P., Chen, J., Chang, T., King, Y., & Lin, C. (2014). A new SAW-like self-recover of interface states in nitride-based memory cell. In *Technical Digest of IEEE International Electron Device Meeting* (pp. 494–497). San Francisco: IEEE.

Chapter 11
Non-Data-Storage Applications

Abstract In this chapter, we describe several non-data-storage applications of the floating-gate memory made in the logic process. We first review digital-to-analog converters that use floating-gate transistors as analog storage for trimming of current sources. Adaptive floating-gate comparator and several applications are briefly described. RFID systems application is then reviewed where many of the novel devices and circuits, including native devices, bandgap-engineered device, and logic NVMs, played a critical role and enabled low cost and low power RFID chips. Use of floating-gate transistor in neutral networks is then discussed. The use of floating-gate devices for multi-input tunable biosensors will be discussed in Chap. 12, while the transmission line-based waveform-shaping structures in the logic process will be investigated in Chap. 13.

Keywords Nonvolatile memory • NVM • Embedded NVM • DAC • Current source • Comparator • Adaptive Floating-Gate Comparator (AFGC) • Trimming • RFID • Neuromorphic computing

11.1 Introduction

As is evident from the previous chapters, the primary use of non-logic devices such as floating-gate transistors has been to store nonvolatile data. In this chapter, we discuss non-data-storage applications of these devices built in standard CMOS processes, including their use in analog and digital circuits, for example, and in adaptive processing/neuromorphic computing. This chapter will focus mainly on information processing applications, while the biosensing and waveform-shaping applications will be in Chaps. 12 and 13.

We have already seen the floating-gate devices being used as analog circuit elements, e.g., the variable capacitor discussed in Chap. 5. It is worth noting that the charge stored on the floating gate and its representation by the threshold voltage, or the current-carrying capability of a transistor, is inherently an analog quantity (i.e., the charge is a continuous value, instead of discrete values, as we are still dealing with many electrons on the floating gate). It is therefore natural to ask if we can use floating-gate transistors to store analog quantities, such as the audio signal in an audio recording chip, directly without the usual A-to-D and D-to-A conversions.

Y. Ma, E. Kan, *Non-logic Devices in Logic Processes*,
DOI 10.1007/978-3-319-48339-9_11

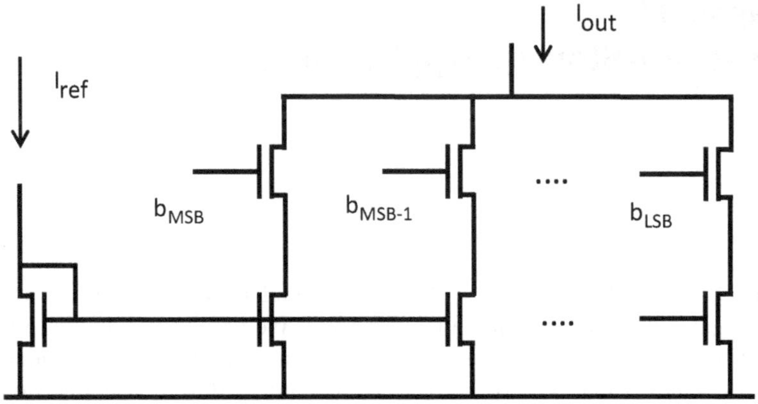

Fig. 11.1 Digital trimmable current source. The binary inputs b_{lsb} to b_{msb} are provided by an embedded digital NVM block

Indeed, such proposals have existed since the 1970s, see, e.g. (Rako, 2009). This article also mentioned several other analog applications, including the neuromorphic computing application which we will discuss in Sect. 11.5.

Today we do not see as many analog applications of the floating-gate transistors as the digital ones. Why is this the case and what is the advantage and disadvantage of analog vs. digital storage using floating-gate transistors? We use the example of digital and analog implementations of a trimmable current source for illustration.

Figure 11.1 illustrates a typical digitally trimmable current source, where a series of current mirrors are controlled by a number of digitally stored bits.

Assuming the current mirrors are in the ratio of 2^n:1, the output current is then:

$$I_{\mathrm{out}} = I_{\mathrm{ref}} \sum_{i=\mathrm{LSB}}^{\mathrm{MSB}} b_i 2^i \tag{11.1}$$

The precision of the current mirrors is a direct function of the number of bits used. For example, for a 4-bit design, LSB = 0, MSB = 3, the width of the FETs in the current mirror would be in the ratio of 8:1, 4:1, 2:1, and 1:1, to give 16 output current levels.

The digital settings can be stored a NVM and rapidly switched. The accuracy of the current is determined by the layout and manufacturing process variations. Once a chip is processed, it can be calibrated and the resulting data stored in the digital form in the NVM. The advantage of the digital solution is that it is more tolerant to the drift of the threshold voltage. Data loss can be corrected by redundancy and error correction discussed in the previous chapters.

Figure 11.2 shows an analog implementation of a current source using a floating-gate transistor. By adjusting the threshold voltage, the floating gate controls how much current from the current mirror can be supplied to the output.

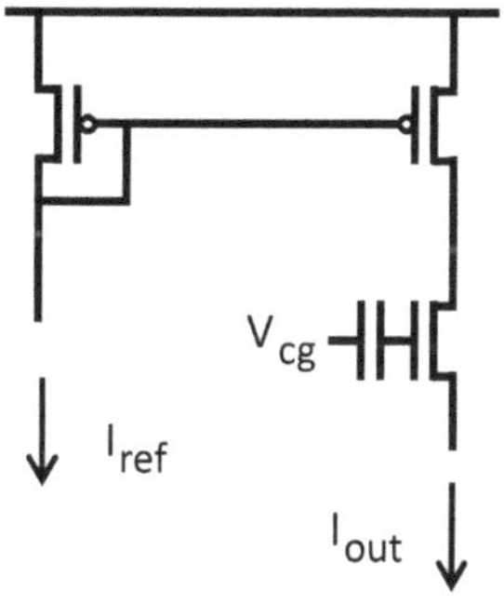

Fig. 11.2 Analog implementation of the current source using a floating-gate transistor

The analog solution is simpler and needs a much smaller area. One memory cell can store the same amount of information equivalent to up to 8 bits of digital storage.

A main drawback of the analog solution is the difficulty with calibration and the charge loss-induced current drift. Programming a floating-gate memory cell to a desired analog voltage is not trivial. As we have seen in Chap. 8, the programming and tunneling speed of floating-gate memory cells varies from cell to cell, due to defects in the oxide and layout dependences. So each cell needs to be calibrated individually. Also due to charge loss, the cell needs to be calibrated periodically.

One of the commonly used techniques, which has been widely used in programming the multilevel cells in NAND flash memory, is the incremental step pulse programming (ISPP) method. In this method, a series of pulses are applied to the floating gate (Suh et al., 1995). After each pulse, the resulting current or threshold voltage is checked against a reference. The same principle can be applied to the erase step, discussed previously, as well.

Perhaps one of the most important features of analog data process is its power efficiency. In general, analog signal processing uses less power and has lower latency than its digital counterpart. Although digital signal processing has become ubiquitous, it usually suffers from significant power penalties. A classic analog processing example is the human brain, which consumes about 10 W of power yet can perform image recognition functions faster than computers with teraflops speed and consuming thousands of watts. For digital processing, we must first digitize the signal, and the fast clock rates in the digital processing cores use a lot of power. Further, we often must convert the signal back to the analog domain. A digital system can perform a 64-point FFT (fast Fourier transform) using 520 mW of power, whereas an analog signal-processing chip uses 13 mW. The analog die is also one-fifth the size of the digital die (Rako, 2009). It is anticipated that as power consumption becomes an ever important design constraint, more analog solutions may be adopted.

However, another interesting trend in the analog and RF applications has also been true in the last 20 years. Whenever precision is needed for more than 4 bits (i.e., more than 16 levels or the requirements of two significant numbers in the floating point number) and power consumption is not the constraining factor for the functional module, digital parts have been widely applied to feedback and calibration of analog and RF circuits, such as phase-lock loops and reconfigurable filters.

In the following section, we discuss a couple of analog trimming methods that illustrate the power and area savings.

11.2 DAC with Floating-Gate Trimming

Digital-to-analog converters (DACs) produce a quantized (in discrete steps) analog output in response to a binary digital input code. DACs are commonly used to interface digital systems with the real physical world, which is mostly analog in nature. There are many DAC architectures, including resistor ladders, charge sharing, and current steering. Due to its inherent high speed, current-steering DACs are commonly used in high-speed applications such as in communications systems.

A typical current source-based DAC using the digital current source from Fig. 11.1 is shown in Fig. 11.3. The n-bit current source is binary weighted in this example.

The total current, I_{DAC}, is given by Eq. 11.1. Because of the operational amplifier, the current going through the resistor is the same as I_{DAC}. As a result, output voltage is

$$V_{out} = I_{DAC}R \sim \sum_{i=0}^{n-1} b_i 2^i \tag{11.2}$$

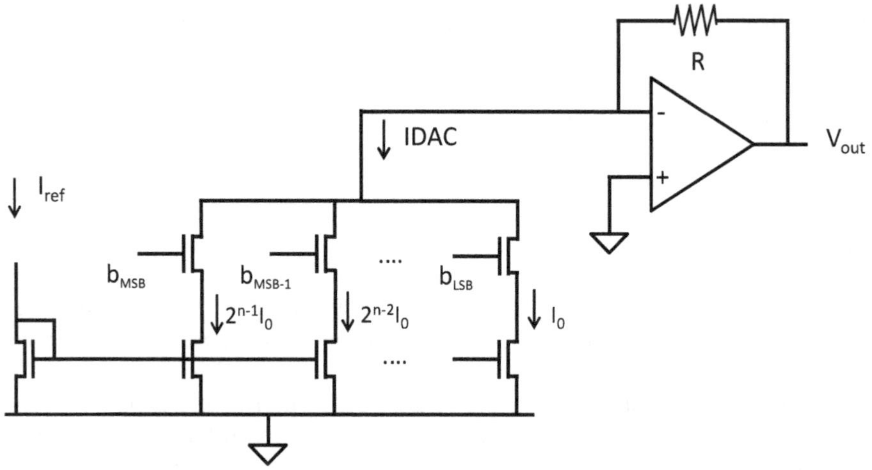

Fig. 11.3 A typical current-steering DAC using digitally trimmed current mirrors

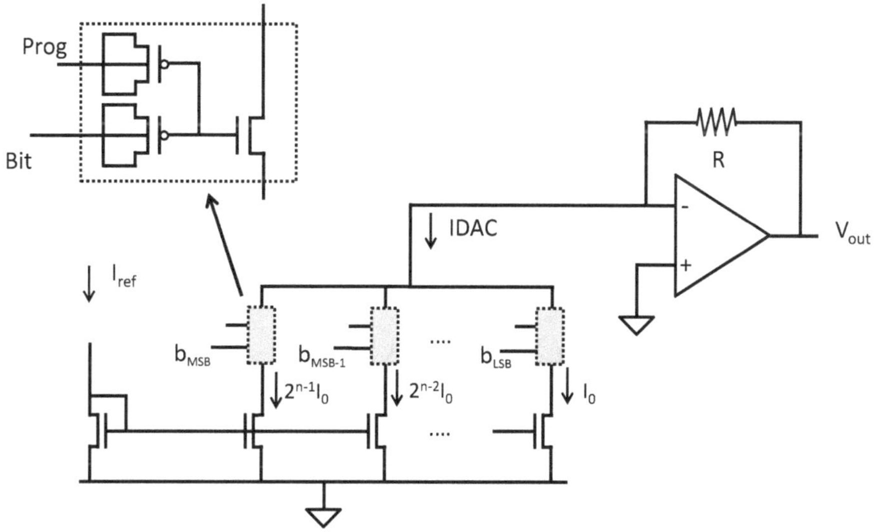

Fig. 11.4 Current-steering DAC using floating-gate devices for precision trimmable current sources

There are issues with this implementation. For good linearity, it is required that the current sources be matched well. The usual practice is to use large transistors for better matching. To make things worse, the transistor size scales with 2^n. These large transistors can dramatically increase the area, hence the cost of the chip. In addition, larger areas increase the parasitic capacitance as well and can have a large impact on the speed of the DAC.

Analog approaches use floating-gate transistors to increase the drive current and to match for the current sources. Instead of increasing the size of the W/L to deliver the desired current, these approaches use the floating gate to adjust the threshold voltage of the current-carrying FET, resulting in an equivalent increase in drive current. One digital-to-analog converter using programmable floating-gate-based current source is described in US Patent No: 7,408,397 (Serrano, Kucic, & Hasler, 2008), as shown in Fig. 11.4.

In this design, each of the current mirrors is controlled by a floating-gate device consisting of three transistors. One transistor is in series with the current mirror and controls the current flow. One transistor is the programming node, where its n-well can be biased to high voltage to change the charge on the floating gate. The third transistor acts as the control gate node that takes the digital input.

Another, more sophisticated approach is shown in Fig. 11.5 (Hyde et al., 2003) which is based on the current sources that can be trimmed using pFET floating gates. In this design, a master reference cell drives many slave trim cells. The floating-gate cells are pFET based and can use either FN/FN or FN/HCI to trim the charge of the floating gate.

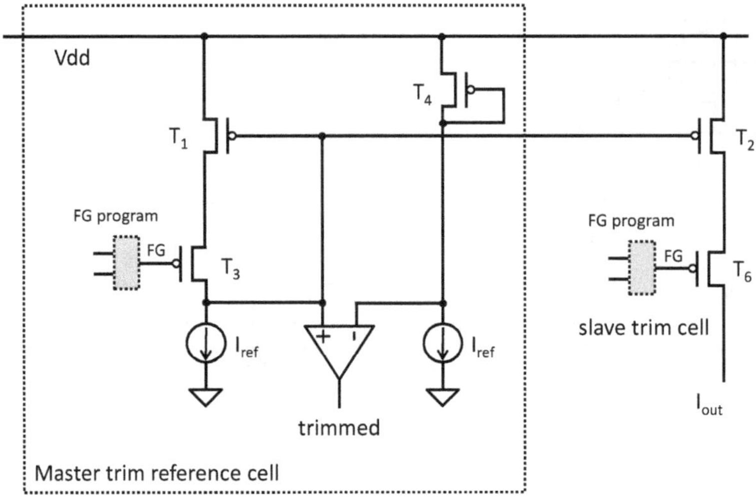

Fig. 11.5 Current source with a master and a slave cell. A DAC may use one master cell and a number of slave cells

The master cell pFET T1 is first calibrated to deliver a current equal to the reference current I_{ref}. The transistors T1 and T3 form a feedback loop to compensate for temperature variation and device aging: T3's transconductance will change its drain voltage and will cause the bias on T1 to change in the opposite direction, thereby maintaining a constant current.

The gate voltage of the T2 of the slave cell is equal to the gate voltage for T1 and maintaining the current of the slave cell as well.

This current source, one master reference cell and 41 slave trim cells, is used in a 14-bit, 250 M samples/s current-steering DAC (Hyde et al., 2003) and fabricated using a standard CMOS processes. In the test chip, the floating-gate transistors all used 3.3 V I/O device with the 7 nm oxide. The tunneling signal is applied to all the floating gates to remove the charge. The trim is performed by hot carrier injection on individual floating gates. Because of the injection efficiency of the pFET occurs near 1/3 to ½ V_{ds} as described in Chap. 2, the drain only needs to go to around −1.5 V (for a total $V_{ds} = 3.3 - (-1.5) = 4.8$ V) and could be supplied by a small, single-stage charge pump at each floating gate.

11.3 Adaptive Floating-Gate Comparator and Applications

Figure 11.6 is a floating-gate-based comparator Wong (2007) and Wong, Abshire, and Cohen (2004) that compares differential analog signals supplied at the inputs. One of the issues with voltage comparators is the mismatch of transistors that can cause a voltage offset. The floating-gate-based comparator is capable of

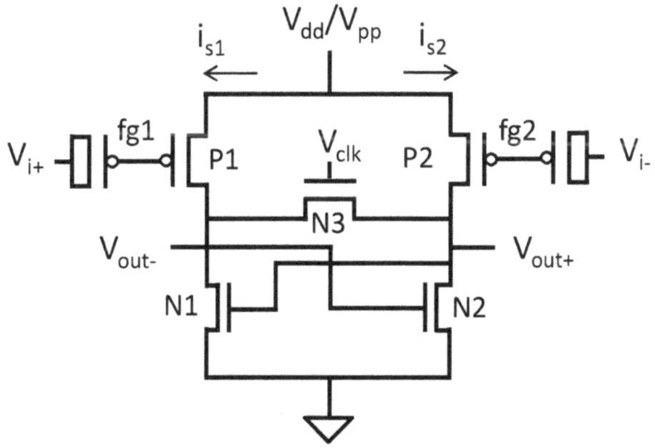

Fig. 11.6 A floating-gate comparator, adapted from Wong (2007)

programming a precise offset voltage automatically. The clock signal, V_{clk}, and the equalization transistor N3 are used to reset the comparator. The comparing phase begins when V_{clk} is set to low.

The offset adjustment can be done in circuit by setting the input differential voltage ($V_{i+} - V_{i-} = 0$). Due to mismatch the output will be unbalanced, e.g., $V_{out+} > V_{out-}$. By bringing V_{dd} to a high voltage, e.g., V_{pp}, and raising the input common mode voltage such that the gate-to-drain voltage V_{gd} is sufficiently high, the impact ionization can cause injection into both floating gates. But since source-to-drain voltage of P1 is greater than that of P2, the injection current into floating-gate FG1 will be greater than the injection to FG2. As a result of the difference in the change of the floating-gate voltages, the current i_{s1} will increase and i_{s2} will decrease, causing the V_{out+} and V_{out-} to approach each other. The charge difference on the floating gates will compensate for the device size mismatches.

This comparator has been used in a number of designs, including a fully functional 6-bit flash analog-to-digital converter. A 750MS/s 6-bit flash ADC (Wong, Cohen, & Abshire, 2007) with 63 high speed adaptive floating-gate comparator (AFGCs) was demonstrated. The online learning feature of the AFGC enables manual and automatic in-circuit programming of reference levels, completely eliminating both resistor ladders and comparator offset problems. Monotonic output, with integrated nonlinearity (INL) and differential nonlinearity (DNL) of 0.24 LSB and 0.79 LSB, respective, was obtained. Finally, since the device mismatch can be completely cancelled, this comparator may be used as a true random number generator (Xu, Wong, Horiuchi, & Abshire, 2006).

11.4 RFID Applications

Radiofrequency Identification (RFID) has a long history—it can be dated to the use of aircraft identification of friend or foe (IFF) using radio waves during World War II. This type of RFID is a so-called active RFID system where the RFID tag is

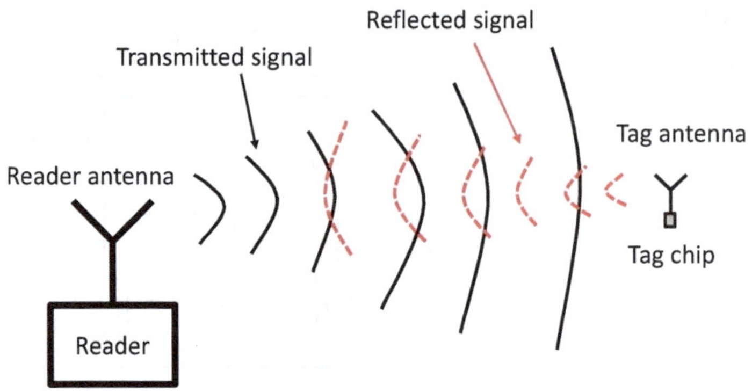

Fig. 11.7 Interaction of a RFID transmitter and a passive tag

connected to a power source such as a battery. Active RFID has found many other applications including long-range animal tagging to trace the movement of wild animals. But due to the high cost, its application in general has been limited.

With IC scaling and the consequent reduction in energy consumption, an RFID tag can now be powered solely by the RF signal it receives. Such RFID tags do not include an energy storage device and are called passive tags. Some passive RFID tags can also harvest energy from the environment, such as WiFi signals.

Passive RFID tags consist of only a chip and a printed antenna. With no battery, the overall tag cost can be very low, to the tune of a few US pennies. This low cost enables the explosive growth in the use of passive RFID tags. In particular, tens of billions of tags in supply chain applications have been found in, among other applications, retail inventory.

The advantage of passive RFID, as compared to other identification technology such as barcodes, is that it has relative long range (up to tens of meters); no line of sight is necessary for a reader to talk to the tag; very high read speed enables reading thousands of tags per second; information stored in an IC can provide much higher information capacity and data security.

Figure 11.7 illustrates how passive RFID works. RFID systems include RFID tags and at least one RFID reader. RFID reader, through its antenna, transmits RF waves. Tag antenna picks up the RF signal, harvest the energy, and power up the chip. Communication between the chip and reader occurs by detecting the modulated baseband signal transmitted from the reader. The tag processes and interprets the command communicated by the reader and formulates its response. The tag then reflects back a portion of the incoming RF wave while adding a modulating component, for example, by modulating the impedance of the tag antenna. Depending upon the command sent by the reader, the reflected RF wave may further encode data stored internally in the tag chip, such as an ID number. The response is demodulated and decoded by the reader, which thereby identifies, counts, or otherwise interacts with the RFID tag.

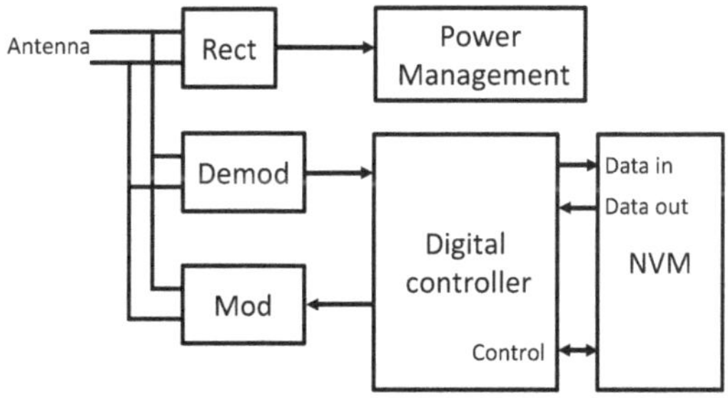

Fig. 11.8 Simplified block diagram of an RFID tag chip where Rect stands for rectifier circuit, Demod is the demodulator circuit, and Mod is the modulator circuits

The range of a passive RFID system is usually dominated by the power that can be delivered to the tag and how much power the tag IC consumes, which is called tag sensitivity limit. In general, the range is given by the Frii's equation, see e.g. (Glidden et al., 2004),

$$R = \frac{\lambda}{4\pi}\sqrt{\frac{P_{\mathrm{t}}G_{\mathrm{tag}}}{P_{\mathrm{tag}}}G_{\mathrm{ant}}} \tag{11.3}$$

where λ is the wavelength of the carrier, P_{t} is the transmitted power of the reader, P_{tag} is the power consumption of the tag at the antenna, and G_{tag} and G_{ant} are the tag and reader antenna gain, respectively.

P_t is usually determined by government regulations. For example in the 915 MHz range, US regulations only allow a total transmitting power of 1 W and a directional gain of 6 dB. G_{tag} is usually constrained by the tag antenna size and form factor. So for longer operation range, P_{tag} needs to be as small as possible, usually a few microwatts.

The simplified block diagram of a typical passive RFID tag chip is shown in Fig. 11.8 (Glidden et al., 2004).

In operation, an RF wave received via antenna port goes through a rectifier, Rect in Fig. 11.8. The DC voltage generated by the Rect circuits is delivered to a power management unit (PMU), which in turn generates the supply voltage for all other circuits.

Harvesting sufficient power from the RF wave to power a tag IC is perhaps the most challenging part of the RFID tag IC and antenna design. Since the voltage amplitude of the reader transmitted RF signal is in the range of approximately 200 mV, and a typical supply voltage is about 1 V or less, a highly efficient rectifier design is needed. Additionally, for high-voltage operations such as programming and erasing nonvolatile memory in the RFID tag, as discussed in Chaps. 8–10, a boosted voltage as high as 12 V may be needed.

The received signal also goes through a demodulator to extract the information from the carrier RF wave. A digital block (also called tag controller) acts as the brain of the RFID chip. It receives the demodulated, digital signal from demodulator, and decodes its meaning. For example, it interprets the command received from the reader and acts accordingly. The digital block will perform other functions like managing security and error correction checks; interface with the embedded NVM to read and write data to the NVM.

Because the limited number of command that needs to be supported for the RFID protocol standards, the tag controller does not have a general purpose microcontroller. The core of the controller is usually a state machine.

A critical component of the chip is the embedded NVM. The NVM stores essential info such as the electronic product code (EPC), security passwords, chip configuration bits, tag manufacturer information, and may have additional user memory. For RFID tags used in supply chain applications, the typical size of the NVM is about 256–512b. Because of the small size and the need for the lowest cost, logic NVMs discussed in Chaps. 8–10 have been the choice for a number of the UHF RFID tags on the market.

For example, an OTP memory based on antifuse technology discussed in Chap. 9, called XLPM (super low-power permanent memory) has been used in RFID chips from Kiloway.[1] The technology is billed as the lowest cost as well as the most secure NVMs for RFID. The drawback is that the code in the chips can be programmed only once and high power consumption is needed during the programming.

For multiple-times programmable memory, other RFID vendors have been using the MTP LNVMs discussed in Chap. 10. For example, the first GEN2 standard RFID tag uses the NVM discussed in Sect. 10.3 (Glidden et al., 2004). As discussed in Chap. 8, for higher memory density LNVM may not be cost competitive to embedded EEPROM technology employing dedicated process. Some RFID tag chips with higher NVM capacity use dedicated embedded flash process.

Two of the most important concerns with a tag IC are the cost, which is the main driver for the wide adoption of RFID technology, and the chip power consumption, which determines the readable range of a tag.

For lower cost, it makes sense to use the basic CMOS process with the lowest cost, e.g., standard logic process. For example, the first Gen2 RFID chip was designed in a standard 0.25 μm CMOS process (Glidden et al., 2004). Since embedded NVM processes usually costs up to 50 % more than the baseline CMOS process, logic NVM discussed in Chap. 10 was used. The embedded NVM uses the 3.3 V IO transistors with its ~7 nm gate oxide to achieve 10-year retention life. The NVM consumes extremely low power (900 fJ/bit read, <400 nA standby current) and has relatively fast write times (<10 ms per row) as well as good, up to 100,000 cycle, write endurance.

In general, the tag power is related to the power consumption of the chip by

$$P_{tag} = \frac{I_{chip} V_{chip}}{\eta} \tag{11.4}$$

[1] http://www.kiloway.com/archive.php?aid-4.html

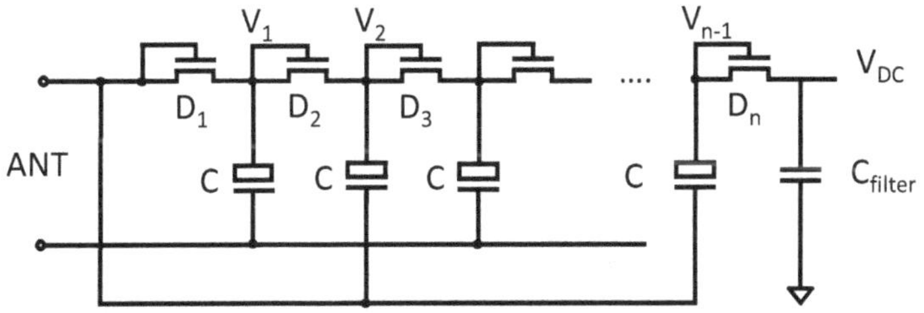

Fig. 11.9 An RFID rectifier based on Dickson charge pump concept

where $I_{\text{chip}}V_{\text{chip}}$ is the power consumption of the chip, and power conversion efficiency η is a product of both tag antenna matching efficiency as well as the rectifier efficiency.

The rectifier is important for determining the power conversion efficiency. The commonly used rectifier in the RFID tags resembles the charge pumps, e.g., the Dickson charge pump (Dickson, 1976), discussed in Chap. 8. Figure 11.9 shows a basic rectifier design that uses the antenna input as both the power and the clock signals. The only difference between the RFID rectifier and the high-voltage generators is the choice of the diode turn-on voltages or the diode-connected transistor threshold. The power impinged at the antenna is often very weak. Even after the resonator with high impedance, the voltage magnitude is often very small, which makes the rectifier performance critically depend on the turn-on voltage loss and reverse-bias leakage.

For this type of rectifier, there are many choices for the diode, including:

1. Diode-connected NMOS as shown in Figure 11.9;
2. Diode-connected native NMOS;
3. Schottky diode discussed in Chap. 7.

As we discussed in Chap. 8, the Dickson type of charge pump suffers from in efficiency where either the diode forward voltage drop, or in the case of MOSFET, the threshold voltage reduces the efficiency. Schottky diodes can be an improvement over the standard p–n diode because of its low forward voltage (Jamali, Cole, Ranasinghe, & Zhu, 2004).

11.4.1 Tag Sensitivity and the Charge Pump Efficiency

The present operation range of passive RFID tags around 10 m is limited by tag sensitivity around −20 dBm with the transmitter at 36 dBm EiRP (equivalent isotropic radiated power) (Boaventura & Varvalho, 2016, Bolic, Simplot-Ryl, & Stojmenovic, 2010), i.e., the battery-free tag can only wake up its ID modulation and sensor circuits when sufficient RF power has impinged on its antenna. The reader receiver sensitivity around −75 dBm is often sufficient, as high reader sensitivity is also subject to more tag collision. In comparison, a typical Wi-Fi data link has receiver sensitivity around −110 dBm. If the tag can operate around −40 dBm impinging power (100 nW or 2.2 mV before the LC resonator), the operational range can be readily extended to 100 m with the receiver sensitivity around −110 dBm. This would greatly enhance the functions of IoT for indoor proximity intelligence. Another possibility is to use the TV band broadcasting power at 90 dBm around 500 MHz in a dual-band system, where the energy harvester is on a different frequency from the backscattering communication link. The tag with −18 dBm sensitivity can be woken up at a distance up to 10 km (Parks, Sample, Zhao, & Smith, 2013) with mild indoor RF attenuation!

The RF-to-DC converter by the Dickson charge pump in Fig. 11.10 exploits the nonlinearity around the turn-on voltage of the rectifying diodes, or equivalently the threshold voltage (V_{th}) of the diode-connected transistors. If the current–voltage (I–V) relation around the RF input voltage (V_{RFin}) is linear, no DC-balanced AC power can be converted to useful DC power. The diode and transistor nonlinearity around V_{th} is normalized by kT/q, where k is the Boltzmann constant, T is the temperature, and q is the elemental charge, which sets the fundamental limit of V_{RFin} at the output of the LC resonator. As V_{RFin} is DC balanced from the receiving antenna and LC resonator, V_{th} of the initial stages of the charge pump need to be close to 0 when V_{RFin} is close to kT/q. After the voltages at the intermediate storage capacitor have reached larger magnitude, the later stages of the charge pump should have slightly larger V_{th} to reduce the reverse leakage to achieve the maximum conversion efficiency (Barnett, Liu, & Lazar, 2009). However, similar to our previous discussions on transistor variations, most logic CMOS technologies have **process variations** of V_{th} on the order of kT/q, as well as limited **availability of zero-threshold transistors** or **zero-turn-on diodes**. The former will cause tags to have different wake-up sensitivity which may result in stronger tags overshadowing weaker tags in close proximity. The latter will restrict the technology choices and often raise the manufacturing cost (Marshall, Morys, & Durgin, 2015). We can use the electronically adaptive-threshold devices discussed in previous chapters here to improve the charge pump design. Each tag can be *programmed* for the diode turn-on voltages at all of the charge pump stages for optimal conversion efficiency at low V_{RFin} through built-in control circuits during the ID program phase. Hence, not only the tag sensitivity can be improved, but also the sensitivity variation can be greatly reduced.

A single-poly tunable-threshold diode-connected transistor based on the single-poly embedded EEPROM, similar to the NVM in Fig. 11.8, is shown in Fig. 11.10. During the program mode when the user-defined ID code is programmed, a high voltage is set to the large poly-coupled capacitor, and ground is set to the source–drain and small diode-connected capacitor to inject electrons to the poly gate, which results in V_{th} increase. We can set the voltage to remove electrons from the poly gate to decrease V_{th} as well. Note that the program mode only needs to be performed once before the wafer is diced or during a period when the tag is very close to the RF source so that the high-voltage generation for EEPROM programming can be achieved. The coupling ratio design of the multiple-gate EEPROM is similar to

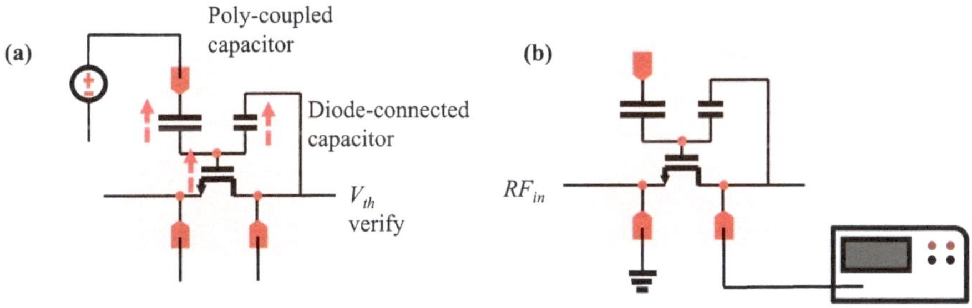

Fig. 11.10 The tunable-threshold diode-connected transistor for the rectifying diode in Dickson charge pumps: (**a**) During the program mode, a high voltage is set to the large poly-coupled capacitor to inject electrons to the poly gate, which results in V_{th} increase; (**b**) during the operational mode, the poly-coupled capacitor is left floated through the tri-state buffer to minimize the parasitic capacitance effect

previous NVM structures. Note that the eventual V_{th} can be very accurate by the program-and-verify scheme.

During the operational mode in Fig. 11.10b, the poly-coupled capacitor is left floated through the tri-state buffer to minimize the parasitic capacitance effect. The threshold voltage of the rectifying diode-connected transistor has already been set in the program mode and remains nonvolatile. Note that V_{th} in various stages can be different from detailed program pulse control to optimize the RF-to-DC conversion efficiency. In general, the higher stages will have slightly higher V_{th} as the threshold voltage loss will be a smaller penalty when the stage voltage is large, while the leakage current has to be reduced from the larger cross voltage.

A simulation study is performed to evaluate the V_{th} magnitude and variation effects on charge pump performance, based on a mixed-signal 180 nm CMOS technology, which offers four other transistor options for implementing the low turn-on diodes: low-threshold 1.8 V (N_LV_18_MM), low-threshold 3.3 V (N_LV_33_MM), zero-threshold 1.8 V (N_ZERO_18_MM), and zero-threshold 3.3 V (N_ZERO_33_MM). We then simulate a six-stage Dickson charge pump with $V_{\text{RFin}} = 250$ mV, which is boosted by the high-impedance resonator after the tag antenna. Here the transistor size W/L of the channel region is 25 and the storage capacitor $C = 2$ pF. As the storage capacitor dominates the layout area, all implementations have similar total cell size. The maximum output voltages for various diode implementations at $f = 1$ GHz and $I_{\text{load}} = 2$ μA are summarized in Table 11.1. It can be observed that the zero-threshold transistors (N_ZERO_18_MM and N_ZERO_33_MM) perform much better than transistors with V_{th} around 165 and 314 mV (N_LV_18_MM and N_LV_33_MM). Furthermore, the tunable-threshold transistor outperforms all other options significantly when the threshold voltages of the six stages are set according to the third row of Table 11.2, which indicates the effects of possible V_{th} variation.

Table 11.1 Charge pump output voltages (mV) by different transistors available in the 180 nm mixed-signal technology

$V_{RFin} = 250$ mV, $2\,ns = 6$, $I_{load} = 2$ μA, $W/L = 25$, $C = 2$ pF

	N_ZERO_33_MM	N_ZERO_18_MM	N_LV_33_MM	N_LV_18_MM	Our design
Description	Zero V_{th} with highest voltage at 3.3 V	Zero V_{th} with highest voltage at 1.8 V	Low V_{th} with highest voltage at 3.3 V	Low V_{th} with highest voltage at 3.3 V	Single-poly embedded EEPROM design
V_{th} (mV)	62.4	−193	314	165	Tunable during the program stage
V_{out} (mV)	994	1140	522	535	1620

Table 11.2 V_{th} (mV) in different charge pump stages and the resulting maximum output voltage V_{out} (mV)

$V_{RFin}=250$ mV, $2ns=6$, $I_{load}=2$ μA, $W/L=25$, $C=2$ pF						
$V_{th,1}$	$V_{th,2}$	$V_{th,3}$	$V_{th,4}$	$V_{th,5}$	$V_{th,6}$	V_{out}
0	30	30	60	130	180	1620
180	130	60	30	30	0	1102
0	0	0	0	0	0	1473

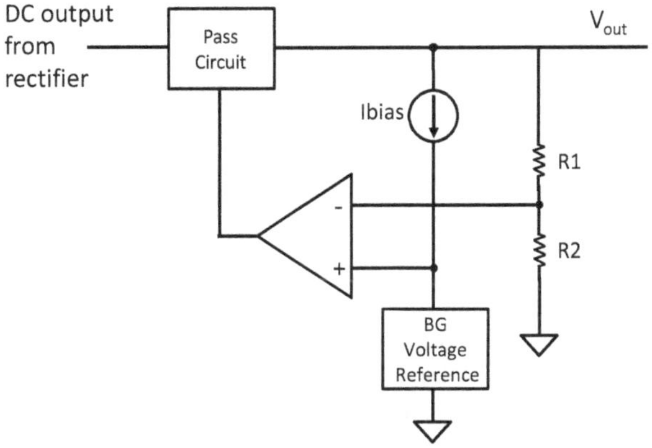

Fig. 11.11 Use of bandgap reference in a voltage regulator

11.4.2 Voltage Reference

The other critical element of a PMU is the voltage reference. Indeed the bandgap voltage reference circuits discussed in Chap. 4 are very useful in RFID circuits because of their low power and small area. A use of the BG reference in a voltage regulator is shown in Fig. 11.11, with the output voltage equal to

$$V_{out} = V_{ref}\left(1 + \frac{R_1}{R_2}\right) \tag{11.5}$$

Because of the low-cost requirement, RFID tag chip designs use many arts of non-logic device design using standard logic CMOS processes, including LNVM, native FETs, bandgap devices, and Schottky diodes.

11.5 Neuromorphic Computing with Floating-Gate Devices

With semiconductor scaling facing increasingly difficult challenges, there have been intense efforts to develop computing architectures that mimic how brain does computing. One of the main motivations is the power efficiency of the brain. Human brains consume only about 10 W of power while performing sophisticated cognitive and learning tasks that can take a supercomputer consuming megawatts of power to perform.

Human brain contains about 100 billion neurons and, between these neurons, up to quadrillion connections, called synapses. Each neutron is connected to many, up to tens of thousands, other neutrons through synapses, forming neutral networks. Neurons can be excited (firing) electrically by the collective actions of its connected neutrons. Models for neutron firing consist of a neutron receiving a stimulation, the action potential, from its upstream neutrons. When the total stimulation received by a neuron exceeds a threshold voltage, the receiving neuron will fire and cause its own action potential to be sent to its downstream neutrons. Strength of each synapse can be modeled by its synaptic weight. Neurons in the brain act in concert as a massively parallel system and can adapt to stimuli over time by strengthening and weakening connections, i.e., changing of synaptic weight through long-term learning.

One way to mimic brain-like computation is to through software. The field of artificial intelligence mostly uses this approach. Since this approach uses computer with conventional architectures, they have not come close to the human brain in terms of performance and power efficiency. For example, the much heralded AlphaGo which is the first computer program that defeated a human professional Go player utilized over 2000 CPUs and 300 GPUs,[2] consumes about 200 kW of power, to compete with one human player that consumes about 20 W of brain power. Thus for this particular task, human brain is about 10,000 times more power efficient. Google did announce that a new "tensor processor" was designed for AlphaGo with a tenfold increase in power efficiency. Analysis of computation efficiency has found that human brains are about one million times more energy efficient than general-purpose computers (Mead, 1990).

To design hardware that mimic brains, or neuromorphic computing, transistors have been used to model synaptic connections. For example, using SRAM cells as synapses, one chip with over 5.4 billion transistors emulating one million programmable spiking neurons and 256 million configurable synapses has been demonstrated (Merolla, Arthur, Alvarez-Icaza, & Modha, 2014). Researchers have also been looking at using analog floating gates as the building blocks for silicon that mimics the neural networks in brains (Lee, Sheu, & Yang, 1991), starting with the work of Mead on artificial neutral network (ANN) (Mead, 1990). This is based on a key attribute of the brain—that brains have capacity for long-term learning and memory.

The following is a simplistic model for a multiple layer artificial neutron network implementable with floating-gate transistors.

To build the model, we recognize that synaptic signals can be excitatory (causing an increase in firing rate for the downstream neutrons) or inhibitory (causing a decrease in firing rate in downstream neutrons), i.e., the synaptic weight can be negative. To implement this bipolar behavior, Fig. 11.11 shows an N input, M output ANN. Each input neuron is represented by a circle and each output neuron

[2] https://en.wikipedia.org/wiki/AlphaGo.

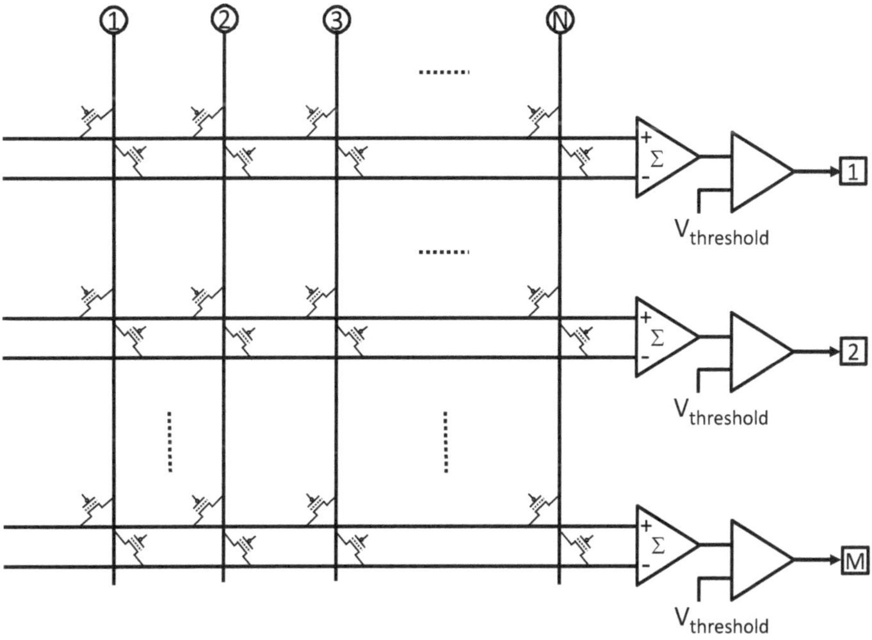

Fig. 11.12 An artificial neutral network implemented using floating-gate pFET as synaptic weight. Cycles represent input neutrons while squares represent output neutrons. The control gate signals for the floating-gate transistors are not shown

is by a square. Each synaptic connection in Fig. 11.12 can be implemented with a pFET floating gate, *synapse transistors* (Diorio, 1997). Using pFET-based floating gate allows bidirectional memory updates, and learn from an input signal without interrupting the ongoing computation.

The charge on the floating gate, as represented by the threshold voltage of the floating-gate transistor, determines how much contribution each input neuron has on an output neuron. The contributions from all input neurons are summed according to each neutron's weight. The weight is added for an excitatory signal and subtracted for an inhibitory signal. The total sum is then compared to a threshold to determine if the output neuron is excited or not. Multiple layers of the neutral network can be built using this model to achieve "deep learning" neutral network. The threshold voltage of each floating-gate transistor can be set by training of the ANN.

Mead and collaborators have used floating-gate devices to build the adaptive synapses nervous systems with memory and learning. More recently, there have been a lot of efforts in using emerging NVM memories such as phase changing memory (PCM) (Eryilmaz et al., 2014), resistive memory (RRAM), or memristors (Prezioso et al., 2015) in neuromorphic circuits.

Finally, we note that we need at least one device, e.g., floating-gate transistor or resistive memory elements, to emulate one synapse. Since there are over one quadrillion synapses in a human brain, it will be very challenging to be able to build a device to simulate the brain as a whole. The current effort has mostly focused on emulating a particular function of the brain, e.g., visual pattern recognition. The neutral networks are then used as accelerators and integrated to work with more general-purpose computers.

References

Barnett, R. E., Liu, J., & Lazar, S. (2009). A RF to DC voltage conversion model for multi-stage rectifiers in UHF RFID transponders. *IEEE Journal of Solid-State Circuits, 44*(2), 354–370.

Boaventura, A., & Varvalho, N. (2016). Measurement of sensitivity improvement in RFID tags. In *IEEE MTT 2016 International Microwave Symposium (IMS)*, San Francisco, CA, May 22–27, 2016.

Bolic, M., Simplot-Ryl, D., & Stojmenovic, I. (2010). *RFID systems: Research trends and challenges*. New York: Wiley.

Dickson, F. (1976). On-chip high-voltage generation in MNOS integrated circuits using an improved voltage multiplier technique. *IEEE Journal of Solid-State Circuits, 11*(3), 374–378.

Diorio, C. (1997). *Neurally inspired silicon learning: From synapse transistors to learning arrays*. Caltech, Ph.D. thesis, California Institute of Technology.

Eryilmaz, S. B., Kuzum, D., Jeyasingh, R., Kim, S., BrightSky, M., Lam, C., et al. (2014). Brain-like associative learning using a nanoscale non-volatile phase change synaptic device array. *Frontiers in Neuroscience, 8*, 205.

Glidden, R., Bockorick, C., Cooper, S., Diorio, C., Dressler, D., Gutnik, V., et al. (2004). Desgin of ultra-low-cost UHF RFID tags for supply chain applications. *IEEE Communications Magazine*, pp. 140–151.

Hyde, J., Humes, T., Diorio, C., Thomas, M., & Figueroa, M. (2003). A 300-MS/s 14-bit digital to analog converter in logic CMOS. *IEEE Journal of Solid-State Circuits, 38*, 734–740.

Jamali, B., Cole, P., Ranasinghe, D., & Zhu, Z. (2004). Design and optimisation of Schottky diodes in CMOS technology with application to passive RFID system. In *Proceedings of SPIE Smart Structures Devices and Systems II* (Vol. 5649).

Lee, B., Sheu, B., & Yang, H. (1991). Analog floating-gate synapses for general-purpose VLSI neural computation. *IEEE Transactions on Circuits and Systems, 38*, 654–658.

Marshall, B. R., Morys, M. M., & Durgin, G. D. (2015). Parametric analysis and design guidelines of RF-to-DC Dickson charge pumps for RFID energy harvesting. In *IEEE International RFID Conference* (pp. 32–39).

Mead, C. (1990). Neuromorphic electronic systems. *Proceedings of the IEEE, 78*, 1629–1636.

Merolla, P., Arthur, J. V., Alvarez-Icaza, R., & Modha, D. S. (2014). A million spiking-neuron integrated circuit with a scalable communication network and interface. *Science, 345*, 668–673.

Parks, A. N., Sample, A. P., Zhao, Y., & Smith, J. R. (2013). A wireless sensing platform utilizing ambient RF energy. In *IEEE Topical Conference on Wireless Sensors and Sensor Networks (RWW)*.

Prezioso, M., Merrikh-Bayat, F., Hoskins, B. D., Adam, G. C., Likharev, K. K., & Strukov, D. B. (2015). Training and operation of an integrated neuromorphic network based on metal-oxide memristors. *Nature, 521*, 61–64.

Rako, P. (2009, December 15). Analog floating-gate technology comes into its own. *Electronic Design News*. Retrieved from http://www.edn.com/design/analog/4312517/Analog-floating-gate-technology-comes-into-its-own.

Serrano, G., Kucic, M., & Hasler, P. (2008). *Digital-to-analog converter with programmable floating gate*. US Patent 7,408,397, Washington, DC.

Suh, K. D., Suh, B.-H., Um, Y.-H., Kim, J.-K., Choi, Y.-J., Koh, Y.-N., et al. (1995). A 3.3V 32Mb NAND flash memory with incremental step pulse programming scheme. *IEEE Journal of Solid-State Circuits, 30*(11), 1149–1156.

Wong, Y. (2007). *Adaptation in standard CMOS Processes with floating gate structures and techniques*. Ph.D. Thesis, University of Maryland.

Wong, Y., Abshire, P., & Cohen, M. (2004). Floating gate comparator with automatic offset manipulation capability. In *Proceeding of IEEE ISCAS* (pp. 529–532). IEEE.

Wong, Y., Cohen, M., & Abshire, P. (2007). On-line histogram equalization for flash ADC. In *Proceedings of IEEE ISCAS*. IEEE.

Xu, P., Wong, Y., Horiuchi, T., & Abshire, P. (2006). Compact floating-gate true random number generator. *Electronics Letters, 42*, 1–2.

Chapter 12
CMOS Biosensors

Abstract We will start from the general requirements in chemical and biosensors and then narrow down the specific advantages of CMOS implementation in view of signal transduction from the biological to the electronic domains. As the amperometric sensing on CMOS is similar to most popular mixed-signal designs, we will only focus on field-effect sensors with a polarizable electrode or interface. A general device based on the neuromorphic principles in the previous chapters will be presented for its structure, operation, and circuit models. Variations in device implementation will be examined under the unified neuromorphic circuit model. The interface between the electrode and the buffer media will be modeled with an electrical network. We then present sample measurements in different buffer media and examine the current difficulties in realistic operations with long-term reliability. We will conclude at future challenges and outlook for the biological interface to the CMOS world.

Keywords Biosensor • CνMOS • Signal transduction • Sensitivity • Specificity • Sampling bandwidth • Neuromorphic MOS • Reference electrode • Electrical double layer • Microelectrode array

12.1 Introduction

The vision of Internet of Things (IoT) has forecast thousands to millions of sensors in the local area network to provide ambient intelligence. We will first consider the general requirements in electronic sensors before we introduce what to aim for in CMOS-based biosensors. Electronic sensors convert a specific ambient signal to an electrical signal of voltage or current. Although most of the "signals" in the environment are continuous values represented by floating points, often the final sensor output will be in digital form to facilitate further processing and communication. Therefore, the first stage of a sensor system is often sampling where the signal is collected, held, and reset for the next sampling point, and the last stage is the analog-to-digital (AD) conversion. From the system point of view, we need to define carefully what the target information is at the sensor output, and any other factor that can change the sensor output will be categorized as interference or noise, whether the unwanted signal is systematic or random. Typical noises in sensors

Y. Ma, E. Kan, *Non-logic Devices in Logic Processes*,
DOI 10.1007/978-3-319-48339-9_12

include white noises like thermal and shot noises (white noises mean the noise frequency spectrum has nearly uniform distribution), flicker noise that originates from discrete particle impinging at interface, and quantization noise from the AD converter. From this point of view, electronic sensors share many common properties with multi-access radio receivers.

With the signal and noise definition in mind, sensing is NOT merely a "Eureka" event (I find it!), but the sensor performance has clear evaluation metrics:

- **Sensitivity**: The minimal quantity of the target analyte that can be detected. This is related to the false-negative statistics, i.e., the target exists but cannot be correctly detected.
- **Specificity**: The signal-to-noise ratio (SNR) when the target analyte is compared to noises and interferers. For example, a pH sensor output should be specific to acidity and not influenced by salinity. This is related to the false-positive statistics, i.e., the target does not exist but detection is positive due to other factors.
- **Operational range**: The ratio of maximum and minimal target values that can be distinguished. The maximum is often set by sensor saturation and the minimum by sensitivity above. This is similar to the concept of the dynamic range in radio receivers.
- **Bandwidth** (Hz): The sampling rate that the target analyte can be reliably collected.
- **Reproducibility**: The consistency of the constant or cyclic sensor output over repeated measurements. This will affect the calibration method as well.
- **Long-term stability**: As an example in biosensors, the stability can be characterized by the lack of slow drift due to surface fouling or aging as a result of ion contamination and coating material fatigue/plasticity.
- **Power consumption (W or Joule)**: The power consumed during the continuous sampling or the energy of each sampling.
- **Manufacturing cost**: The cost of the sensor system from sampling to final electrical output.

As an example, a biosensor may be deployed to measure the DNA segment concentration from a specific pathogen lysis for diagnostic purposes (Yager, Domingo, & Gerdes, 2008). The full procedure may involve several sequential steps (Allain, 2000):

1. **Sample preparation**: from patient blood to pathogen lysis,
2. **Signal amplification**: generation of more copies of targets if necessary, e.g., polymerase chain reaction (PCR),
3. **Signal transduction**: the sensor sampling the concentration through optical, mass, or electrical methods,
4. **Signal readout**: conversion to a digitized output.

Here the signal is either the number of the original pathogen cell counts as a diagnostic measure or simply a yes–no answer for infection exposure. The interferer can come from other nonspecific DNA segments, proteins, or molecules of similar molecular weights. The noise can come from salinity variation, readout background, and amplification noise during PCR. Alternatively, we may hope to measure the immunization response to such pathogen invasion directly. The signal is then the antibody biomarker concentration in the patient sample. Although the purpose of diagnostic is similar, immunological sensors have different noise and interferers to their sensitivity and specificity metrics.

In a conventional pathology lab, the four steps earlier involve standard biological and biomedical equipment and protocols such as incubator, autoclave, centrifuge, and microscopes. Not only the equipment is relatively large (though with mass production not very expensive), but also regulation on operator training is stiff. A conventional biosensing procedure needs significant time (such as 1–2 days) if incubation is needed in the sample preparation and signal amplification steps. With the advances in **Lab-on-a-Chip** (**LoC**), all four steps can be integrated with the help of microfluidics to finish the entire diagnostic process. However, Steps 1, 2, and 4 are out of the scope of this book, and we will only focus on signal transduction as the sensor function later.

12.2 Integrated Sensors: Amperometric or Field Effect

To convert chemical and biological signals from the target in electrolytes or biomedia to the electrical domain, three types of electrochemical sensors are commonly adapted (Janata, 2003): **Amperometric** sensors for the measurement of current with low impedance electrodes, **potentiometric** sensors for the measurement of voltage with very high impedance and hence nonamperometric (no current flow), and **chemiresistive** sensors for the measurement of resistance when the target penetrates into the sensing media. Various chemical sensors have been made along the way for over 40 years (Gardner, Varadan, & Awadelkarim, 2001; Kress-Rogers, 1997; Madou & Morrison, 1989) with different approaches, including chemiresistors, chemicapacitors (Steiner et al., 1995), microelectrodes (Kovacs et al., 1994; Lonergan et al., 1996), microcalorimeters, surface-acoustic-wave (SAW) devices (Wenzel & White, 1988), chemomechanical sensors (Lang et al., 1999), and field-effect-transistor (FET)-based sensors (Colapicchioni, Barbaro, Porcelli, & Giannini, 1991; van der Spiegel, Lauks, Chan, & Babic, 1983). Some of the sensors such as chemiresistors and SAW-based sensors are difficult to integrate with CMOS, while some sensors such as microelectrode arrays and FET-based sensors are ready for CMOS integration.

Chemical and biological sensors integrated with CMOS transistors have fundamental advantages in operation and cost. Existing CMOS technology provides ready analog-to-digital converter for sampling, fast computational speed for signal processing, large data-storage capacity, accurate control circuitry, standard wired (e.g., USB) and wireless (e.g., Bluetooth) communication support, and high reliability on the affordable mass production. The sensing electrodes, amperometric or nonamperometric, can be very small if needed by the sensing application, although most of the time the choice of sensing gate area is governed by the noise consideration instead of what can be achieved in production. Many of these factors can be used to improve the sensor performance defined earlier.

For amperometric **microelectrode arrays (MEA)**, the sensing principle is to measure an analyte-related current, either a natural current such as the ionic current in the neuron action potential cycles or an induced current by applying a test voltage such as the cyclic voltammetry. To complete a DC current path in the buffer fluid, the electrode has to induce a redox reaction at the surface for the continuity of electronic and ionic currents. This can be the target analyte such as oxidation of glucose or some balance provided by the electrode such as the Ag/AgCl reference electrode.

Note that buffer ion impinging will cause AC noises or transient IV characteristics of field-effect sensors, but not DC current. When the surface potential is higher than 1.2 V, water electrolysis and bubble formation are possible hazards in saline when a non-negligible DC current exists. This makes field-effect sensing attractive, especially for in vivo scenarios. The sensitivity of the MEA (Berdondini et al., 2005, 2006) is similar to standard measurements of electrical currents in a large DC range, e.g., from pA to mA by source-measuring units (SMU), or in a large AC range, e.g, from 0 to −90 dBm by radio receivers. Isolation of the leakage current path is similar to the practice in digital or analog circuit design, as the frequency of interest is often much lower than 10 MHz and all devices are in quasi-static modes. Therefore, we will limit our attention in this book to FET-based chemical and biosensors.

12.2.1 Floating-Gate-Based Devices as Sensors

The ISFET (ion-sensitive field-effect transistor), introduced in 1970 by Piet Bergveld is the first example of CMOS-based chemical and molecular sensors (Bergveld, 1970). An FET with a gate open to solution for signal transduction and the drain current as the readout is shown in Fig. 12.1a. "Ion sensitive" indicates that charged species in solution contribute to the operation of the device. The original ISFET has the gate oxide or the poly1 gate directly exposed to the electrolyte, while a possible modification is to have the gate coupling through the interconnect structure, and use the top metal (or another functional coating on the top metal) as the sensing gate. This is shown in Fig. 12.1b. This is often referred as the extended-gate ISFET or floating-gate ISFET. Note that the extended gate will assume one potential in the quasi-static operation determined by all of its nearby structures, but the capacitive coupling to the channel will determine the output current. This will become clear in the neuron MOS equations in the next section.

The concept of adapting the "floating gate" for sensing purpose has been explored further experimentally (Bousse, Shott, & Meindl, 1988; Smith, Huber, & Janata, 1984; van der Spiegel et al., 1983). For example, the extended gate structure in (van der Spiegel et al., 1983) is used as a signal line with one end coated with the chemically sensitive membrane, and the polysilicon layer on top of the floating gate serves as a guard/shield for the signal line. The floating gate in Smith

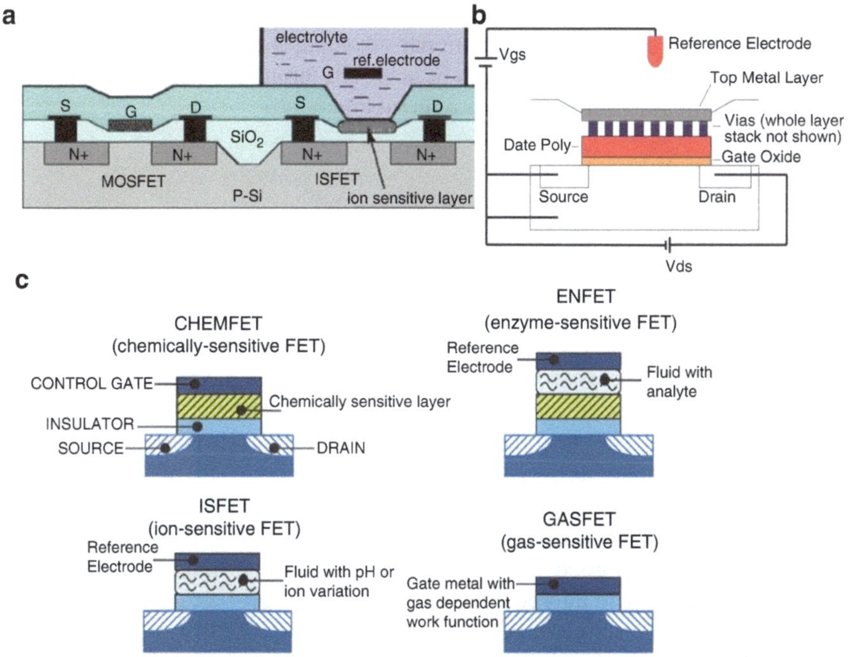

Fig. 12.1 (**a**) The original ISFET structure compared to a standard MOSFET on the *left*. The poly gate is exposed to solution allowing ions in the electrolyte to influence the surface potential (Bergveld, 1970). (**b**) A possible structural variation of ISFET with sensing gate or selective sensing layer coated on sensing gate to be placed on the top metal. (**c**) Possible sensing material variation of FET-based sensors

et al. (1984) and Bousse et al. (1988) is covered with Si_3N_4, which is adapted as the sensing layer in direct contact with the fluid buffer. To unify the operational principle and circuit models in ISFET-based sensors, we will use the relatively complex structure and circuit elements of **chemoreceptive neuron MOS** (**CνMOS**) transistors (Shen, Liu, Lee, Minch, & Kan, 2003) as an illustration, and then see the different approximations of sensing gate, control gate, and channel in various possible implementations.

As the field effect sensing is nonamperometric, we will first assume the sensing signal will come from capacitive coupling to the extended floating gate. If the extended gate is directly driven from a sensing potential, we can just use the approximation of infinite sensing capacitance. Furthermore, the total amount of charge at the floating gate can be configured by electron injection mechanism introduced in previous chapters for embedded flash, which can further enhance the signature analyses and the potential fluidic actuation (Jayant, Rajwade, Pollack, & Kan, 2010; Shen, Liu, Jacquot, Minch, & Kan, 2004). CνMOS transistors take the combined advantages of ISFET-based sensors (Bergveld, 1970), chemicapacitors (Steiner et al., 1995), and neuron MOS transistors (νMOS) (Shibata & Ohmi, 1992), and can be integrated into standard CMOS with a simple postprocessing of sensing surface functionalization and fluidic isolation. The extended floating gate in CνMOS transistors inherits the operational advantages of the nonlinear I–V in FET-based sensors: high sensitivity can be achieved through current differentiation and large dynamic range from voltage referencing.

12.2.2 Generalized CνMOS FET-Based Sensor Structure

Fabrication of CνMOS devices consists of two main steps: the physical design of the foundry-fabricated MOSFETs with the extended floating-gate structure where the sensing gate will be exposed in the contact pad step. The second step is to create specific sensing surface if desirable and the fluidic sample isolation in the postprocessing step. The floating-gate structure operates according to similar neuromorphic principles as discussed in Chap. 8. Figure 12.2a illustrates the CνMOS device structure in a possible two-poly process. Poly2 was chosen here as the sensing and control gates, because in the specific foundry process considered the interpoly oxide gives good capacitive coupling, and the design rules give overall compact device area. The interpoly oxide between the Poly1 floating gate and the Poly2 sensing and control gates has thickness of 58 nm, which allows for a larger range of voltage exploration for various target analytes.

From the equivalent circuits in Fig. 12.2b, we can see that the extended floating gate is one circuit node with a potential V_{FG} determined from the various capacitance input, identical to the coupling ratio consideration in Chap. 8. The potential of a floating-conductor structure will be determined from all of its capacitor coupling, normalized by the total capacitance seen from that floating structure. In the extreme case when one capacitor is larger than the sum of all of the other capacitors, the floating gate will basically follow the potential of that big capacitor, which can be seen as a virtual voltage reference. In Fig. 12.2b we have C_{CG}, C_{SG1}, C_{SG2} and the transistor channel coupling to the extended floating gate. Also, if we fix the output drain bias (often set at a relatively low value of 50–100 mV to keep the transistor in

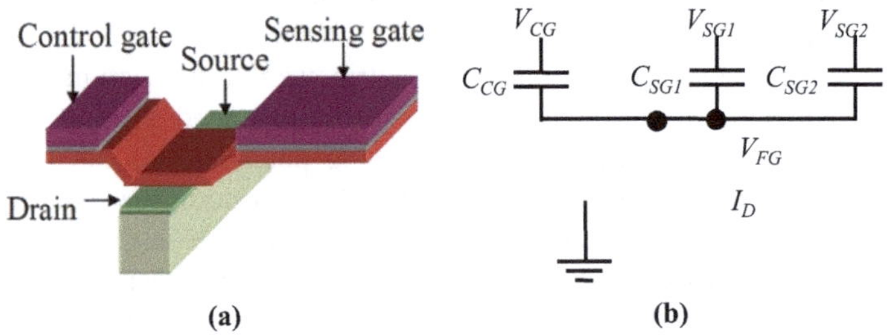

Fig. 12.2 (**a**) A 3D illustration for the multiple-input design of the CνMOS sensor with a modular isolation structure. An internal control gate is also shown. The sensing and control gates were implemented in poly2 in a two-poly process. (**b**) The equivalent circuits for one control gate and two sensing gates

the linear range and to avoid errors introduced from the drain coupling), the readout current I_D is entirely determined by V_{FG}. We can have at least three potentiometric operation modes: (a) fixed V_{CG} and use I_D to derive V_{SG}, (b) obtain $V_{CG} - I_D$ sweeps to derive V_{SG}, (c) fixed I_D and use the compensation V_{CG} to derive V_{SG}. Operation (a) has the simplest setup, but V_{SG} may drive I_D into deep subthreshold and therefore difficult to read. Operation (b) is robust in range, but can have larger power consumption during the V_{CG} sweep with unnecessary bulk heating. Operation (c) needs a feedback circuit, but otherwise robust in the differential sensing scheme. Note that we can achieve a capacitive amplification scheme readily when $C_{SG}/C_{CG} \gg 1$ in operation (c). If V_{SG} has a known small range as set by the fluidic reference electrode, the control gate can be omitted, and the CνMOS sensor becomes an extended-gate ISFET.

12.2.3 Structural Variations

For the sensing purposes, as long as the circuits can be represented by Fig. 12.2b, we can have various implementations for CνMOS, and consider other FET-based sensors as equivalent cases of CνMOS. Note that C_{CG} and C_{SG} are defined by the respective coupling area to the floating gates, and they can be implemented by PMOS capacitors as in the embedded flash in Chap. 8 or even Metal 1 to poly, if the coupling area is sufficiently large. The other terminal of the capacitor away from the floating gate can be routed through the interconnect structure to the top metal layer to become the sensing gate, which is similar to the strategy used in Fig. 12.1b. If the sensing gate potential is directly driven by the electrolyte, then there is no penalty in the long wire routing, as long as the sampling frequency is not too high (lower than the inverse of the total RC time constant of that routing line). If the sensing gate potential is affected by the electrolyte through field effects, the routing line is then another "floating gate," where the potential is also determined by all of its capacitive input or following the largest capacitive coupling.

From the capacitive coupling point of view, the original ISFET in Fig. 12.1a can be seen as the asymptotic cases where $C_{CG} = 0$ and $C_{SG} \to \infty$ in the CνMOS structure. Another possible implementation can be using the substrate bias as the control gate, as the channel is "capacitively" coupled to the bulk bias in the reverse bias, which is also known as "the body effect." Many non-CMOS sensor structures such as nanowires can also be modeled by the CνMOS capacitive network as shown in Fig. 12.3a, b, where the target analyte is capacitively coupled to the nanowire channel, and the back gate served as the control gate (Stern et al., 2008). Other possible structural variations in the control and sensing gate placement are shown in Fig. 12.3c–e, which have different design trade-offs. For technology that does not have high-voltage well isolation but with two poly layers, CG and SG can be implemented in poly 2 or metal 1 as in Fig. 12.3c. As the inter-poly and metal-to-poly oxides are relatively thick, the area penalty will be large, and the charge injection voltage will need to be high. The thick oxide can tolerate a large voltage

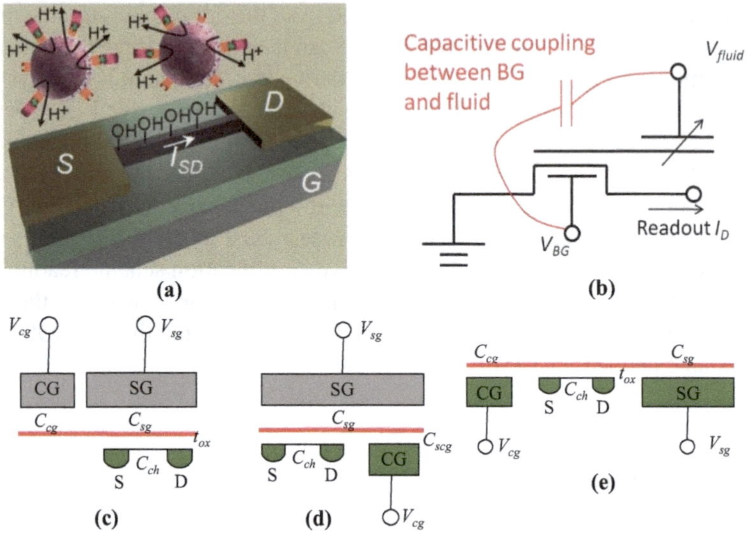

Fig. 12.3 (**a**) Nanowire-based sensor with surface functionalization (Stern, Steenblock, Reed, & Fahmy, 2008). (**b**) Equivalent circuit model similar to CνMOS capacitive network. (**c–e**) Structural variations in control and sensing gate placement

variation, especially from electrostatic charge. As well isolation is not required, many logic processes can directly implement this structure. If high-voltage well isolation or silicon-on-insulator (SOI) is available, then the inverted structures or source–drain–well shorted PMOS in Fig. 12.3d, e can be used.

There is no fundamental difference for the sensing transistor to be in PMOS or NMOS. NMOS is preferred when the source–drain–bulk-shorted PMOS transistor is used as the control gate and sensing gate coupling, as these PMOS capacitors will sit in their own well to achieve the readout transistor isolation. When nonvolatile charge will be tunneled in, the choice of PMOS and NMOS should follow the discussion in Chap. 2 for the preferred injection mode.

12.2.4 Fluidic and Electrical Isolation

To isolate the electrical connections to the chip and to prevent alkali ion diffusion into the chip, an electrolyte chamber or microfluidic channel is often formed to confine the fluid delivery to the sensing area. For structural integrity, a stiction layer between the chip top surface and the fluidic part is important, especially when pressure is used to deliver the fluidic sample. For prevention of ion contamination, the surface that touches the electrolyte will need to be noble metal such as Au and Pt. Aluminum can also be used when dense alumina has been formed on the surface. Metals have very small permeability for most ions, if corrosion is not a concern. The next choice will be silicon nitride, where the alkali ion permeability is

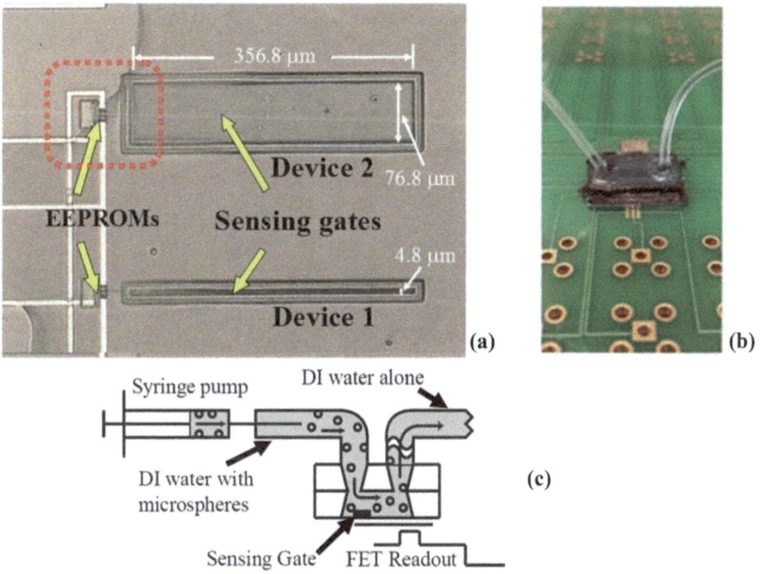

Fig. 12.4 (**a**) A microscopic image of two CνMOS transistors with different sensing-gate dimensions. The large sensing gate dimension is to accommodate the flicker noise from the buffer media. (**b**) Fluidic delivery with PCB integration. (**c**) A fluidic channel driven by syringe pump to deliver analytes to sensing gates

sufficiently low for years of operation. Note that this layer is to prevent ion diffusion, and often a functional layer such as self-assembled monolayer or antibody will be anchored on this layer for specificity.

In the previous two-poly example, stacked vias were opened to expose the poly2 to microfluidic chambers, as shown in Fig. 12.4. The microfluidic chamber is made by silicone elastomer, cured from a master wafer with DRIE patterns, because the elastomer has good stiction property to the spin-on oxide in the IC packaging. The sensing gate is made large to reduce the flicker noise and achieve a capacitive amplification. An alternative example of 2×2 sensor pixels by CνMOS is shown in Fig. 12.5 where the pixel area is encapsulated in a fluidic isolation chamber to avoid ion contamination to other parts of the chip.

As an example, the silicone elastomer mixture is composed of 20 parts of the elastomer base with 1 part of curing agent (Sylgard 184, Dow Corning). When the mixture is applied onto the master silicon substrate, a de-airing process is performed in a vacuum chamber for 30 min to get rid of tiny air bubbles during the curing process so as to reduce the surface roughness of the elastomer film, which degrades the stiction. The pattern is transferred from the master silicon substrate to the insulation elastomer layer by further curing the mixture on a hotplate set at 150 °C for 15 min. The encapsulated microfluidic channels for control fluid delivery are then shaped by bonding the patterned elastomer layer on top of the 4.6 mm × 4.7 mm CMOS chip. There are many other ways to make the electrolyte chamber and microfluidic channels, but that is outside the scope of this book.

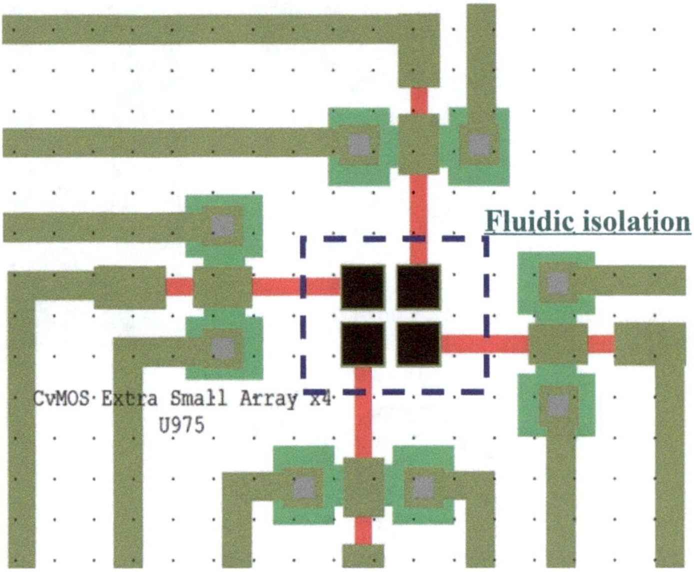

Fig. 12.5 A layout example of the 2 × 2 CνMOS sensor with the four sensing gate at the center for ease of microfluidic isolation

12.3 Circuit Models for CνMOS FET-Based Sensors

12.3.1 Neuron MOS Equations

The neuron MOS equation was first proposed by Shibata and Ohmi (1992) for neuromorphic devices. The characteristic feature of the νMOS transistor is that such a device can be constructed with any number of control-gate electrodes which couple into the floating gate through capacitive coupling (Burns & Powlus, 1996). The floating-gate voltage is established through charge sharing or capacitive voltage division, as a weighted sum of the voltages that are applied to the control gates. The weight on each control gate is directly proportional to the capacitance of that control gate and is normalized by the total capacitance of the floating gate. CνMOS shares the same capacitive network considered in the coupling ratio in Chap. 8, with the sensing gate as one of the inputs.

Capacitive-divider models (Minch, 1997) are adapted from νMOS. As discussed in Chapter 5, e.g. Eq. 5.1, the floating-gate voltage V_{FG} can be expressed as a weighted sum through charge sharing or capacitive voltage division, including those from all control and sensing gates labeled from 1 to X:

$$V_{\mathrm{FG}} = \frac{Q}{C_{\mathrm{T}}} + \frac{C_{\mathrm{gs}} V_{\mathrm{S}}}{C_{\mathrm{T}}} + \frac{C_{\mathrm{gd}} V_{\mathrm{D}}}{C_{\mathrm{T}}} + \sum_{j=1}^{X} \frac{C_j V_j}{C_{\mathrm{T}}} \tag{12.1}$$

where the total capacitance C_{T} can be written as

$$C_{\mathrm{T}} = \left(C_{\mathrm{ox}} \| C_{\mathrm{dep}}\right) + C_{\mathrm{b}} + C_{\mathrm{gs}} + C_{\mathrm{gd}} + \sum_{j=1}^{X} C_j \tag{12.2}$$

Q is the total static charge stored on the floating gate. C_j and V_j are the capacitance and voltage at the jth input gate. The capacitances C_{ox}, C_{dep}, C_{gs}, and C_{gd} follow the same definitions as in conventional MOS transistors. C_{ox} is the gate oxide capacitance, and C_{dep} is the depletion-layer capacitance in the silicon substrate under the gate area. C_{gs} and C_{gd} are the capacitances from the floating gate to the overlapped source and drain region, respectively. C_{b} is the capacitance from the floating gate to the bulk. Q is estimated to be only around pC level to create several volts of the threshold-voltage shift in CνMOS transistors. All potentials are bulk referenced.

Once we know V_{FG}, the output current I_{D} in the subthreshold region can be expressed as

$$V_{\mathrm{th}} = V_{\mathrm{th0}} - \frac{Q}{C_{\mathrm{T}}} - \frac{C_{\mathrm{gs}} V_{\mathrm{S}}}{C_{\mathrm{T}}} - \frac{C_{\mathrm{gd}} V_{\mathrm{D}}}{C_{\mathrm{T}}} \tag{12.3}$$

$$I_{\mathrm{D}} = \mu_{\mathrm{n}} C_{\mathrm{ox}} \frac{W}{L} \exp\left(\frac{\kappa (V_{\mathrm{FG}} - V_{\mathrm{th}})}{U_{\mathrm{T}}}\right) \cdot \left(\exp\left(-\frac{V_{\mathrm{S}}}{U_{\mathrm{T}}}\right) - \exp\left(-\frac{V_{\mathrm{D}}}{U_{\mathrm{T}}}\right)\right) \tag{12.4}$$

where V_{th0} is the threshold voltage of the intrinsic transistor. The body coefficient κ is defined as $\kappa = C_{\mathrm{ox}}/(C_{\mathrm{ox}} + C_{\mathrm{dep}})$ in the intrinsic transistor and U_{T} is thermal voltage. V_{S} and V_{D} are the voltages at the source and drain. Note that as the floating gate is not directly driven, Eq. 12.3 is the additional correction on the threshold voltage resulting from the coupling between the source–drain and the floating gate. If there is any additional drain-induced barrier lowering (DIBL) by V_{DS}, V_{th} needs to be further corrected.

When V_{FG} is higher than V_{th}, the transistor enters the above-threshold region, and is "turned on." When $V_{\mathrm{DS}} = V_{\mathrm{D}} - V_{\mathrm{S}} < V_{\mathrm{FG}} - V_{\mathrm{th}}$, I_{D} is in the linear region and can be expressed by

$$I_{\mathrm{D}} = \mu_{\mathrm{n}} C_{\mathrm{ox}} \frac{W}{L} \left((V_{\mathrm{FG}} - V_{\mathrm{th}}) V_{\mathrm{DS}} - \frac{V_{\mathrm{DS}}^2}{2}\right) \tag{12.5}$$

Just like the expression in regular MOSFET, there is a discontinuity at V_{th} for $I_{\mathrm{D}}(V_{\mathrm{FG}})$ modeled by Eqs. 12.4 and 12.5. This is just because the simplification during the model derivation step. The actual measurement is surely continuous. The discontinuity can be remedied by the **EKV** (Enz–Krummenacher–Vittoz) model, which is based on a smoothing function that asymptotically approaches exponential and quadratic at the two extremes of subthreshold and above-threshold linear:

$$
\begin{aligned}
I_\mathrm{D} = {} & I_\mathrm{S}\log^2\left(1+\exp\left(\frac{\kappa(V_\mathrm{FG}-V_\mathrm{th})-V_\mathrm{S}}{2V_\mathrm{T}}\right)\right) \\
& - I_\mathrm{S}\log^2\left(1+\exp\left(\frac{\kappa(V_\mathrm{FG}-V_\mathrm{th})-V_\mathrm{D}}{2V_\mathrm{T}}\right)\right)
\end{aligned}
\tag{12.6}
$$

The EKV model, although the functional form is complex for intuition, is $C-\infty$ (continuous for a function and all of its derivatives to the infinite order) and includes the body effect. Therefore, the EKV model is the more appropriate form in circuit simulation. Remember that all biases in EKV are reference to the bulk potential.

For FET-based sensor operations, when we do not need a voltage self-gain in the underlying transistor, we often use the linear or subthreshold region to save power consumption, as excessive self-heating can be problematic to the electrolyte solution. I_D is very small in the subthreshold region and can have a limited bandwidth for fast sampling. Therefore, we often use V_CG or a feedback circuits to bring the readout to the linear region.

We can define the subthreshold slope S_CG seen from the control gate V_CG ramp as

$$
S_\mathrm{CG} \equiv \left(\frac{\partial \log_{10}\left|\frac{I_\mathrm{D}}{I_\mathrm{th}}\right|}{\partial V_\mathrm{CG}}\right)^{-1} = \frac{U_\mathrm{T}\ln 10}{\kappa}\cdot\frac{C_\mathrm{T}}{C_\mathrm{CG}}
\tag{12.7}
$$

The subthreshold slope S_CG (in mV) stands for the change of V_CG that can cause a decade of change in I_D. A smaller S_CG corresponds to a steeper slope in the *I–V* characteristics. In CνMOS devices, C_CG represents a constant capacitance depending on the dimension of the control gate, and therefore Eq. 12.7 indicates the subthreshold slope is directly related to C_T, assuming the body coefficient κ only has little or no variation during the subthreshold operation.

As expressed in Eq. 12.2, the total capacitance C_T includes all capacitance as seen from the extended floating gate. For the CνMOS device with a single sensing gate, the schematic of capacitive loads can be illustrated as in Fig. 12.6. Here of particular interest is the sensing-gate capacitance C_SG, which contributes to the change of the total capacitance and hence the subthreshold-slope S_CG variation. C_SG consists of the capacitance formed by inter-poly oxide and the capacitance by the solid–liquid interaction, i.e., the **electrical double-layer** (EDL) capacitance C_EDL (Adamson, 1982; Bard & Faulkner, 2001). V_SG is now the surface potential at the sensing gate. When the fluid has an Ag/AgCl nonpolarizable electrode to set up the fluidic bulk potential at V_B, V_SG can still be different from V_B depending on the ionic profile in the EDL, which can in turn depend on the surface charge, salinity, and steric effects (Kilic, Bazant, & Ajdari, 2007; Liu, Huber, Tabard-Cossa, & Dutton, 2010). The difference between V_SG and V_B may not be constant either, as the ion profile can change with V_{SG}. If we opt to use a polarizable electrode such as Au, Pt, or carbon fiber, we have to treat additionally the surface potential and ion profile on that polarizable electrode surface. We will look at more details of EDL modeling in the next section.

12.3.2 *The Role of Fluidic Polarizable and Reference Electrodes*

To set V_{FG} in the correct operating point, we need knowledge of the fluid bulk potential φ_{bulk}, as shown in Fig. 12.6. Note that the standard Ag/AgCl reference electrode can carry a continuous faradic current (nonpolarizable) and uses the chemical balance between Ag^+ and Cl^- to set a constant electrochemical potential for Cl^-, and for Cl^- **only**. The reference electrode is nonpolarizable where a faradic current can pass with negligible overpotenial and without polarization. This has involved several assumptions (Huang & Huang, 2002; Wong & White, 1989):

1. The buffer media have dominant Cl^- as the an-ion.
2. The surface areas of Ag and AgCl nanocrystals are sufficiently large to allow instantaneous (nonrate limiting) redox operation.
3. The overpotential, i.e., the difference between the potential of the half reduction reaction and the potential at which the Ag-to-AgCl redox event is experimentally observed, is negligible.
4. Silver ions are toxic and hazardous to many living cells, and therefore the Ag/AgCl electrode cannot be applied in vivo, or when in vitro cellular vitality is important for the sensing.

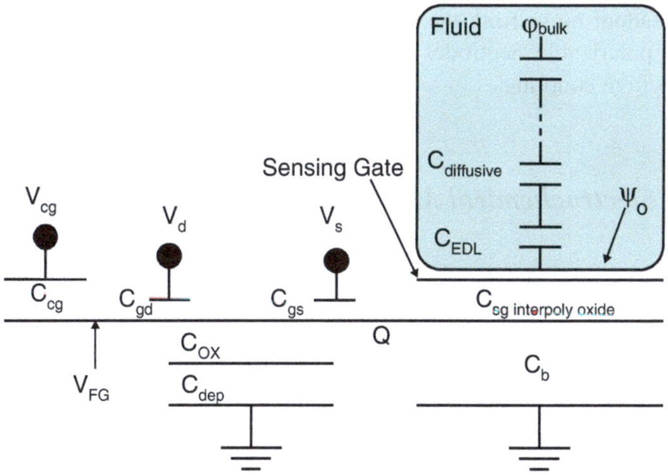

Fig. 12.6 The capacitive network includes capacitors contributed by the control gate, sensing gate, and the parasitic capacitors from the MOS transistor

In addition to these questionable assumptions in realistic experiments, Ag ions are toxic and cannot be used for in vivo applications. Therefore, although Ag/AgCl reference electrode has still been regarded as the gold standard, many efforts had been applied to decrease the dependence on the reference electrode (Wong & White, 1989) or reduce its invasive impacts (Huang & Huang, 2002).

For electrodes of in vivo measurements, a polarizable (nonfaradic) electrode such as Pt and carbon fiber is often used to give a reference electrostatic voltage to the body fluid. For FET-based sensing purposes, this reference voltage is often the same voltage at the transistor source electrode. Remember that when there is no direct amperometric electrode in touch with the buffer fluid, the fluidic potential will mainly follow the voltage of the largest capacitance, just like the floating-gate equation in Eqs. 12.1 and 12.2. That is, with all polarizable contacts, the buffer behaves like a (leaky) floating gate. Therefore, either we still need a sizable polarizable electrode in the buffer, or we can design a large capacitive coupling from the sensing chip.

Many electrochemical experiments rely on the use of the Ag/AgCl reference electrode for repeatability. Note that the only chemical balance achieved here is the chlorine ion where there is a continuous supply and drain of Cl^- during the continuous operation and where the constant bulk an-ion concentration can be maintained. All other cat-ions and an-ions, including even Cl^- when Ag/AgCl is depleted or not used, have a constant total dose within the fluidic volume regardless of the variation in surface potentials. For example, if Na^+ is pushed away from the surface by a positive surface potential, it would has to pile up in some other places.

Both polarizable and reference electrodes are effective in removing the static electron charge in the electrolyte. Static charge, often due to tribological interaction with the container walls, is well studied in fuels and gasoline to avoid the instantaneous combustion from electrostatic discharge or spark generation. Although such concerns are not needed for most electrolytic buffers, static charge can affect the FET readout by shifting the fluidic potential with respect to the FET source. Therefore, polarizable electrodes can still help stabilize the readout more than pure large capacitive coupling.

12.3.3 Electrochemical Models for the Fluidic Interface

At the metallic sensing gate of a CνMOS transistor, image charge is induced upon the solid–liquid interaction, and an electrical double layer (EDL) is formed in the liquid side, as illustrated in Fig. 12.7. In the Gouy–Chapman–Stern (GCS) interface model, we have a fixed layer of adsorbed ions (including protons) at the solid–liquid interface which is determined by the chemical functional group on the surface. For example, to consider proton adsorption, the charged hydroxyl group (OH^-) is most critical. The charge centroid of this interface layer defines the **inner Helmholtz plane** (**IHP**). We can then have a densely packed ion layer when the surface potential is sufficiently high (Kilic et al., 2007), whose outer boundary defines an

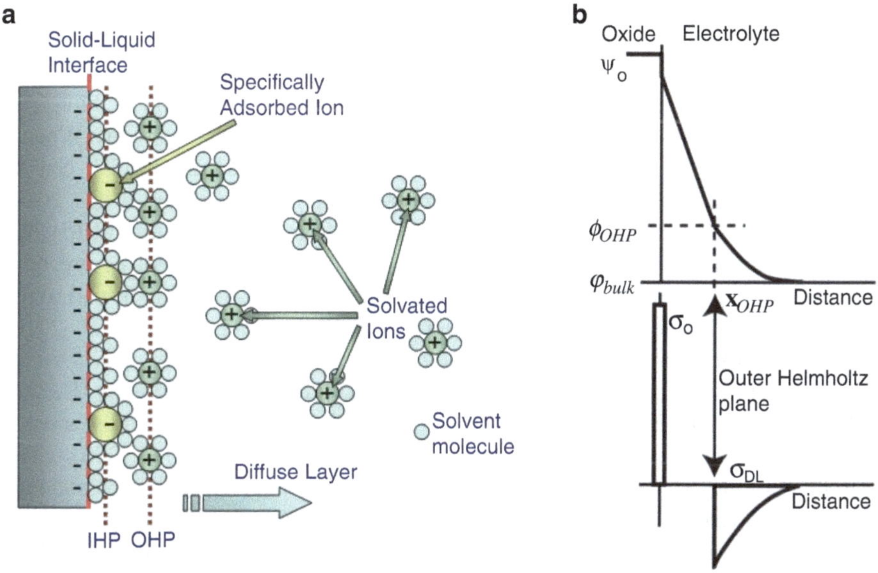

Fig. 12.7 The illustration of an electrical double layer. (**a**) The Outer Helmholtz Plane (OHP) represents the closest plane at which nonspecifically adsorbed ions can arrive. (**b**) Potential and net charge distribution for the GCS model

outer Helmholtz plane (**OHP**). Due to the **steric hindrance** (caused by the size of the hydrated ions), the densely packed ion layer can be more than a monolayer but not covalently bonded. Outside OHP, we then have a **diffusive layer** of free-moving ions (the layer that can be driven in electro-osmotic drive).

According to GCS model (Bard & Faulkner, 2001), the EDL capacitance per unit area can be expressed as

$$\frac{1}{C_{\mathrm{EDL}}} = \frac{1}{C_{\mathrm{H}}} + \frac{1}{C_{\mathrm{DIF}}} \tag{12.8}$$

where C_{H} is the Helmholtz-plane capacitance per unit area and can be written as

$$C_{\mathrm{H}} = \frac{\varepsilon_{\mathrm{buf}}\varepsilon_0}{x_{\mathrm{OHP}}} \tag{12.9}$$

and the diffuse capacitance per unit area C_{DIF} can be estimated by

$$C_{\mathrm{DIF}} = \sqrt{\frac{2\varepsilon_{\mathrm{buf}}\varepsilon_0 z^2 q^2 n}{k_{\mathrm{B}}T}}\cosh\left(\frac{zq\phi_{\mathrm{OHP}}}{2k_{\mathrm{B}}T}\right) \tag{12.10}$$

where $\varepsilon_{\mathrm{buf}}$ is the relative permittivity of the solution; ε_0, the permittivity in vacuum; z, the magnitude of charge on the ions; n, the number concentration of the potential-determining ions (Hunter, 2001) in the liquid bulk; k_B, the Boltzmann constant; T, the temperature; x_{OHP}, the thickness of the condensed ion layer (Kilic et al., 2007; Storey & Bazant, 2012); and ϕ_{OHP} is the potential at the OHP. The OHP can be envisioned as the closest plane at certain distance x_{OHP}, where the solvated ions can best approach due to the finite size of those nonspecifically adsorbed ions. When a densely packed ion layer is formed under high surface potential which can be physically observed by **atomic-force microscopy** (AFM) (Siretanu et al., 2014) or reflection ellipsometry (Wang, Zhao, Duits, Mugele, & Siretanu, 2015), the surface is said to reach its **steric limit**, and further complication of descreening (i.e., ε_{buf} becomes smaller) and charge reversal (i.e., counter-ions may be part of the steric layer) (Landheer, Aers, McKinnon, Deen, & Ranuarez, 2005; Liu et al., 2010) can modify the expression in Eq. 10.8.

The values of ϕ_{OHP} usually range from tens to hundreds of millivolts (Hunter, 2001), and they indicate both the charge condition at the solid–liquid interface and the closest distance the solvated ions can reach (Bard & Faulkner, 2001). The structure of the electrical double layer can affect the interfacial interaction and vice versa, considering the ions in the steric layer is not covalently bonded. Particularly in poly-electrolyte with multiple cat-ions such as Na^+, K^+, Mg^{2+}, and Ca^{2+}, which are parts of the standard composition in biological buffer solution such as **phosphate buffered saline** (PBS). Change of ionic composition in the steric layer can be VERY low (on the order of seconds to minutes) due to the dense packing and had been long suspected as one of the reasons for the slow drift in surface potential.

The free-moving ion species can only approach the boundary at OHP, and the ϕ_{OHP} is one of the important factors determining the contribution of the diffuse-layer capacitance C_{DIF}, as in Eq. 12.10, which is often smaller than C_H in series and hence set the observable capacitance. Although the relative permittivity of the buffer ε_{buf} is known to be potential dependent as well due to local polarization level (Liu et al., 2010), it is assumed to be constant in our extraction of ϕ_{OHP} for the ease of calculation in the conceptual demonstration.

Note that the sensing gate potential V_{SG} in Fig. 12.2 will be the surface potentialΨ_0 denoted in Fig. 12.7b, which will be a function of the charge composition in the condensed layer and the interface adsorbed charge density σ_0. At high surface potential, as σ_0 is not directly measurable and cannot be pinned down easily (depending on the surface functional group and prior history of acid/salt exposure), we do not have an accurate evaluation of the charge composition in the condensed layer, which renders Ψ_0 hard to predict even we know the bulk concentration. The Gauss law at the interface can be used to evaluate the electric field in the oxide, but we still do not have sufficient information to distinguish Ψ_0 unambiguously. Therefore, differential sensing or V_{CG} compensation to fix V_{FG} is a better way to evaluate the fluid ionic conditions than just using I_D to evaluate a variable V_{FG} under different ionic distribution.

12.3.4 Deposition of a Functional Layer in Postprocessing

Selectivity and sensitivity broadly comprise two most important figures of merit for a sensor, with selectivity often accomplished through careful selection of functional materials in the case of chemical and biological sensing. For chemical sensing, one option is the deposition of a compound such as ionomers or **self-assembled monolayers** (SAMs) that are selective for one or a group of chemicals (Albert et al., 2000; Freund & Lewis, 1995). Another option is the use of a filter to allow only one type of analyte to reach a broadly sensitive sensor interface. For example, various kinds of insulators have a predictable voltage response to the presentation of ions (van Hal, Eijkel, & Bergveld, 1996). Materials such as SiO_2, Si_3N_4, and Ta_2O_5 can be grown or deposited on the interface and have predictable Nernst responses of -45, -54, and -58 mV/dec, respectively, with -60 mV/dec being ideal. Ta_2O_5 functional coating has been used in acidity ISFET sensors for many years.

For biological detection, surface selectivity is usually accomplished by depositing a layer intended to grab a molecule that has high selectivity for the analyte of interest, such as the use of gold/thiol or (3-aminopropyl) triethoxysilane (APTS) chemistry to bind a specific antibody with a given antigen or protein as the target analyte (Jacquot, Muñoz, Branch, & Kan, 2008; Ulman, 1996).

There are many techniques for applying thin functional coating layers to a sensor interface. Techniques for deposition during fabrication include low pressure vapor deposition (LPCVD), plasma enhanced chemical vapor deposition (PECVD), molecular vapor deposition (MCVD), sputtering, and evaporation, which is often not at the CMOS designer's disposal. Postprocessing steps on the sensing gate surface of the standard CMOS chip include molecular printing, dip coating, dip pen, spin on coating, sol–gel process, SAMs, and aerosol spray. Adhesion and integrity of these postprocessing films are critical for the long-term reliability of the CMOS sensors.

Last but not least, the selectivity through functional groups with very high affinity does cause much confusion in the literature, especially in the case of low bulk concentration and fast sampling when the surface cannot reach equilibrium within the sampling time. Consider the following surface adsorption reaction equation:

$$A_{(1)} + *_{(s)} \underset{k_r}{\overset{k_f}{\Leftrightarrow}} A^*_{(s)} \tag{12.11}$$

where $A_{(1)}$ is the target analyte in the fluid, * is the surface adsorption site defined by the functional group, $A^*_{(s)}$ is the adsorbed target on the surface, and k_f and k_r are the forward and reverse reaction constants. High affinity of A to the functional group implies k_f is very large. The transient response for $A^*_{(s)}$ can be expressed as

$$A^*_{(s)}(t = t_1) = A^*_{(s)}(t = t_0) + \int_{t_0}^{t_1} \left(k_f \left[A_{(l)}\right] - k_r \cdot A^*_{(s)}(t) \right) dt \qquad (12.12)$$

When $A_{(l)}$ has sufficient supply in the bulk fluid, the bulk concentration $[A_{(l)}]$ can be regarded as constant, or we will enter into the source depletion mode and Eq. 12.11 cannot readily achieve chemical balance.

The sensor output is a function of the adsorbed ion density and will only be stable when $A^*_{(s)}$ has reached its steady state, which implies:

$$\left[A_{(l)}\right] = \frac{k_r}{k_f} \cdot A^*_{(s)} \qquad (12.13)$$

Note that $[A_{(l)}]$ can only be reliably determined from $A^*_{(s)}$ when the chemical balance of Eq. 12.11 has been achieved. Otherwise $A^*_{(s)}(t = t_1)$, and hence the translated $[A_{(l)}]$, will depend on $A^*_{(s)}(t = t_0)$ and the sampling period $(t_1 - t_0)$. For very large k_f, or equivalently highly selective adsorption with very small k_r, chemical balance of Eq. 12.11 can take a long time to achieve especially when $[A_{(l)}]$ is low.

12.3.5 Transient Models

For transient analyses, consider the time-varying voltage signal at the control gate ($\tilde{v}_{CG}$) can be expressed as

$$\tilde{v}_{CG} = V_{CG0} + V_{CG} \sin \omega t \qquad (12.14)$$

where V_{CG0} is the offset DC control-gate voltage and V_{CG} and ω are the magnitude and the angular frequency of the applied AC voltage signal. The sinusoidal AC input has two advantages over DC-only input: (1) AC responses can be treated with simpler small-signal linear models; (2) the output can be filtered at the inquisitive frequency to reduce the noises, especially away from the low-frequency Flicker noise. However, when the test signal period is shorter than the time required to achieve the chemical equilibrium, the surface adsorption of the analyte will stop responding to the AC signal, similar to the cutoff frequency in transistor operations.

Assume the constant drain voltage V_D of the MOS transistor is biased at a small value and $V_S = 0$, the above-threshold transient drain current i_D is in the linear region and can be expressed by

$$i_D = \mu_n C_{ox} \frac{W}{L} V_D \left[\left(\frac{Q + C_{CG} V_{CG0} + C_{CG} V_{CG} \sin \ \omega t}{C_T} - V_{th0} \right) + \left(\frac{C_{gd}}{C_T} - \frac{1}{2} \right) V_D \right] \tag{12.15}$$

The induced charge at the solid–liquid interface and the according total net charge Q affects the drain current, which is also influenced by the variation of the total capacitance upon the formation of electrical double layer. In addition, the change of the interface charge may create fluctuation in the potential at the Outer Helmholtz Plane ϕ_{OHP} in the liquid bulk, and cause C_{SG} to change accordingly with the frequency of the AC signal. This effect further influences the total capacitance C_T, the C_{CG}/C_T ratio on both the swinging amplitude and the offset voltage, and hence the drain current.

The transient measurement is useful in the step response and monotone excitation and sensing. For example, i_D can go through a bandpass filter or a lock-in amplifier to investigate only the part with the corresponding frequency, which can be viewed as a generalized impedance spectroscopy with flexibility in the operating point and excitation magnitude. When larger molecules are involved in the EDL, the frequency domain information (magnitude and phase) is valuable to improve specificity (Jayant et al., 2013b).

12.4 Sample Sensor Measurements

The capacitive load and static charge introduced onto the floating gate cause specific changes in both subthreshold slopes and threshold voltages. The total amount of nonvolatile charge Q on the floating gate further influences the interaction with the fluid at the interface. The drain current from the control-gate voltage sweep then serves as a signature for the fluid and its solute contents. High sensitivity is derived from the high transconductance gain of the MOS transistor and a large dynamic range is achieved concurrently from the large control-gate voltage V_{CG} sweep.

After measuring the full-swing I–V curves for each type of sample fluid in the encapsulated channel, we can also opt to tunnel electrons onto and off the floating gate of the CνMOS device by a fixed bias at the control gate, 30 V for tunneling in and -30 V for tunneling out. The I–V characteristics of various sample fluids are separately recorded after the electron-tunneling operations. The drain voltage is biased at 0.5 V for all the measurements to prevent hot carrier injection. The I–V traces in bare devices and when the microfluidic chamber is filled with deionized water, acetone, and saline are shown in Fig. 12.8. For deionized (DI) water and acetone, the subthreshold slopes change significantly due to the dominant diffuse capacitors in the fluid. The different I–V behavior of water and acetone on polysilicon primarily stems from the smaller ionic strength of acetone. The pH value for DI water is around 5.5 and around 7 for acetone, which is used to estimate

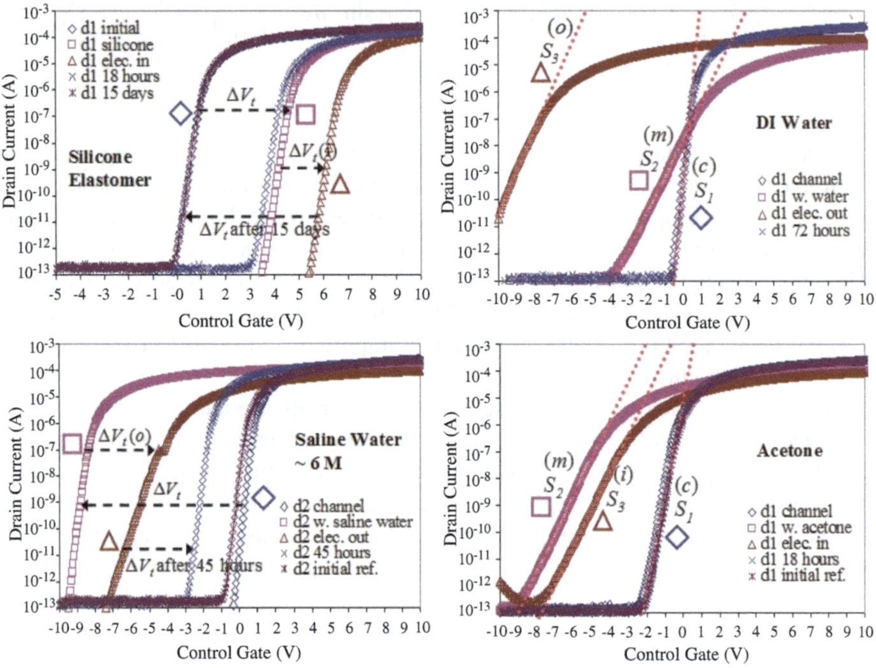

Fig. 12.8 Saturated saline water was delivered to the channel, and then the electrons were tunneled out of the floating gate. Note the distinctive shift of the threshold voltage

the number concentration n in Eq. 12.10. We assume the sampling time is sufficient to build the proton or cat-ion balance at the surface.

The charge interaction further suggests the feasibility of an actuation scheme, in which the amount or distribution of the interfacial charge can be modified in correspondence with the charge tunneled from the control gate into the floating gate. This type of actuation can be utilized in modifying the surface hydrophobicity (e.g., electrowetting (Lee & Kim, 2000; Matsumoto & Colgate, 1990), or in electroosmotic (Kohashi & Kurosawa, 1991; Manz et al., 1994) applications).

The generic SiO_2 sensing surface has distinctive response to salinity and acidity from both the contribution of Ψ_0 (which affects the threshold voltage V_{th}) and C_{DIF} (which affects the subthreshold slope S_{CG}). Sample IV sweeps are shown in Figs. 12.9 and 12.10. Note that due to the complicated ion composition in EDL, it is better to use to entire sweep to determine the salinity and acidity than a direct interpretation of I_D.

We also monitor the I–V relationship at the transient mode on CνMOS transistors with sample fluids in the microfluidic channels. An AC signal with 5-V amplitude is applied at the control gate, with the frequency ranging from 1 Hz to 100 kHz, as shown in Fig. 12.11. The monitored drain current is amplified and converted to the voltage signal through a transimpedance amplifier with the sensitivity set at 50 μA/V, and the output voltage is then examined on an oscilloscope. At

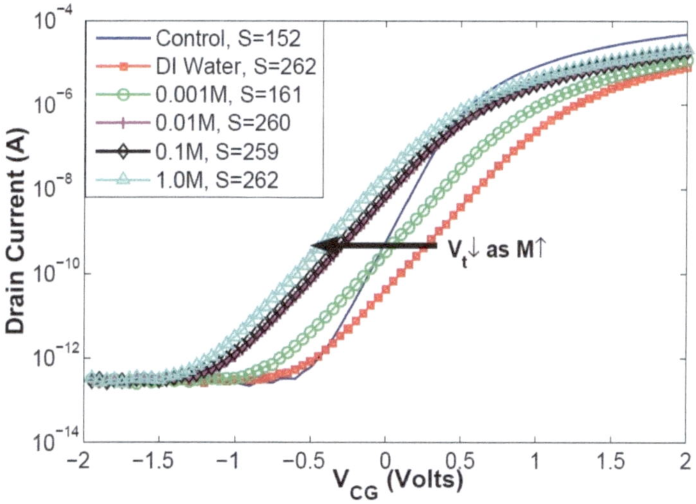

Fig. 12.9 Varying salinity on CνMOS sensor with SiO_2 as the sensing gate coating

Fig. 12.10 Varying acidity on CνMOS sensor with SiO_2 as the sensing gate coating. The solution is prepared by mix of CH_3COOH and NH_4OH

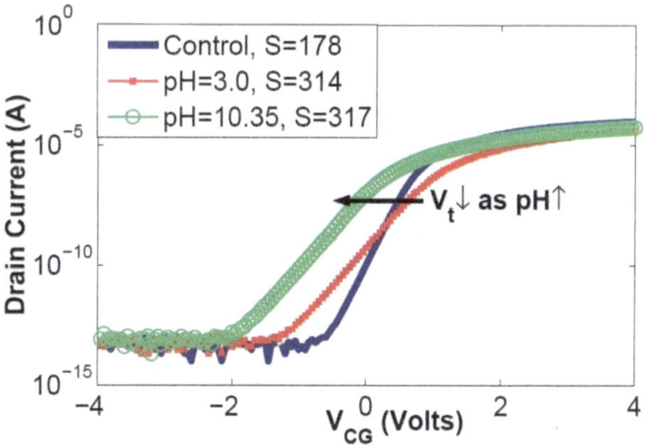

frequencies of 1 Hz–10 kHz the quasi-static behavior has been valid for the control devices as well as the devices with water and saline. Above 10 kHz, significant deviation has been observed, which corresponds to the regime when the ions stop responding to the AC excitation.

12.5 Summary and Further Reading

Although the fluidic interface has still many complicated electrochemical effects to be resolved before the FET-based sensors can be predictively modeled, there are several sensing modality such as V_{CG} compensation and general

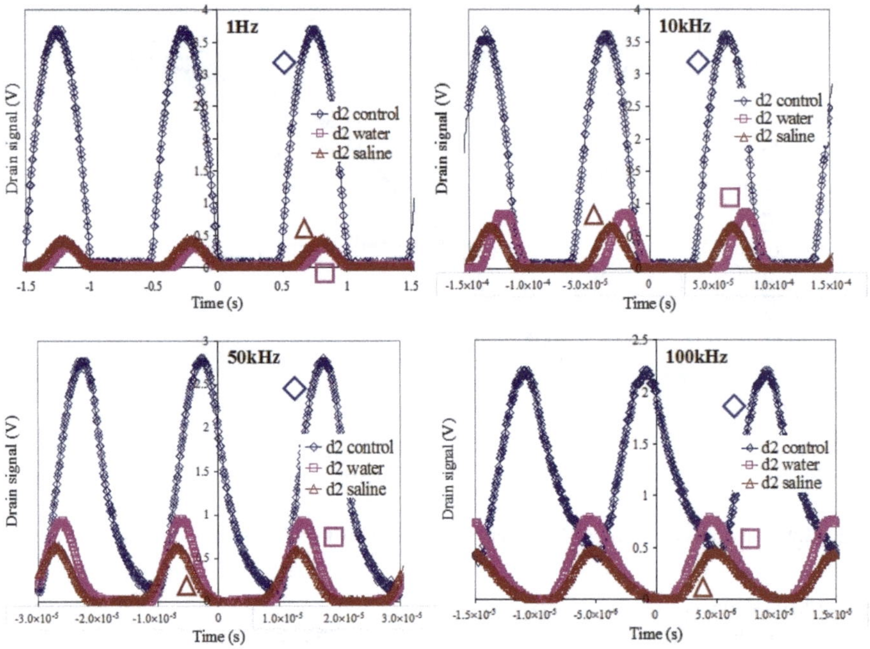

Fig. 12.11 Saturated saline water was delivered to the channel, and then the electrons were tunneled out of the floating gate. Note the distinctive shift of the threshold voltage

impedospectrocopy that can be used today for reliable sensing (Gordon et al., 2015). Research on interface condition and how it can affect sensor reading for acidity, salinity, molecules, action potentials, and vesicle release (Jayant et al., 2013a, 2013b; Jayant, Auluck, Rodriguez, Cao, & Kan, 2014) will still need more efforts for broader sensing applications and controllable actuation. The electronic interface to molecules and cells holds many promises to investigate the secrets of life itself, as at the present time only the electronic devices are up to the task requirements in view of the feasible geometrical scale, the computing/communication system support, and the cost structure. Using the deep integration with CMOS, as we finally have such precise, affordable, and powerful tools after the invention of microscopes, we are at the door of a new research paradigm.

References

Adamson, W. (1982). *Physical chemistry of surfaces* (4th ed.). New York: Wiley.

Albert, K. J., Lewis, N. S., Schauer, C. L., Sotzing, G. A., Stitzel, S. E., Vaid, T. P., et al. (2000). Cross-reactive chemical sensor arrays. *Chemical Reviews, 100*(7), 2595–2626.

Allain, J. P. (2000). *Clinical & Laboratory Haematology, 22*, 1–10.

Bard, J., & Faulkner, L. R. (2001). Double-layer structure and adsorption, Chap. 13. In *Electrochemical methods: Fundamentals and applications* (2nd ed., pp. 534–577). New York: Wiley.

Berdondini, L., Chiappalone, M., van der Wal, P. D., Imfeld, K., de Rooij, N. F., Koudelka-Hep, M., et al. (2006). A microelectrode array (MEA) integrated with clustering structures for investigating in vitro neurodynamics in confined interconnected sub-populations of neurons. *Sensors and Actuators B: Chemical, 114*(1), 530–541.

Berdondini, L., van der Wal, P. D., Guenat, O., de Rooij, N. F., Koudelka-Hep, M., Seitz, P., et al. (2005). High-density electrode array for imaging in vitro electrophysiological activity. *Biosensors and Bioelectronics, 21*, 167–174.

Bergveld, P. (1970). Development of an ion-sensitive solid-state device for neurophysiological measurements. *IEEE Transactions on Biomedical Engineering, 17*(1), 70–71.

Bousse, L., Shott, J., & Meindl, J. D. (1988). A process for the combined fabrication of ion sensors and CMOS circuits. *IEEE Electron Device Letters, 9*(1), 44–46.

Burns, J. R., & Powlus, R. A. (1996, July 12). Threshold circuit utilizing field effect transistors. U.S. Patent #3,260,863 (filed March 19, 1964).

Colapicchioni, C., Barbaro, A., Porcelli, F., & Giannini, I. (1991). Immunoenzymatic assay using CHEMFET devices. *Sensors and Actuators B: Chemical, 4*, 245–250.

Freund, M. S., & Lewis, N. S. (1995). A chemically diverse conducting polymer-based "electronic nose". *Proceedings of the National Academy of Sciences, 92*(7), 2652–2656.

Gardner, J. W., Varadan, V. K., & Awadelkarim, O. O. (2001). *Microsensors, MEMS, and smart devices*. New York: Wiley.

Gordon P. H., Jayant K., Cao Y., Auluck K., Phelps, J. B., & Kan, E. C. (2015). Critical assessment on modeling and design of non-Faradic CMOS electrochemical sensing. *IEEE Sensor Journal*, art. 2445292.

Huang, I. Y., & Huang, R. S. (2002). Fabrication and characterization of a new planar solid-state reference electrode for ISFET sensors. *Thin Solid Films, 406*(1–2), 255–261.

Hunter, R. J. (2001). *Foundations of colloid science* (2nd ed.). New York: Oxford University Press.

Jacquot, B. C., Muñoz, N. L., Branch, D. W., & Kan, E. C. (2008). Non-faradic electrochemical detection of protein interactions by integrated neuromorphic CMOS sensors. *Biosensors and Bioelectronics, 23*(10), 1503.

Janata, J. (2003). Electrochemical microsensors. *Proceedings of the IEEE, 91*(6), 864–869.

Jayant, K., Auluck, K., Funke, M., Anwar, S., Phelps, J. B., Gordon, P. H., et al. (2013a). Programmable ion sensitive transistor interfaces I: Electrochemical gating. *Physical Review E, 88*(1), 012801.

Jayant, K., Auluck, K., Funke, M., Anwar, S., Phelps, J. B., Gordon, P. H., et al. (2013b). Programmable ion sensitive transistor interfaces II: Biomolecular sensing and manipulation. *Physical Review E, 88*(1), 012802.

Jayant, K., Auluck, K., Rodriguez, S., Cao, Y., & Kan, E. C. (2014). Programmable ion-sensitive transistor interfaces III. Design considerations, signal generation and sensitivity enhancement. *Physical Review E, 89*(5), 052817.

Jayant, K., Rajwade, S., Pollack, L., & Kan, E. C. (2010) Controlled adsorption and desorption of DNA on CMOS—Towards a bi-directional bio-electronic interface. *Biosensors*, Galsgow, UK.

Kilic, M. S., Bazant, M. Z., & Ajdari, A. (2007). Steric effects in the dynamics of electrolytes at large applied voltages. I. Double-layer charging. *Physical Review E, 75*, 021502.

Kohashi, T., & Kurosawa, T. (1991). Electroosmotic display device. *IEEE Transactions on Electron Devices, ED-38*(9), 2064–2069.

Kovacs, G. T. A., Storment, C. W., Halks-Miller, M., Belczynski, C. R., Jr., Santina, C. C. D., Lewis, E. R., et al. (1994). Silicon-substrate microelectrode arrays for parallel recording of neural activity in peripheral and cranial nerves. *IEEE Transactions on Biomedical Engineering, 41*(6), 567–577.

Kress-Rogers, E. (1997). *Handbook of biosensors and electronic nose: Medicine, food, and the environment*. Boca Raton: CRC Press.

Landheer, D., Aers, G., McKinnon, W. R., Deen, M. J., & Ranuarez, J. C. (2005). Model for the field effect from layers of biological macromolecules on the gates of metal-oxide-semiconductor transistors. *Journal of Applied Physics, 98*, 044701.

Lang, H. P., Baller, M. K., Battiston, F. M., Fritz, J., Berger, R., Ramseyer, J.-P., et al. (1999). The nanomechanical NOSE. In *Proceedings of the IEEE Micro Electro Mechanical Systems (MEMS '99)*, pp. 9–13.

Lee, J., & Kim, C.-J. (2000). Surface-tension-driven microactuation based on continuous electrowetting. *Journal of Microelectromechanical Systems, 9*(2), 171–180.

Liu, Y., Huber, D. E., Tabard-Cossa, V., & Dutton, R. W. (2010). Descreening of field effect in electrically gated nanopores. *Applied Physics Letters, 97*, 143109.

Lonergan, M. C., Severin, E. J., Doleman, B. J., Beaber, S. A., Grubbs, R. H., & Lewis, N. S. (1996). Array-based sensing using chemically sensitive, carbon black-polymer resistors. *Chemistry of Materials, 8*, 2298–2312.

Madou, M. J., & Morrison, S. R. (1989). *Chemical sensing with solid state devices*. San Diego: Academic.

Manz, A., Effenhauser, C. S., Burggraf, N., Harrison, D. J., Seiler, K., & Fluri, K. (1994). Electroosmotic pumping and electrophoretic separations for miniaturized chemical analysis systems. *Journal of Micromechanics and Microengineering, 4*, 257–265.

Matsumoto, H., & Colgate, J. E. (1990) Preliminary investigation of micropumping based on electrical control of interfacial tension. In *Proceedings of the IEEE Micro Electro Mechanical Systems (MEMS)*, pp. 105–110.

Minch, B. A. (1997). *The subthreshold floating-gate MOS transistor, Chap. 5*. Ph.D. Dissertation, California Institute of Technology, Pasadena, CA, pp. 127–158.

Shen, Y. N., Liu, Z., Jacquot, B. C., Minch, B. A., & Kan, E. C. (2004). Integration of chemical sensing and electrowetting actuation on chemoreceptive neuron MOS transistors (CνMOS). *Sensors and Actuators B, 102*(1), 35–43.

Shen, N. Y., Liu, Z., Lee, C., Minch, B. A., & Kan, E. C. (2003). Charge-based chemical sensors: A neuromorphic approach with chemoreceptive neuron MOS (CνMOS) transistors. *IEEE Transactions on Electron Devices, ED-50*(10), 2171–2178.

Shibata, T., & Ohmi, T. (1992). A functional MOS transistor featuring gate-level weighted sum and threshold operations. *IEEE Transactions on Electron Devices, ED-39*(6), 1444–1455.

Siretanu, I., Ebeling, D., Andersson, M. P., Stipp, S. L. S., Philipse, A., & Stuart, M. C. (2014). Direct observation of ionic structure at solid-liquid interfaces: A deep look into the Stern Layer. *Scientific Reports, 4*, 4956.

Smith, R., Huber, R. J., & Janata, J. (1984). Electrostatically protected ion sensitive field effect transistors. *Sensors and Actuators, 5*(2), 127–136.

Steiner, F. -P., Hierlemann, A., Cornila, C., Noetzel, G., Bachtold, M., Korvink, J. G., et al. (1995). Polymer coated capacitive microintegrated gas sensor. In *Proceedings of the 8th International Conference on Solid-State Sensors and Actuators (Transducers '95)*, pp. 814–817.

Stern, E., Steenblock, E. R., Reed, M. A., & Fahmy, T. M. (2008). Label-free electronic detection of the antigen-specific T-cell immune response. *Nano Letters, 8*(10), 3310–3314.

Storey, B. D., & Bazant, M. Z. (2012). Effects of electrostatic correlations on electrokinetic phenomena. *Physical Review E, 86*, 056303.

Ulman, A. (1996). Formation and structure of self-assembled monolayers. *Chemical Reviews, 96* (4), 1533–1554.

van der Spiegel, J., Lauks, I., Chan, P., & Babic, D. (1983). The extended gate chemically sensitive field effect transistor as multi-species microprobe. *Sensors and Actuators, 4*, 291–298.

van Hal, R. E. G., Eijkel, J. C. T., & Bergveld, P. (1996). General model to describe the electrostatic potential at electrolyte oxide interfaces. *Advances in Colloid and Interface Science, 69*, 31–62.

Wang, L., Zhao, C., Duits, M. H. G., Mugele, F., & Siretanu, I. (2015). Detection of ion adsorption at solid–liquid interfaces using internal reflection ellipsometry. *Sensors and Actuators B: Chemical, 210*, 649–655.

Wenzel, S. W., & White, R. M. (1988). A multisensor employing an ultrasonic lamb-wave oscillator. *IEEE Transactions on Electron Devices, ED-35*(6), 735–743.

Wong, H. S., & White, M. H. (1989). A CMOS-integrated ISFET-operational amplifier chemical sensor employing differential sensing. *IEEE Transactions on Electron Devices, 36*(3), 479–487.

Yager, P., Domingo, G. J., & Gerdes, J. (2008). *Annual Review of Biomedical Engineering, 10*, 107–144.

Chapter 13
Waveform Shaping Structures and Transmission Lines on CMOS

Abstract Most of the CMOS mixed-signal circuits, including digital, analog, and RF applications, have layout sizes much smaller than the wavelength of design interest (or equivalently the operating frequency is much lower than the speed of light divided by the layout size), and are hence treated as lumped circuit elements or scattering blocks in conventional microwave circuits. These lumped circuits do not need to consider the coupling between the electric and magnetic fields governed by the Maxwell equations, but only the electrostatic Poisson equation with the displacement current would be sufficient. On the other hand, this chapter will introduce how to design distributive circuit structures in logic CMOS processes. Distributive circuits means the physical size of the component under consideration is comparable to the wavelength of interest, the electrostatic picture is insufficient, and the electromagnetic wave propagation needs to be considered for the module characteristics. As wave propagation is part of the design considerations, we will investigate the waveform shaping characteristics for these distributive modules, whether the shaping is a desirable feature or an unwanted distortion. As there are many excellent texts on the CMOS RF circuit modules such as on-chip inductors and resonators, we will focus only on distributive structures such as on-chip waveguides and transmission lines. We will then illustrate the design and characteristics of both semidiscrete and lumped-element transmission lines, together with varactor loading to make functionalities available in nonlinear transmission lines (NLTL). We will finally investigate the layout dependence of line folding and floating-metal isolation structures in practical waveguide and transmission line structures.

Keywords Transmission lines • Nonlinear transmission line (NLTL) • MMIC • Coplanar waveguide, CPW • Coplanar stripline, CPS • Waveform shaping circuits • Distributive circuits • Wave guides • Lumped elements

13.1 Introduction

Microwave monolithic integrated circuits (MMIC) have been an important part of high-frequency applications for many years. The main applications include radar, communication above 10 GHz, sensors in harsh environment, and high-speed

Y. Ma, E. Kan, *Non-logic Devices in Logic Processes*,
DOI 10.1007/978-3-319-48339-9_13

instrumentation. However, most MMIC capabilities were not implemented in CMOS due to both material limitation and historical design expertise. With the advance of CMOS beyond 22 nm and f_{max} over the 100 GHz range (where the wavelength is less than 3 mm, smaller than many of the die size), many of the MMIC designs are now possible in standard CMOS and the distributive effects need to be considered for on-chip global interconnect structures. This electromagnetic feature presents both new opportunities and challenges. For RF and microwave components on CMOS such as inductors and resonators, there are already many excellent articles and books available (Lee, 2004; Razavi, 2011; Rouphael, 2014), especially for components that have layout size much smaller than the wavelength and can be well characterized by a scattering matrix representation, even though they are meant to be used in a signal path where electromagnetic wave propagation is the main function.

In the spirit of this book for hybrid implementation with logic CMOS components, we will thus focus only on realization of distributive waveform shaping circuits in this chapter, as these structures illustrate the design principles to accommodate the high-frequency circuit modules on the logic CMOS platform, without emphasizing (in danger of diversion) the mixed-signal circuit functions that have been well introduced in other texts. We will start from the distributive transmission lines and waveguides on CMOS. We will then compare its behavior with its close cousins on CMOS: semidiscrete transmission lines with capacitor or varactor loading, and discrete transmission lines built by lumped-element LC ladders. Performance in terms of bandwidth, pulse shaping, and waveform dispersion will be presented. Applications in data interconnect and harmonic generation will be used for design examples. We will also investigate the effectiveness of floating-metal isolation structures.

Before we go into specific distributive structures, do keep in mind that this is a deviation to the usual circuit models that many IC designers are familiar with. SPICE, which is universally used for all kinds of digital, analog, and RF circuit designs, means "Simulation Program with an Integrated Circuits Emphasis." The IC emphasis actually implied that the circuit element in simulation is small enough that the wave propagation effect does not need to be considered. Transmission lines are included as a SPICE module, but mainly modeled as a delay line, instead of any physical waveform shaping behavior. The closest module one can implement in SPICE is actually an LC ladder. This point will become clear in the later sections.

13.2 CMOS Scaling to Distributive Effects in Digital Links

Digital ICs communicate and process information with "quasi-static" voltage signals representing 1 or 0 s (Rabaey, 2002). Local memories such as the NVM introduced in previous chapters may use the different threshold voltage values from charge storage or current magnitude to represent data, but eventually in the static logic blocks, the information will be transformed back to the quasi-static voltage

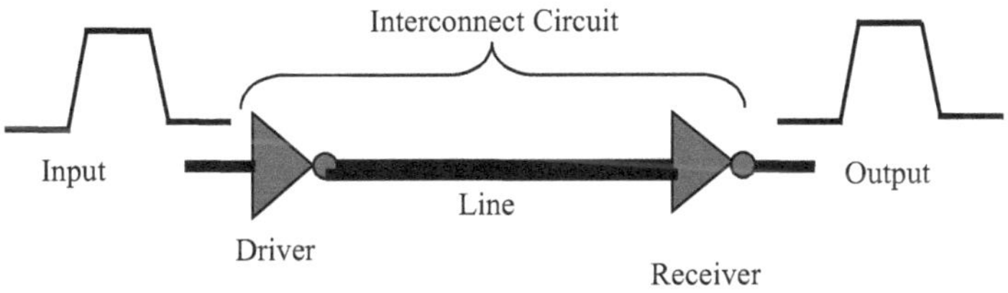

Fig. 13.1 A digital interconnect transports digital data from the driver to receiver by a quasi-static line

representations, or equivalently the charge condition on the node capacitance. This viewpoint has an important implication to implement the charge conservation law in SPICE.

When the quasi-static digital data representation needs to be transferred to the other part of the die or off the chip, typically it would involve sending and receiving a series of trapezoidal-shape voltage signals in time. The conductive paths carrying the signals are known as "interconnects," which include the entire electrical pathway from the transistors sending a signal to the transistors receiving the signal (Celik, Pileggi, & Odabasioglu, 2002; Hall, Hall, & McCall, 2000), as shown in Fig. 13.1. The whole interconnect line is charged or discharged during the signal transmission process (Young, 2001), or equivalently the forward and backward wave propagation has reached a steady state. The line can be viewed by its resistance and capacitance only, without considering the inductive effect. Digital IC performance is characterized by the circuit speed and power consumption, which had been device centric before 0.25 μm technology (Cheng, Lillis, Lin, & Chang, 2000). The effects of interconnects have been simplified as parasitic resistors and capacitors with some simple design guidelines, such as Elmore Delay (Elmore, 1948; Rabaey, 2002) where the propagation is treated as a broadband amplifier in the analog sense.

In deep submicron technologies, the traditional treatment is still true for the **local interconnects**, which connect transistors within a logic module and scale with the transistor dimension. However, regional, global, and off-chip interconnects, as presented in Table 13.1, have the same function of transferring the quasi-static digital information but can now be comparable to the wavelength of interest, if repeaters are not added to break the length. Continuous CMOS scaling with faster and faster device switching and about constant die size has made this distributive effect come into effects in shorter line length.

A simplified, yet informative propagation delay comparison, presented in Table 13.2, exemplifies this transition (Davis et al., 2001).

A significant manifestation of the distributive effects arises in the form of timing closure problem in today's CMOS design methodology, where the logic synthesis (optimization) tools often only account for logic gate loading prior to physical synthesis without considering the distributive effects of line delay and waveform shaping. On-chip communications may have to explicitly consider the distributive effects for interconnect-centric design methodology (Cong, 2001).

Table 13.1 A typical interconnect circuit and interconnect line classification

Line	Classification	Characteristics	% on typical digital chip
Short	Local	<500 μm ($L \ll \lambda/4$)	90
Medium	Regional	~0.5–3 mm ($L < \lambda/2$)	9
Long	Global	~3–15 mm ($L \sim \lambda$)	1
Off-chip	Board	Up to 1.0 m	N/A

L is the line length and λ is the wavelength corresponding to the significant frequency defined in (Cheng et al., 2000)

Table 13.2 Interconnect and transistor delay scaling properties (Davis et al., 2001)

Technology	MOSFET switching delay ($t_d = CV/I$) (ps)	Pulse spatial length (signal velocity: 1.5×10^8 m/s) (mm)	Intrinsic delay of minimum scaled 1 mm interconnect ($t_d = RC$) (ps)
1.0 μm (Al, SiO_2)	~20	~3	~5
0.1 μm (Al, SiO_2)	~5	~0.75	~30
35 nm (Cu, low k)	~2.5	~0.375	~250

Every conductor has a capacitance, inductance, and resistance, all of which can be eventually frequency dependent. At high enough frequency, none of these things is negligible for long interconnects. Thus, a global interconnect line is no longer a simple RC wire but a distributed parasitic element that is coupled to everything around it. The signal is not contained entirely in the conductor itself but is a combination of all the local electric and magnetic fields around the conductor. The signals on one interconnect will affect and be affected by the signals on another, commonly called **crosstalk**. The occurrence of these signals on different interconnects is independent and may force the designers to live with some random variations. Furthermore, at high frequencies, complex interactions occur between different parts of the same interconnect, such as the packages, connectors, vias, and bends. All the distributive or transmission line effects tend to distort waveforms that can give the digital designers a completely different view of high-speed digital signals, in addition to the conventional trapezoidal transient waveform and clear signal eye diagrams (Hall et al., 2000). Microwave design techniques, such as scattering parameter analysis (Pozar, 1998), are then needed to deal with the distributive details of the waveform shaping to ensure the timing, noise, and power consumption requirements.

Faster signal transmission is necessary to meet the exponentially decreasing timing budget (Hall et al., 2000). The signal transmission velocity is limited by the RC/RLC delay (Rabaey, 2002) of the line with the local speed of light as the hard limit (Davis et al., 2001). Traditionally, repeaters have been inserted into RC or RLC lines (Rabaey, 2002) to effectively reduce the signal propagation delay. But repeaters add their own propagation delays in addition to some other issues, such as power consumption discussed later. Moreover, the distorted waveforms caused by transmission line effects and crosstalk result in timing uncertainties and synchronization difficulties. Consequently, timing closure is becoming very difficult. An asynchronous IC does not need clock signals (Rabaey, 2002), but a fast signal transmission is still essential for high performance.

Couplings (crosstalk) through electric (mutual capacitance) and magnetic fields (mutual inductance) bring one of the most difficult aspects of high-speed on-chip interconnect and off-chip links: there are a large number of interdependent variables that can affect the outcome of a physical design. The mutual capacitance coupling is a short-range near-field effect. As a result, many design tools can now extract full-chip coupling capacitances in reasonable efficiency. Increasing line spacing is an effective approach to control the related crosstalk. The mutual magnetic field coupling, however, is a long-range near-field effect, which makes its shielding, extraction, and modeling very difficult (Devgan, Ji, & Dai, 2000). The magnetic field coupling is actually used in **near-field communication** (NFC), a standard protocol for up to 10 cm wireless communication for smart cards (Bolic, Simplot-Ryl, & Stojmenvoic, 2010).

Many other factors, such as simultaneous switching noise, charge sharing, leakage current, and voltage scaling, make noise issue worse (Young, 2001). Various methods are being developed to tackle noise problems, including repeater insertions, interconnect terminations, reduced voltage swings, slew rate controls, differential receiver with a reference voltage, fully differential drivers and receivers (Hall et al., 2000; Young, 2001). Even so, noises remain one of the most difficult design considerations.

Last but not least, the dynamic power consumed by the global interconnect lines and the repeaters scales up with the operating frequencies (Banerjee & Mehrotra, 2001). Global interconnect, including global clock distribution network, already accounts for ~20–40 % of the on-chip power consumption. This power consumption constraint has set the limit to the global clock frequency in the last 10 years.

3D integrations stack processed wafers together with vertical connections such as through-silicon vias (TSV), which can somewhat alleviate the long interconnect line limitations, besides its other possible manufacturing and cost advantages in the integration of disparate signals (digital, analog, and RF) and technologies (SOI, SiGe, GaAs, GaN, MEMS, etc.), especially in different layers. When the overall die size is reduced, the distributive effects in the long interconnect are also relieved. It is a very attractive way for system-on-chip (SoC) development. In terms of digital circuits, 3D integration does not require additional design innovations, but heat dissipation and reliability (Liu, Zhang, Datta, & Tiwari, 2002) remain the prominent issues in most 3D fabrication technologies. One notable exception is the stacked NVM layers, which has a guaranteed low access rate, as in the 3D V-NAND chip (Lapedus, 2016). Although V-NAND has used 48–64 layers of 3D stacking (i.e., the stacking manufacturing technology is available with high registration accuracy), the gain in reducing the interconnect length and distributive coupling will also gradually die out after 3–6 layers of wafer stacking.

13.3 CMOS Distributive Transmission Lines and Waveguides

Distributive effects can be a design nuisance for digital interconnect in terms of noise and interference, as described earlier, but can also be exploited to build functional modules for information propagation (Chang, Yue, & Wong, 2002; Wang, Pei, & Kan, 2004), clock distribution (O'Mahony, Yue, Horowitz, & Wong, 2003), crosstalk reduction (Kim, Ni, & Kan, 2006), waveform shaping (Afshari & Hajimiri, 2005), frequency conversion (Lyon, Yu, & Kan, 2010), and broadband harmonic generation (Yu, Lyon, & Kan, 2010). Notice that we focus on "electromagnetic wave propagation" structures and did not include lumped-element RF inductors here as there are many review articles and textbooks (Razavi, 2011). The fundamental structure for the earlier application is the on-chip transmission line and wave guides, linear or nonlinear, which will be the focus later.

Low-loss, electrically long, and high-impedance transmission lines can be constructed on CMOS substrate, despite of several design challenges. The most common types of on-chip transmission line are microstrip, **coplanar waveguide** (CPW), and **coplanar stripline** (CPS), which are shown in Fig. 13.2 with electric field lines for their fundamental modes. Both coplanar lines control their characteristic impedance through the combination of conductor width and spacing, both parameters available to an IC designer. The impedance of a microstrip is controlled by the conductor height (around 20 μm in typical CMOS) and width, making high impedance lines very narrow. The CPS line offers the highest possible characteristic impedance Z_0, about 100–150 Ω on mixed-signal IC. The maximum possible impedance for both coplanar lines is limited by the presence of the substrate ground plane, which causes a hybrid microstrip-coplanar mode when the conductor spacing is on the order of the conductor height (Simons, 2001).

Losses are a major issue for all on-chip transmission lines. Loss, especially its associated frequency dependence, can cause signal reduction, increasing noise and waveform dispersion. The main mechanisms are conductor losses from resistance and skin effect, dielectric losses, and radiation losses. Radiation losses are often

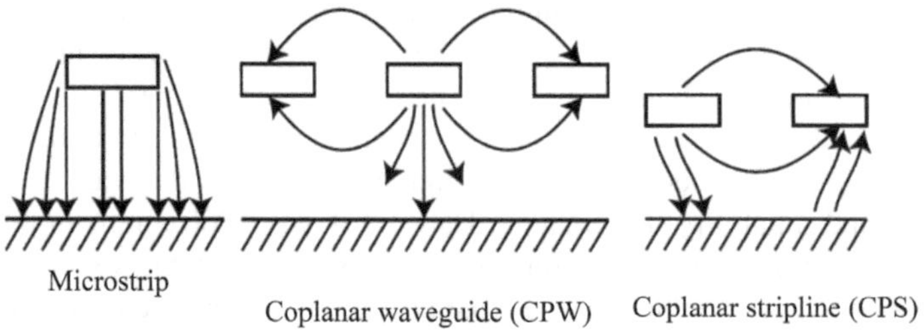

Fig. 13.2 Common transmission line structures, microstrip, coplanar waveguide (CPW), and coplanar stripline (CPS), on CMOS with electric field configuration for the dominant mode of propagation. CPW and CPS lines have a hybrid microstrip mode due to the proximity of the substrate

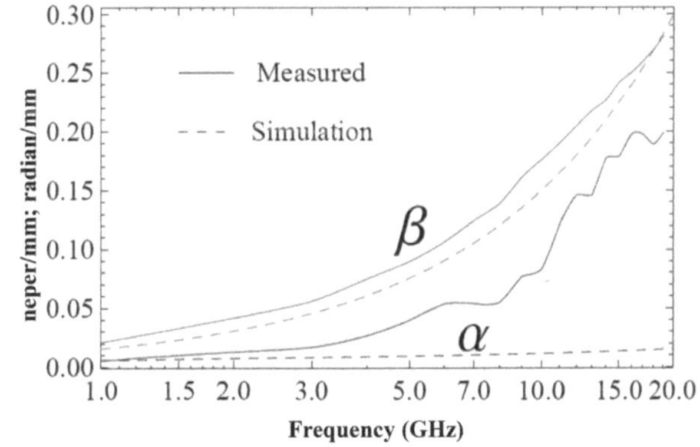

Fig. 13.3 The attenuation coefficient α in neper/mm and the phase constant β in radian/mm for a 100 Ω CPS transmission line in the 0.13 μm CMOS process. Attenuation is much higher than prediction in HFSS while measured β matches closely. A neper is a standard unit to describe exponential decay and is about 8.7 dB

small when the dimensions of the transmission line are smaller than the wavelength. This is the opposite side of efficient antenna design and was explored in the conventional PCB designs to avoid leaking the local oscillator signature. Radiative coupling to dielectric slab modes is unlikely because the dielectric contrast at the Inter-metal dielectric-air interface is small (4.1:1 for SiO_2, compared to 13.1:1 for a conventional GaAs MMIC) and the presence of metal fills in the upper metal layers tends to disrupt the slab mode. Dielectric loss is often the main concern due to the proximity and conductivity of the silicon substrate. The electric and magnetic fields from the transmission line can penetrate into the substrate, causing conduction and eddy currents which dissipate energy. These losses become more severe as the conductivity of the silicon increases and are difficult to model because the three-dimensional doping concentration is not known by the IC designer. A modern mixed-signal IC process starts with 1–2 Ω-cm wafers, with the conductivity increased further by various dopant implants. To prevent substrate losses, transmission lines are placed on the top metal layers, farthest from the substrate, with metal shielding on lower layers, and lightly doped substrate regions underneath. We will discuss the shielding structures in a later section.

We will use a CPS design to illustrate the transmission line design difficulties. Simulation is performed in ANSYS HFSS to predict the propagation and attenuation constants, as shown in Fig. 13.3.

The design goal of the CPS line is to achieve a high Z_0 and low loss here (Lyon et al., 2010). A conductor spacing of 25–50 μm, larger than the conductor to substrate spacing, was used. While good results had been reported for CPW lines and inductors with floating metal shields (Cheung & Long, 2006), the CPS line on a commercial silicon chip still showed significant high-frequency attenuation beyond that predicted in HFSS. Ultimately, the metal shielding proved somewhat ineffective at higher frequency because the thickness of the available shield layers was less than the skin depth, allowing fields to penetrate the substrate. The wide-spacing CPS lines thus had large substrate losses, as well as conduction losses in the thin floating shield.

13.4 Capacitive Loading for Semidiscrete Transmission Lines and Nonlinear Transmission Lines

We can reduce Z_0 for the transmission lines by adding additional capacitance periodically or in a tapered manner. This is not only a way to tune Z_0 (one can use switches and capacitors to dynamically tune Z_0 to the desirable value), but also introduces a design method for making **nonlinear transmission lines** (NLTL). The segment length to add additional capacitance will introduce a **Bragg frequency** f_{Bragg}, above which no traveling wave mode exists and the transmission line behaves like a low-pass filter (Afshari & Hajimiri, 2005; Wang et al., 2004; Yu et al., 2010).

$$Z_0 = \sqrt{\frac{L_{\mathrm{seg}}}{C_{\mathrm{seg}}}} \tag{13.1}$$

$$f_{\mathrm{Bragg}} = \frac{1}{2\pi\sqrt{L_{\mathrm{seg}}C_{\mathrm{seg}}}} \tag{13.2}$$

where L_{seg} is the distributive segment inductance and C_{seg} is the sum of the distributive segment capacitance plus the load capacitor. You can use the capacitors or varactors introduced in previous chapters, as long as they are fast enough to respond to the transient signal. For MOS varactors with majority carrier responses (accumulation mode), the dielectric relaxation time is usually on the order of ps, which can be comfortably used in waveforms with characteristic frequency below 100 GHz. With this high varactor cutoff frequency, we can manipulate signals faster than the transistor cutoff frequency (Afshari & Hajimiri, 2005; Lyon et al., 2010). However, practical varactors will be limited by the parasitic inductance and resistance connecting them, imposing a self-resonant frequency (due to inductance) and extrinsic cutoff frequency (due to resistance) significantly lower than the intrinsic dielectric relaxation frequency. A sample varactor response in high frequency is shown in Fig. 13.4. The capacitance at higher frequency is often higher, not due to the increase in permittivity (most materials have decreasing permittivity in high frequency), but due to self-resonance with the inductance of the connected wire.

NLTL can be used in pulse sharpening (Afshari & Hajimiri, 2005) to rise/fall time below 5 ps in standard CMOS, frequency conversion (Lyon et al., 2010) to obtain f_{Bragg} oscillation, broadband harmonic generation (Yu et al., 2010) in harmonic RFID (Ma & Kan, 2014), and soliton circuits (Wang et al., 2004) when the nonlinearity nearly cancel the dispersion caused by line loss. The nonlinearity can be additionally controlled by the biases across the varactor, and hence be explored for dynamic waveform shaping (Yu et al., 2010).

To illustrate the characteristics of on-chip NLTLs, sample layout structures were fabricated in TSMC's 0.25 μm mixed-signal process. The varactors are 200fF nMOS capacitors spaced at 675 μm intervals along $Z_0 = 62\ \Omega$ CPW with different

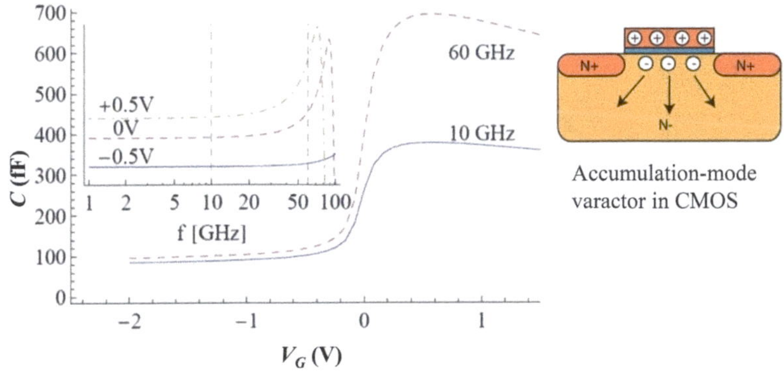

Fig. 13.4 C–V characteristics for a typical MOS varactor in the accumulation mode at 10 and 60 GHz. At 60 GHz, the device has a larger effective capacitance due to the approach to its self-resonance frequency. The inset shows the capacitance–frequency characteristics at different V_G

Fig. 13.5 Die photo showing (*a*) tightly folded, (*b*) loosely folded, and (*c*) straight NLTL. Inner boxes indicate varactor placement. The die size is about 3 mm × 3 mm

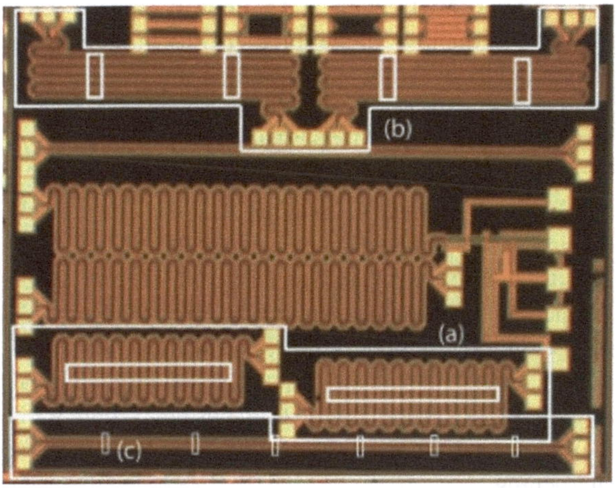

folding. Versions were constructed with and without shielding by floating metal strips 2 μm wide and 2 μm apart placed on the metal layer 1 μm beneath the CPW (referred to here as Ground-backed CPW or GbCPW). Both tight- and loose-folded serpentine patterns are implemented with ten NLTL sections using both conventional CPW and GbCPW. A six-section straight NLTL (no fold) is used for reference. Figure 13.5 shows the die with all NLTLs and varactor locations in white boxes.

The phase constants β in Fig. 13.6a, extracted from S-parameter measurements, show good agreement with the analytical model up to about 25 GHz. Overall, the unfolded and tight-folding lines have lower than expected propagation velocities, while the loose-folding lines are faster than expected. Figure 13.6b shows the extracted attenuation constants along with the analytical model. The cutoff is

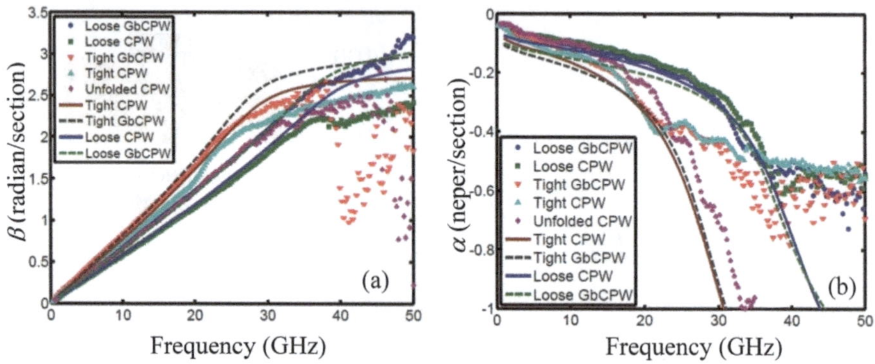

Fig. 13.6 (**a**) Extracted phase constants compared to the analytical model; (**b**) extracted attenuation constants, where the cutoff frequencies are well predicted by varactor losses

accurately predicted by considering varactor loss alone, while transmission line loss dominates at low frequency. The loosely folded NLTLs have slightly higher cutoff frequencies than the model predicts due to their higher than expected propagation velocities. Conversely, the unfolded and tightly folded NLTLs have lower than predicted cutoff frequencies due to their larger transmission line delay. All of the folded lines have relatively low stopband attenuation due to coupled line effects between the folds. These features are similar for semidiscrete transmission lines with constant capacitive loading.

Finally, the per-section edge sharpening in the NLTL is extracted and shown in Fig. 13.7. The tightly folded CPW NLTL and unfolded NLTL show reasonable sharpening performance. Both GbCPW NLTLs have significantly reduced performance due to additional capacitive loading. Finally, we see that the use of folding also increases the parasitic capacitive loading, with the largest degradation in loose-folding NLTLs. Here, coupled line effects create a transmission path that bypasses the varactors, improving the small signal response at the expense of edge sharpening.

13.5 Lumped-Element Transmission Lines and NLTLs

As the substrate and transmission line losses are very high for the semidiscrete transmission lines and NLTLs, lumped-element NLTLs were constructed in the same CMOS technology using spiral inductors. Notice that as shown in Fig. 13.8, the TEM mode wave propagation in a planar waveguide has an equivalent propagating current–voltage form in the lumped-element LC ladder below f_{Bragg}.

The main advantages of spiral inductors over bondwires (Ricketts, Li, Sun, Woo, & Ham, 2007) are the lack of additional postfabrication steps, and the availability of mature, measurement-based foundry models. The improvement of the lumped-

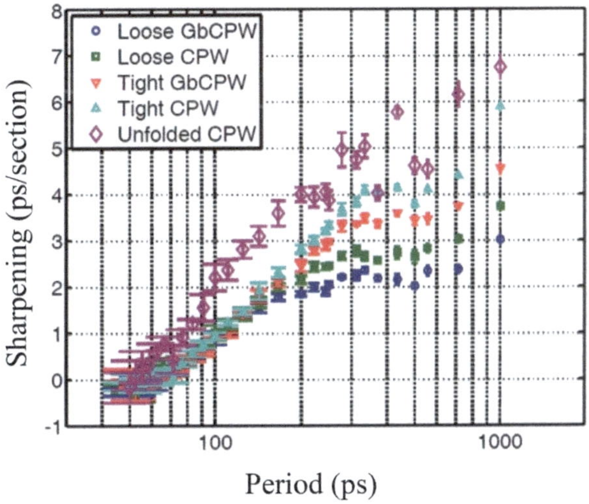

Fig. 13.7 Per-section fall-time sharpening with respect to the period of the pulses

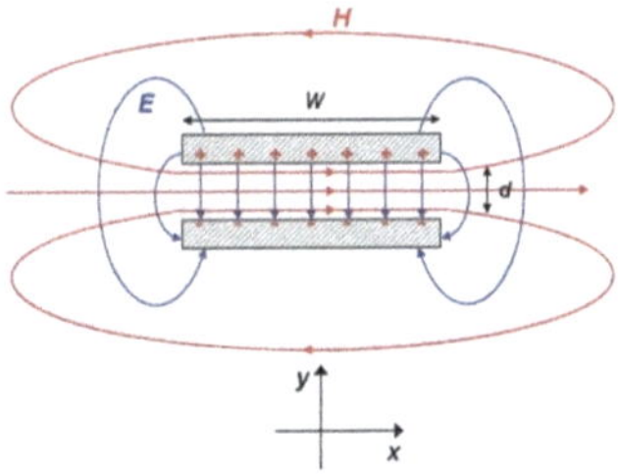

Transmission line variables:

$$\begin{cases} V(z) = V_+ \, e^{-jkz} \\ I(z) = I_+ \, e^{-jkz} = \dfrac{V_+}{Z_o} \, e^{-jkz} \end{cases}$$

Plane wave variables (ignoring fringing fields):

$$\begin{cases} \vec{E}(z) = -\hat{y} \, \dfrac{V_+}{d} \, e^{-jkz} \\ \vec{H}(z) = \hat{x} \, \dfrac{I_+}{W} \, e^{-jkz} \end{cases}$$

Fig. 13.8 Matchup of transmission-line and plane-wave waveforms from Maxwell phaser solutions

element design over the previous semidiscrete designs is evident from S-parameter transmission measurements taken from both structures. Figure 13.9 compares the magnitude transmission from the semidiscrete design with the 20 GHz lumped-element design. The semidiscrete design has losses which increase rapidly with frequency, up to 15 dB near f_{Bragg} cutoff. By contrast, the lumped-element design has as much as a 6 dB advantage in the 10–15 GHz range and losses no greater than −6 dB near f_{Bragg} cutoff. Additionally, the lumped-element design has a significantly reduced layout area (1 mm by 0.2 mm) and meets pattern density rules quite easily.

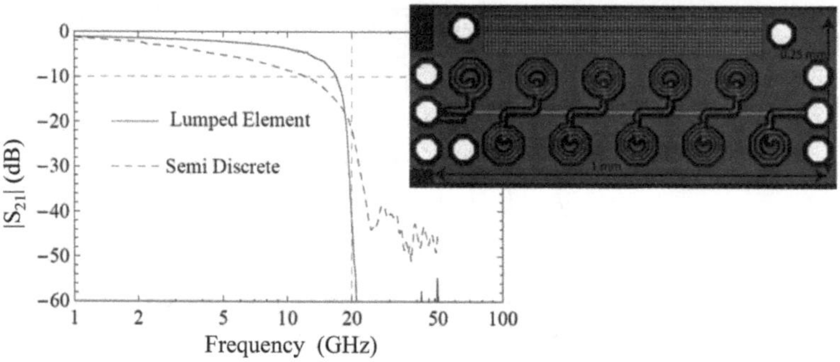

Fig. 13.9 S-parameter transmission for semidiscrete CPS and lumped-element LC ladders on mixed-signal CMOS. For the same f_{Bragg} limit, the lumped-element line has better performance due to smaller loss

13.6 Summary

In this chapter, we first illustrate the scenarios when the distributive effects in long interconnect need to be considered in the design phase. For functional use of the distributive wave propagation effects, we then illustrate how to effectively implement distributive, semidiscrete, and lumped-element transmission lines and NLTL on CMOS substrate for waveform shaping and pulse sharpening. The dependence of the phase constants and attenuation coefficients on layout folding and isolation structures is then investigated. Although transmission lines and NLTL can be readily implemented on CMOS to provide unique capabilities of waveform shaping, frequency conversion, and noise isolation, they have not been broadly adopted in commercial ICs due to the strong competition from repeater-inserted lines and less design library experience.

References

Afshari, E., & Hajimiri, A. (2005). Nonlinear transmission lines for pulse shaping in silicon. *IEEE Journal of Solid-State Circuits, 40*(3), 744–752.

Banerjee, K. M., & Mehrotra, A. (2001). Global (interconnect) warming. *IEEE Circuits and Devices Magazine, 17*, 16–32.

Bolic, M., Simplot-Ryl, D., & Stojmenvoic, I. (2010). *RFID systems: Research trends and challenges*. New York: Wiley.

Celik, M., Pileggi, L., & Odabasioglu, A. (2002). *IC interconnect analysis*. New York: Kluwer Academic.

Chang, R. T., Yue, C. P., & Wong, S. S. (2002) Near speed-of-light on-chip electrical interconnect. In *Technical Digests, VLSI Circuits Symposium*, pp. 18–21.

Cheng, C., Lillis, J., Lin, S., & Chang, N. (2000). *Interconnect analysis and synthesis*. New York: Wiley.

Cheung, T., & Long, J. (2006). Shielded passive devices for silicon based monolithic microwave and millimeter-wave integrated circuits. *IEEE Journal of Solid-State Circuits, 41*(5), 1183–1200.

Cong, J. (2001). An interconnect-centric design flow for nanometer technologies. *Proceedings of the IEEE, 89*, 505–528.

Davis, J. A., Venkatesan, R., Kaloyeros, A., Beylansky, M., Souri, S. J., Banerjee, K., et al. (2001). Interconnect limits on gigascale integration (GSI) in the 21st century. *Proceedings of the IEEE, 89*, 305–324.

Devgan, A., Ji, H., & Dai, W. (2000). How to efficiently capture on_chip inductance effects: Introducing a new circuit element K. Presented at *IEEE International Conference on Computer-Aided Design.*

Elmore, W. C. (1948). The transient response of damped linear networks with particular regard to wideband amplifiers. *Journal of Applied Physics, 19*, 55–63.

Hall, S. H., Hall, G. W., & McCall, J. A. (2000). *High-speed digital system design, a handbook of interconnect theory and design practices.* New York: Wiley.

Kim, J., Ni, W., & Kan, E. C. (2006). Crosstalk reduction with nonlinear transmission lines for high-speed VLSI system. In *IEEE Custom Integrated Circuits Conference (CICC)*, San Jose, CA, September 11–13, 2006, Paper No. 29.6.

Lapedus, M. (May 2016). What's next for NAND? *Semiconductor engineering, manufacturing, design and test.*

Lee, T. H. (2004). *The design of CMOS radio frequency integrated circuits* (2nd ed.). Cambridge: Cambridge University Press.

Liu, C., Zhang, J., Datta, A. K., & Tiwari, S. (2002). Heating effects of clock drivers in bulk, SOI, and 3-D CMOS. *IEEE Electron Device Letters, 23*, 716–718.

Lyon, K. G., Yu, F., & Kan, E. C. (2010). A UWB-IR transmitter using frequency conversion in nonlinear transmission lines with 16pJ/pulse energy consumption. *IEEE Transactions on Microwave Theory and Techniques, 58*(12), 3617–3625.

Ma, Y., & Kan, E. C. (2014). Accurate indoor ranging by broadband harmonic generation in passive NLTL backscatter tags. *IEEE Transactions on Microwave Theory and Techniques, 62* (5), 1249–1261.

O'Mahony, F., Yue, C. P., Horowitz, M. A., & Wong, S. S. (2003). A 10-GHz global clock distribution using coupled standing-wave oscillators. *IEEE Journal of Solid-State Circuits, 38* (11), 1813–1820.

Pozar, D. M. (1998). *Microwave engineering* (2nd ed.). New York: Wiley.

Rabaey, J. M. (2002). *Digital integrated circuits: A design perspective* (2nd ed.). Upper Saddle River: Prentice-Hall.

Razavi, B. (2011). *RF microelectronics* (2nd ed.). Upper Saddle River: Prentice Hall.

Ricketts, D., Li, X., Sun, N., Woo, K., & Ham, D. (2007). On the self-generation of electrical soliton pulses. *IEEE Journal of Solid-State Circuits, 42*, 1657–1663.

Rouphael, T. J. (2014). *Wireless receiver architecture and design: Antennas, RF, synthesizers, mixed signal and digital signal processing.* Oxfordshire: Academic.

Simons, R. N. (2001). *Coplanar waveguide circuits, components, and systems.* New York: Wiley.

Wang, P., Pei, G., & Kan, E. C. (2004). Pulsed wave interconnect. *IEEE Transactions on Very Large Scale Integration (VLSI) Systems, 12*(5), 453–463.

Young, B. (2001). *Digital signal integrity, modeling and simulation with interconnects and packages.* Upper Saddle River: Prentice Hall.

Yu, F., Lyon, K. G., & Kan, E. C. (2010). A novel passive RFID transponder using harmonic generation of nonlinear transmission lines. *IEEE Transactions on Microwave Theory and Techniques, 58*(12), 4121–4127.

Chapter 14
Conclusions and Outlook

Abstract We conclude this book by looking at the potential directions that IC industry is going to continue the Moore's law scaling and the associated impacts to the non-logic devices discussed in previous chapters. We first review the fundamental limits to the continued IC scaling, including those implied by the quantum mechanics and quantum information theory. As we close to those fundamental limits, two main directions that IC industry is moving toward include the use of the vertical dimension and new materials to build new ICs. The later approach has resulted in the development of many emerging memory technologies, including STT-RAM, PCRAM, and RRAM. Both directions could open up new opportunities for designing new non-logic devices. Finally new computing paradigms, including quantum computing, may usher in fundamentally new processes for building these new computing devices.

Keywords Shannon–von Neumann–Landauer limit • Moore's law • More than Moore (MtM) • 3D IC • RRAM • Phase change memory • STT-RAM • finFET • Tri-gate FET • VNAND • 3D NAND • Memory hierarchy • Neuromorphic computing • Spintronics • Quantum computing

14.1 Introduction

As stated in Sect. 2.1, Moore's law scaling is slowing down. Some aspects of the scaling have already stopped while others could stop in the very near future. There are fundamental physical limits that the current CMOS technology will bump into very soon, which will cause semiconductor scaling to stop completely.

One of these limits is the thermodynamic limit, called Landauer limit (also called Shannon–von Neumann–Landauer limit). It deals with the minimal energy needed to flip a bit. In this classical information process regime, the thermodynamic scaling limit is given by

$$E_{\text{bit\_min}} = kT \ln(2), \tag{14.1}$$

which is equal to 17 meV at room temperature (300 K). Today (2016) the minimal switching energy is associated with the charging of a gate to the supply voltage and

Y. Ma, E. Kan, *Non-logic Devices in Logic Processes*,
DOI 10.1007/978-3-319-48339-9_14

can be estimated as following: for a 10 nm by 10 nm gate with a gate oxide that has an oxide thickness that is equivalent to that of the 1.0 nm SiO_2 (or EOT of 1 nm = equivalent oxide thickness) and charged to 0.7 V, the switching energy cost is about CV^2~3.5 aF × 0.7^2~10 eV. The energy is more than two orders of magnitude higher than the Landauer limit. Therefore, there is still a lot of room left for scaling in the energy limit.

Spatial scaling is constrained first by the economics of lithography. For the 10 nm process technology, quad patterning or other advanced lithography needs to be employed at greatly increased cost. It is still not clear what tools can be used in mass production of 7 nm and below technologies. If we change from the current immersion lithography to extreme UV (EUV) sources, then the economical model we presented in Chap. 2 will be even more relevant:

BIP.1 *Focusing IC development effort on using the simplest, most basic, and highest volume process.*

BIP.1 is usually best achieved with incremental improvements of existing process. Major changes such as the transition to EUV lithography and the corresponding changes in mask and photoresist technology represent a major deviation from BIP and will cause dramatic increase in cost, and it would be even more costly to change the process and material to adapt to make the non-logic devices.

The second fundamental limit is the way we view electron conduction as in the electronic band theory. From Anderson's tight-binding model and strong localization (Anderson, 1958), if we have less than ten unit cells in a lattice, then the electron transport should be treated with molecular orbitals (MO) instead of band electrons. While MO works well for information processing in DNAs, we do not have an effective way at the present to perform logic operations using MO. This will set a fundamental spatial limit around 2 nm for the current transistor architecture to work, i.e., band electron flow controllable with a gate.

Note that this limit also does not conflict Eq. 14.1. Assuming that we are able to use an oxide with EOT of 0.7 nm and a supply voltage of 0.3 V in some advanced FET and materials, the energy cost for charging a gate of size 2 nm by 2 nm is then about 110 meV, which is still about 6× of the Landauer limit of Eq. 14.1.

The 2 nm here is just a fundamental limit from the electron localization point of view. Surely other factors, especially from the viewpoints such as system power consumption and process variations including dopant fluctuation and line-edge roughness, may stop the scaling even earlier.

At the time of this writing (2016), technology with 10 nm minimal feature sizes is being rolled out to production. This means that at current rate of progress, we may reach the practical physical limit within approximately 3–4 technology generations or around 7 years.

Also in 2016, the DRAMs are being produced in 20 nm process. It is projected that we may be able to scale the DRAM process to 13–14 nm but unlikely to below 10 nm. The early stop of the scaling is mainly due to the limit set by the storage capacitor requirement to cause sufficient perturbation to a common bit-line bus.

After the scaling of DRAM is stopped, manufacturing will likely continue to use the last process technology but with continued design and process improvement to reduce the cost: by increasing the yield, reducing the memory cell size, and improving the test efficiency. In the meantime, DRAM replacement technology is being actively researched, with phase change memory (PCRAM), magnetic RAM (MRAM), and resistive RAM (RRAM) showing promising results.

Scaling in NAND flash technology has also reached a milestone. Because of the interference from neighboring cells, it is believed that the planar NAND process technology will stop after reaching the 12 nm node. However, as discussed in the next section, 3D NAND technology has already been in production since 2013.

14.2 Continued Scaling with FinFET

3D NAND (see, e.g., Micheloni, 2016 for a review) is a *bona fide* true 3D IC, where the channel of the storage transistor is along the vertical dimension. NAND chips with many layers of transistors, up to 48, are in production in 2016. The channel material of the 3D NAND transistors is polysilicon, which can be deposited after the gate and gate dielectrics in the bulk have been formed.

This type of 3D NAND is possible because the speed requirement for NAND flash is low, and the section that is active and consumes power is always a very small portion in the circuits. Typical read access time is only on the order of 50 μs. Typical data access of a page (4096 bits) or a block (64 pages) is a very small portion of the die. As a result the channel of the memory transistor can be made of low-mobility polysilicon. The state-of-the-art NAND chip (in 2016) consists of a 16 die stack, each die is 40 μm thick and having 48 storage transistor layers (Gibb, 2016).

3D NAND is the only true 3D ICs in production now. The reason similar approaches have not been used to produce 3D logic ICs is due to performance and heat dissipation. High drive and low leakage current requirements on logic transistors mean that the channel of a logic FET has to be made of materials with high mobility and saturation velocity. Only single-crystal channel is now capable of providing high mobility and, at the same time, good junctions with low leakage. Unfortunately low defect single crystals cannot be deposited at the practical temperatures used in present IC production.

One key movement of logic front-end FETs to the third dimension is the introduction of finFET. FinFET was invented around 1998 (Hisamoto et al., 1998) and has been successfully deployed commercially since 2011 (Jan et al., 2012). It is projected that the finFET transistor can be scaled to at least the 7 nm node, to around 2020. Figure 14.1 illustrates the basic structure of a finFET compared to a traditional planar FET. The planar FET has the usual channel length L and width W, shown in the figure. For finFET, the channel length is defined similar to that of the planar FET, but the equivalent channel width is given by $W = 2h + w$, where h is the height of the fin and w is the width of the fin.

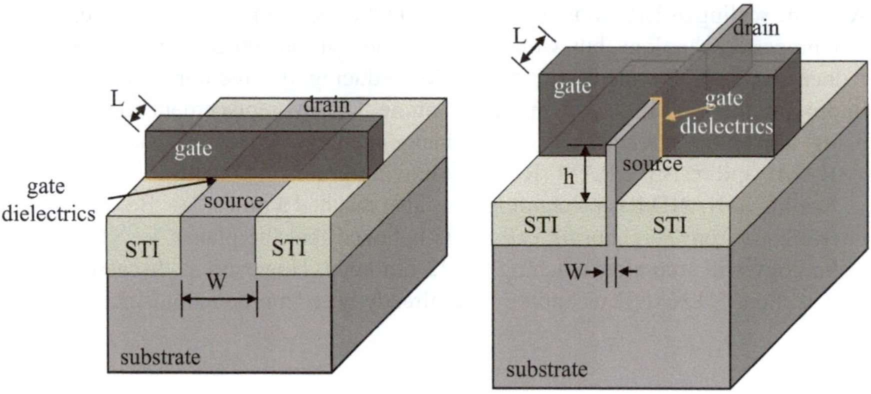

Fig. 14.1 finFET (*right*) and planar FET (*left*) comparison

The main advantage of the finFET over the planar FET is that finFET reduces source to drain leakage current and allows lower supply voltage. For finFET, the gate wraps around the channel on three sides,[1] thus providing a better electrostatic control of the channel. This reduces the short channel effect, or equivalently, lowers the source to drain leakage. The sharp subthreshold slope then enables the threshold voltage to be lowered and the transistor can keep the same small I_{OFF}. As a consequence, the supply voltage can be lowered correspondingly, e.g., to below 0.7 V from about 1.0 V for the planar FET. This results in a large reduction in power consumption. As we noted in Chap. 2, power consumption has become one of the main barriers for the continued scaling. Lower power consumption allows more dense packing of transistors and helps to continue the Moore's law scaling.

Another advantage of the finFET is that the width of the FET is the fin height. FinFET is a 3D device in this sense. In contrast to the planar FET, increasing the width of the finFET does not increase the area of the transistor directly, but through the requirement of the aspect ratio due to the fin etching step.

The front-end process for making finFET is significantly different from the planar CMOS process described in Chap.3. To make the tri-gate type of finFET, we can use a silicon nitride layer as hard mask to etch the bulk silicon and form the fin. This step is very similar to the formation of the active region in the planar process of Fig. 3.5. The gate dielectrics are then deposited, followed by the deposition of the gate. Process alternatives such as the gate-last process to allow high-k gate dielectrics are similar to the planar FET process.

Since the finFET is relatively new and many of the process details are still proprietary, we will not cover device design in this book and will leave the art of non-logic device design from finFET to future editions. The physical and design principles we presented are nevertheless valid, such as the capacitance coupling

[1] This is the reason Intel has named their version of the finFET as tri-gate FET, and TSMC as Ω-FET.

ratio to enable an electron injection structure and the oxide thickness requirement for the floating-gate carrier retention time. FinFET has a higher capacitive coupling from the gate/floating gate to the channel, and the coupling ratio will need to be adjusted accordingly. Also, the electrostatic coupling in the fin or nanowire or nanotube-based FET can be very different from the parallel-plate approximation, and more careful 3D electrostatic models should be employed for coupling ratios (Ganguly, Kan, & Zhang, 2005; Guo, Kan, Ganguly, & Zhang, 2006).

14.3 Emerging Memories

FinFET made it possible to continue the scaling of logic CMOS for a few more generations. It has not solved the scaling problems in other ICs, especially in the memory ICs. Memory is one of the fundamental blocks of integrated circuits, and for many computing architectures, the memory wall (time delay to access the intended memory location) limits the present computing throughput. Figure 14.2 shows the semiconductor memory hierarchy. Colored boxes indicate the degree to which the memory can be made in logic processes. Three classes are listed. The memories in the first class can be made in pure logic processes, examples including SRAM and LNVM. The second class is the memory that can be easily integrated with logic processes with minor modifications, e.g., by adding a few steps or masks. The memory in the third class includes those that needs more drastic changes to the logic process and are often not integrated with logic circuits.

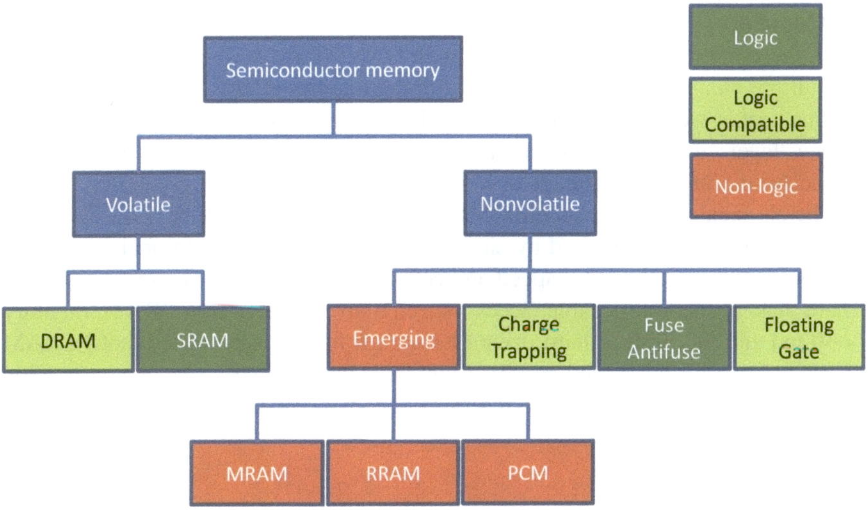

Fig. 14.2 Semiconductor memories hierarchy. *Colored boxes* indicate the degree to which the memory can be made in logic process or not

Current memories, SRAM, DRAM, and NAND flash all face serious scaling challenges. SRAMs are manufactured in standard logic processes and have been produced with the finFET technology successfully.

SRAM is volatile and has also large layout area when compared with DRAM and NAND flash memory cells. The most common SRAM cell has six transistors. For lower voltage and more advanced processes, cells with eight or more transistors have been proposed, mostly to relieve the effect of the process variation.

Even though 3D NAND flash technology has made it possible to continue scaling, NAND flash memory is facing the problems of low write endurance and slow program and read speed.

One of the most significant and promising developments is the progresses made in emerging nonvolatile memories. There are a number of emerging memories that look attractive as candidates for replacing or complementing the existing SRAM, DRAM, or NAND flash memory, including spin transfer torque RAM (STT-RAM), resistive memory (RRAM), and phase change RAM (PCRAM).

Because it is capable of virtually unlimited write cycles, STT-RAM has been studied as a "universal" memory. However, because of its larger cell size, more recent focus has been to use STT-RAM as a SRAM replacement in embedded applications. It has been demonstrated that magnetic tunneling junctions (MTJs) can be manufactured and integrated in advanced logic process (Lin, Wang, Lee, Chen, & Hsu, 2009). With the perpendicular MTJs (Yuesa, 2013), the drive current needed to program the MTJ is decreasing to a feasible range, as is the size of the select transistor. It is believed that the STT-RAM memory cell can then scale as the design rules of logic transistors in advanced process nodes. When this happens, STT-RAM will have the potential to replace SRAM.

PCRAM (Lee, 2010) and RRAM usually have limited write endurance cycles, from 10^6 to 10^{10} cycles. Because of such limitation they cannot replace DRAM or SRAM, which usually need up to 10^{16} cycle in most applications. On the other hand, these memories cannot compete with NAND flash memory in consideration of multiple bits per cell and the potential integration problems in 3D as we cannot confine heat as well as electrons. Nevertheless, these memories have been proposed as new layers in the memory hierarchy, as shown in Fig. 14.3, to exploit their respective merits of low voltage or high program speed.

The classical memory elements (SRAM, DRAM, flash memory, hard disk, CD, and tape) are usually arranged by either speed or cost in the traditional computing memory hierarchy. Note that speed and cost are usually in an inverse relationship. There has been a noticeable access speed gap between the DRAM memory ($<$10 ns) and the flash or hard drive ($>$1 ms). It has been proposed that PCRAM or RRAM can fill this speed gap. It is too early for us to speculate how non-logic devices can be derived from STT-RAM, PCRAM, or RRAM, except for embedded nonvolatile memory.

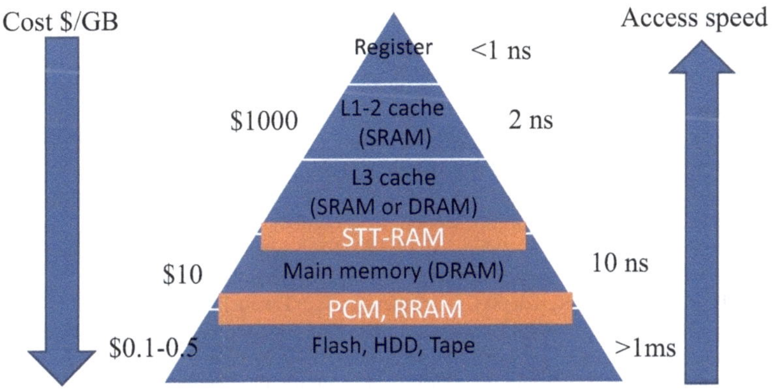

Fig. 14.3 Memory hierarchy (in *blue*) with the possible positions of emerging memories (in *orange*)

14.4 New Computing Paradigms

As we stated in the introduction section, fundamental physical limits dictate that the scaling of computing devices to smaller dimension will end soon. So to continue the economical equivalent of Moore's law, i.e., ~50 % reduction of computing cost every 1–2 years, we may have to seek architectural or paradigm changes in computing.

We briefly discussed neuromorphic computing in Sect. 11.5 which has the potential to provide computing power in special circumstances that mimics human brains in capacity and energy efficiency.

Spintronics, which uses spins instead of electrons for computation (Wolf & Treger, 2006), is another possible approach. Molecular computing including DNA computing (Kahan, Gil, Adar, & Shapiro, 2008) is also being researched heavily.

More fundamentally, quantum computing (Nielsen, 2010), which utilizes some quantum mechanical concepts such as eigenstate superposition and quantum entanglement to perform operations on data, has been shown to be able to solve some of the toughest computing problems, such as large integer factorization, much more quickly than classical computers.

Indeed a large number of quantum computers have been proposed, including systems based on the spins of electrons or photons, superconducting Josephson junctions, trapped ions, and others.

The global information industry will depend upon some of these approaches to succeed to continue its growth. These new computing paradigms may usher in fundamentally new devices and processes for building other required devices in the system level as well.

References

Anderson, P. (1958). Absence of diffusion in certain random lattices. *Physical Review, 109*, 1492–1505.

Ganguly, U., Kan, E., & Zhang, Y. (2005). Carbon nanotube based non-volatile memory with charge storage in metal nanocrystals. *Applied Physics Letters, 87*, 043108.

Gibb, K. (2016, April 6). *First look at Samsung's 48L 3D V-NAND flash*. Retrieved from EE Times http://www.eetimes.com/author.asp?section_id=36&doc_id=1329360

Guo, J., Kan, E., Ganguly, U., & Zhang, Y. (2006). High sensitivity and nonlinearity of carbon nanotube-based charge sensors. *Journal of Applied Physics, 99*, 084301.

Hisamoto, D. ; Lee, W.-C. ; Kedzierski, J. ; Anderson, E. ; Takeuchi, H. ; Asano, K. ; King, T.-J. ; Bokor, J. ; Hu, C. (1998). A folded-channel MOSFET for deep-sub-tenth micron era. In *IEEE International Electron Device Meeting* (pp. 1032–1034). Washington, DC: IEEE.

Jan, C.-H.; Bhattacharya, U.; Brain, R.; Choi, S.- J.; Curello, G.; Gupta, G.; Hafez, W.; Jang, M.; Kang, M.; Komeyli, K.; Leo, T.; Nidhi, N.; Pan, L.; Park, J.; Phoa, K.; Rahman, A.; Staus, C.; Tashiro, H.; Tsai, C.; Vandervoorn, P.; Yang, L.; Yeh, J.-Y. ; Bai, P. (2012). A 22nm SoC platform technology featuring 3-D Tri-gate and high-k/metal gate, optimized for ultra low power, high performance and high density SoC applications. In *IEEE International Electron Device Meeting* (pp. 44–47). Washington, DC: IEEE.

Kahan, M., Gil, B., Adar, R., & Shapiro, E. (2008). Towards molecular computers that operate in a biological environment. *Physica D, 237*, 1165–1172.

Lee, B. Z. (2010). Phase-change technology and the future of main memory. *IEEE Micro, 30*, 131–141.

Lin, C. K., Wang, Y., Lee, K. Z., Chen, W., & Hsu, W. (2009). 45nm low power CMOS logic compatible embedded STT MRAM utilizing a reverse-connection 1T/1MTJ cell. In *IEDM Technical Digest* (pp. 279–282). Washington, DC: IEEE.

Micheloni, R. (2016). *3D Flash memories*. New York: Springer.

Nielsen, M. C. (2010). *Quantum computation and quantum information*. Cambridge: Cambridge University Press.

Wolf, S. A., Chtchelkanova, A. Y., & Treger, D. M. (2006). Spintronics—A retrospective and perspective. *International Journal of Advance Research and Engineering, 50*, 101–110.

Yuesa, S. (2013). Future prospects of MRAM technologies. In *IEDM Technical Digest* (pp. 56–59). Washington, DC: IEEE.